Environmentally Friendly Manufacturing of Food Additives and Functional Ingredients

Environmentally Friendly Manufacturing of Food Additives and Functional Ingredients

Editors

Fuping Lu
Wenjie Sui

MDPI • Basel • Beijing • Wuhan • Barcelona • Belgrade • Manchester • Tokyo • Cluj • Tianjin

Editors
Fuping Lu
Tianjin University of Science
& Technology
China

Wenjie Sui
Tianjin University of Science
& Technology
China

Editorial Office
MDPI
St. Alban-Anlage 66
4052 Basel, Switzerland

This is a reprint of articles from the Special Issue published online in the open access journal *Foods* (ISSN 2304-8158) (available at: https://www.mdpi.com/journal/foods/special_issues/food_additives_ingredients).

For citation purposes, cite each article independently as indicated on the article page online and as indicated below:

LastName, A.A.; LastName, B.B.; LastName, C.C. Article Title. *Journal Name* **Year**, *Volume Number*, Page Range.

ISBN 978-3-0365-6400-5 (Hbk)
ISBN 978-3-0365-6401-2 (PDF)

Contents

About the Editors

Fuping Lu

Fuping Lu (Professor), Ph.D., is president of Tianjin University of Science & Technology, doctorial supervisor, director of State Key Laboratory of Food Nutrition and Safety, and chief expert of China Biotech Fermentation Industry Association. He is awarded the title of the National Candidate for Millions of Talents Project, the National Mid-Aged Expert with Remarkable Contributions, and the First-Level Candidate for Tianjin "131" Creative Human Resource Project. His research fields involve in Food Biotechnology. Meanwhile, he is granted the State Technological Innovation Award, the China Patent Excellence Awards, the China Light Industry Council Science and Technological Innovation and Progress Award, the Tianjin Science and Technology Progress Award, and the Tianjin Patent Gold and Excellence Awards.

Wenjie Sui

Wenjie Sui (Associate Professor), Ph.D., is Associate Professor for Food Science and Engineering at the State Key Laboratory of Food Nutrition and Safety of Tianjin University of Science & Technology, China. She studied the Biochemical Engineering and received her Ph.D. from the Institute of Process Engineering, Chinese Academy of Sciences. Her scientific interests are centered at plant cell wall breaking techniques and biorefinery engineering. To this end, she pursues the principles and applications of multi-physics cell wall breaking technologies, with special attention on their chemical-stress coupling effects and cell wall breaking and utilizing mechanism. For the aim of clean and efficient manufacturing of food additives and functional ingredients, they are further integrated with high-solid conversion and high-efficiency extraction technologies in order to broaden their application in food processing fields.

Preface to "Environmentally Friendly Manufacturing of Food Additives and Functional Ingredients"

Food additives and functional ingredients are naturally or artificially synthesized substances that are added to foodstuffs to improve their color, aroma, and taste, as well as to meet anti-corrosion and other technological purposes. They are considered to be the catalysts and headstones of the modern food industry, affecting every step of food production, processing and storage. Currently, in the highly competitive global market amidst stringent environmental constraint, the eco-friendly and cost-effective production of high-purity food additives has remained a challenge, mainly due to involvement of multiple processing techniques and their associated resource and energy waste, production inefficiency and environmental pollution. The need of the hour is thus innovation of the environmentally friendly manufacturing process of food additives and functional ingredients, involving bioconversion, extraction, separation and purification, stabilization, formulation and compounding, etc., as well as relevant process intensification and integration, all of which ensure solving the above deep-seated problems and manufacturing additive products with accurate functions and reliable safety. The book aims to disseminate research on addressing and discussing theoretical and practical clean production, encompassing environmental and sustainability issues in the production of food additives and functional ingredients.

We are, therefore, pleased to introduce this book on Environmentally Friendly Manufacturing of Food Additives and Functional Ingredients. This book contains 12 articles about various aspects of the clean production of food additives and functional ingredients and their structure characterization as well as functional application. We would like to thank all of the authors for their compelling contributions to this book and the reviewers for their willingness to assess the submitted chapters.

The book should be of interest to students, academics scientists, engineers from industry, and potential investors in the food science and engineering field. We hope you will enjoy reading the chapters presented in the book as much as we enjoyed writing and editing it.

Fuping Lu and Wenjie Sui

Editors

Article

Extraction, Structural Analysis, and Biofunctional Properties of Exopolysaccharide from *Lactiplantibacillus pentosus* B8 Isolated from Sichuan Pickle

Guangyang Jiang [1,2], Ran Li [3], Juan He [4], Li Yang [1,2], Jia Chen [1,2], Zhe Xu [1,2], Bijun Zheng [1,2], Yichen Yang [1,2], Zhongmei Xia [5] and Yongqiang Tian [1,2,*]

[1] College of Biomass Science and Engineering, Sichuan University, Chengdu 610065, China
[2] Key Laboratory of Leather Chemistry and Engineering, Ministry of Education, Sichuan University, Chengdu 610065, China
[3] College of Food Science, Sichuan Agricultural University, Ya'an 625014, China
[4] Key Laboratory of Bio-Resources and Eco-Environment of Ministry of Education, College of Life Sciences, Sichuan University, Chengdu 610065, China
[5] Institute of Biotechnology and Nucleic Technology, Sichuan Academy of Agricultural Sciences, Chengdu 610066, China
* Correspondence: yqtian@scu.edu.cn; Tel.: +86-028-85461102

Citation: Jiang, G.; Li, R.; He, J.; Yang, L.; Chen, J.; Xu, Z.; Zheng, B.; Yang, Y.; Xia, Z.; Tian, Y. Extraction, Structural Analysis, and Biofunctional Properties of Exopolysaccharide from *Lactiplantibacillus pentosus* B8 Isolated from Sichuan Pickle. *Foods* 2022, 11, 2327. https://doi.org/10.3390/foods11152327

Academic Editors: Fuping Lu and Wenjie Sui

Received: 8 July 2022
Accepted: 1 August 2022
Published: 4 August 2022

Publisher's Note: MDPI stays neutral with regard to jurisdictional claims in published maps and institutional affiliations.

Abstract: Two novel exopolysaccharides, named LPB8-0 and LPB8-1, were isolated and purified from *Lactiplantibacillus pentosus* B8. Moreover, their structure and bioactivities were evaluated through chemical and spectral means. The study results demonstrated that LPB8-0 was primarily composed of mannose and glucose and had an average molecular weight of 1.12×10^4 Da, while LPB8-1 was composed of mannose, glucose, and galactose and had an average molecular weight of 1.78×10^5 Da. Their carbohydrate contents were 96.2% ± 1.0% and 99.1% ± 0.5%, respectively. The backbone of LPB8-1 was composed of (1→2)-linked α-D-Man*p* and (1→6)-linked α-D-Man*p*. LPB8-0 and LPB8-1 had semicrystalline structures with good thermal stability (308.3 and 311.7 °C, respectively). SEM results displayed that both LPB8-0 and LPB8-1 had irregular thin-slice shapes and spherical body structures. Additionally, an emulsifying ability assay confirmed that LPB8-0 and LPB8-1 had good emulsifying activity against several edible oils, and this activity was retained under acidic, neutral, and high temperature conditions. Furthermore, an antioxidant assay confirmed that LPB8-1 had stronger scavenging activity than LPB8-0. Overall, these results provide a theoretical basis for the potential application of these two novel exopolysaccharides as natural antioxidants and emulsifiers in the food and pharmaceutical industries.

Keywords: exopolysaccharide; structural analysis; biofunctional properties; *Lactiplantibacillus pentosus*

1. Introduction

Exopolysaccharides (EPSs), produced by various microorganisms (bacteria, fungi, and microalgae) during their growth phases, are high-molecular-weight and structurally diverse groups of natural biomacromolecules [1]. EPSs have recently attracted considerable attention from many researchers because of their potential applications in various industries. Compared with artificial polymers, natural EPSs are almost inexhaustible polymers, as they are not dependent on external environmental conditions.

Lactic acid bacteria (LAB) have been generally recognized as safe (GRAS) by the FDA, and LAB-EPSs have also been recognized as safe agents. Previous studies have reported that LAB-EPSs from different sources possess multiple functional bioactive activities, such as antioxidation, anticancer, immunomodulatory, and prebiotic activities [2–4]. Unlike other polymers, LAB-EPSs have attracted increasing attention because of their unique physicochemical properties and comprehensive applications in some industrial fields, including food additives, pharmaceuticals, and wastewater treatment, as an emulsifying agent, coagulant, thickener, flocculant, etc. Indeed, the bioactive activities and physicochemical properties of EPSs have been verified to be closely associated with their chemical structures and complexity, including their monosaccharide composition, molecular weight (MW), linkage type, and substituent groups [5]. Consequently, a systemic characterization of the structure of various LAB-EPSs is of high significance to exploiting their functional properties and providing a theoretical basis for their future applications in various industries.

So far, numerous LAB-EPS-producing strains have been screened and identified from traditional fermented foods. For instance, LAB-EPSs from *Lactobacillus rhamnosus* ZFM231 [6], *Lactobacillus plantarum* CNPC003 [7], and *Streptococcus thermophilus* ZJUIDS-2-0 [8] have been reported to possess potential bioactivities and functional properties. As a representative of fermented vegetables, Sichuan pickle is a LAB-rich resource. Notably, *Lactiplantibacillus* spp. (gram-positive, facultatively anaerobic, non-motile, and non-spore-forming) plays an indispensable role in pickle fermentation. Additionally, the EPS from *Lactiplantibacillus plantarum* MM89 exhibited excellent immunomodulatory activity and hence could be regarded as a convenient additive or functional immunomodulatory agent for food products [3]. A new EPS-producing strain, *L. pentosus* B8, was recently screened from Sichuan pickle in our laboratory. So far, very few studies have verified the structural characteristics and biofunctional properties of the EPSs from *L. pentosus*.

The present work sought to isolate EPSs from *L. pentosus* B8. Furthermore, the structural characteristics, emulsifying activities, and antioxidant activities of these EPSs were assessed.

2. Materials and Methods

2.1. Bacterial Propagation

An EPS-producing strain was isolated from Sichuan pickle. Based on its morphological and physiological characteristics and 16S rDNA sequence analysis, the strain was identified as *L. pentosus* (GenBank accession number: MW898221) and named as strain B8. This strain was routinely grown on MRS broth at 30 °C and preserved in 30% (v/v) glycerol.

2.2. Crude EPS Extraction and Purification

The EPS of *L. pentosus* B8 was obtained according to a previously reported method, with minor modifications [9]. Briefly, *L. pentosus* B8 was inoculated in MRS medium supplemented with 40 g/L sucrose and incubated at 30.0 ± 0.1 °C for 48 h. Subsequently, the fermented broth was centrifuged ($12,000 \times g$, 10 min), and the supernatant was added to 95% (v/v) cold ethanol. The precipitates were resuspended in ultrapure water, and the protein was removed using an enzyme combined with the Savage method [10]. Briefly, papain (800 U/mL) was mixed with the precipitate solution (pH 6.0) and maintained at 55 °C (water bath) for 100 min, followed by thorough mixing with chloroform and n-butanol (4:1, v/v). Later, this mixture was centrifuged, dialyzed (MW: 8–14 kDa), and lyophilized to obtain the crude EPS. Using an anion-exchange chromatographic column (DEAE-52; GE Healthcare, Stockholm, Sweden), the sample was eluted with ultrapure water and three NaCl concentrations (0.1, 0.3, and 0.5 M NaCl). The fractions were collected from 4 mL aliquots per tube, and then the total sugar content was measured using the phenol–sulfuric acid method. The primary fractions were pooled, dialyzed, and then lyophilized. Afterward, the lyophilized EPS sample was redissolved in ultrapure water at a final concentration of 30 mg/mL. Further EPS purification was performed on Sephacryl S-300 HR and S-400 HR columns (1.6 cm × 90 cm; GE Healthcare, Stockholm, Sweden) and eluted with ultrapure water at a flow rate of 1 mL/min. Finally, the resulting fractions were pooled and freeze-dried for structural characterization and functional evaluation.

2.3. Chemical Composition Characterization Assays

The total sugar content of the EPSs was estimated by the phenol–sulfuric acid method by using glucose as a standard. The protein content was estimated using the bicinchoninic acid (BCA) method. The EPSs (1 mg/mL) were recorded in the range of 200–800 nm by using a UV–vis spectrophotometer (Hitachi, Tokyo, Japan) to evaluate the presence of protein or nucleic acid.

2.4. Determination of MW

The MWs of the EPSs were determined through high-performance size-exclusion chromatography (HPSEC) along with refractive index (RI) (Optilab T-Rex, Wyatt Technology, Santa Barbara, CA, USA) and multiangle laser light scattering (MALLS) detectors (DAWN HELEOS II, Wyatt Technology, Santa Barbara, CA, USA). The EPS samples were dissolved in 0.1 M $NaNO_3$ aqueous solution (5 mg/mL) and filtered through a filter (0.45 μm). The LPB8-0 and LPB8-1 solutions (5 mg/mL, 100 mL) were added to the HPSEC–RI–MALLS system. The EPSs were eluted with 0.1 mol/L $NaNO_3$ solution.

2.5. Monosaccharide Composition

The monosaccharides of LPB8-0 and LPB8-1 were determined through high performance anion-exchange chromatography with pulsed amperometric detection (HPAEC-PAD). The LPB8-0 and LPB8-1 samples (5.0 mg) were hydrolyzed with trifluoroacetic acid (2.0 M) at 121 °C for 2 h. Then, two hydrolysates were evaporated to dryness under an N_2 stream blowing instrument and eluted with methanol. The released monomers and all standards were further measured using a Thermo ICS-5000 ion chromatography system (Thermo Scientific, Waltham, MA, USA) fitted with a Dionex CarboPac PA-20 analytical column and a Dionex ED50A electrochemical detector.

2.6. Linkage Analysis by Methylation

The linkage analysis method of Nasir et al. [11] was used, with minor modifications. In total, 10 mg of the LPB8-1 sample was dissolved in anhydrous dimethyl sulfoxide and methylated with 0.5 mL methyl iodine. After methylation, the methylated sample was hydrolyzed with TFA, reduced with NaBD4, and acetylated to produce the derivative for a 7890A-5977B GC–MS system (Agilent Technologies, Palo Alto, CA, USA) equipped with an HP-5MS capillary column. The program was isothermal at 140 °C; the hold time was 2 min, with a temperature gradient of 3 °C/min up to a final temperature of 230 °C.

2.7. FTIR and NMR Spectroscopy Analysis

FTIR spectroscopy (Thermo Fisher Scientific, Waltham, MA, USA) was used to determine the major chemical groups of the EPSs. The IR spectra were recorded from 600 to 4000 cm^{-1}. Then, a 40 mg/mL solution of the sample was prepared with 99.9% D2O (500 μL) as the solvent. The 1D and 2D NMR data were recorded using a Bruker spectrometer (600 MHZ, Bruker, Karlsruhe, Switzerland).

2.8. Thermal Stability Evaluation

The thermal properties of the EPSs (LPB8-0 and LPB8-1) were measured by a thermogravimetric analyzer (209F3, NETZSCH, Free State of Bavaria, Germany). Approximately 5 mg of each of the EPS samples were placed in a standard aluminum pan and heated between 50 and 800 °C in a nitrogen atmosphere (10 °C/min).

2.9. Examination of X-ray Diffractometry, Particle Size, and Zeta Potential

The X-ray diffraction (XRD) data of LPB8-0 and LPB8-1 were analyzed using an X-ray diffractometer (D8 ADVANCE Bruker, Germany). The LPB8-0 and LPB8-1 samples were recorded at 2θ angles from 5 to 85° with a scanning rate of 10°/min. The size distributions and zeta potentials of the EPSs (0.5% *w/v*) were measured using a NanoPlus Zeta Potential and Particle Size Analyzer (ZEN5600, Malvern Instruments, Malvern, UK) at 25 °C. The LPB8-0 and LPB8-1 samples were dissolved in ultrapure water.

2.10. Scanning Electron Microscopy (SEM) and Atomic Force Microscopy (AFM)

The morphological characteristics of LPB8-0 and LPB8-1 samples were observed using a Apreo 2C SEM (Thermo Fisher Scientific, Waltham, MA, USA) at a voltage of 15.0 kV with magnification of 1000× and 5000×, respectively. Then, 10 μg/mL of EPS solution was deposited onto the surface of freshly mica plate and was allowed to air-dry. The tapping mode was used to acquire the topographies of LPB8-0 and LPB8-1 by using an atomic force microscope (TESPA-V2, Bruker, Billerica, MA, USA), and the AFM image size was selected as 8 μm × 8 μm.

2.11. Emulsifying Activities of the EPSs

The emulsifying activities of LPB8-0 and LPB8-1 were measured according to our previously reported methods [12]. Briefly, 3 mL of several edible oils (soybean oil, coconut oil, olive oil, corn oil, rap oil, peanut oil, palm oil, and sunflower oil) were added to 2 mL of the EPS solutions at a concentration of 1 mg/mL. Then, each sample was stirred for 5 min. Later, the effects of different EPS concentrations (0.1, 0.5, 1.0, 1.5, and 2.0 mg/mL), temperatures (25, 40, 60, 80, and 100 °C), and pH values (4.0, 6.0, 8.0, 10.0, and 12.0) on emulsion stability were assessed. The emulsifying activity (EA, %) was analyzed after 1, 24, and 168 h using the following equation:

$$\text{EA } (\%) = (\text{emulsion layer height}/\text{total height}) \times 100$$

2.12. Analysis of the Antioxidant Activities of LPB8-0 and LPB8-1 In Vitro

The antioxidant activities (DPPH radical scavenging ability, ABTS$^+$ free radical scavenging ability, and hydroxyl free radical scavenging ability) of LPB8-0 and LPB8-1 were evaluated according to the previously reported method [13].

2.12.1. DPPH Scavenging Activity

Briefly, approximately 100 μL of LPB8-0 and LPB8-1 sample solution of different concentrations were immersed in 10 mL DPPH solution. Then, the mixture was incubated for 30 min in darkness. The absorbance was recorded at 517 nm on a UV–visible spectrophotometer (PerkinElmer, Waltham, MA, USA). The DPPH radical scavenging ability of samples was calculated using the following formula:

$$\text{scavenging ability } (\%) = (1 - A_{sample})/A_{blank} \times 100$$

where A_{sample} and A_{blank} represent the absorbances of the test and control samples, respectively.

2.12.2. ABTS Radical Scavenging Activity

For ABTS radical scavenging activity, equal volumes of ABTS solution (7 mmol/L) and potassium persulfate solution (2.45 mmol/L) were mixed and placed at room temperature overnight. Then, the stock solution was added to various concentrations of LPB8-0 and LPB8-1 samples. The reaction mixture was reacted at 37 °C for 30 min. The absorbance of the samples was measured at 732 nm, and the ABTS radical scavenging activity was determined as follows:

$$\text{scavenging ability } (\%) = (1 - A_{sample})/A_{blank} \times 100$$

where A_{sample} is the absorbance of a sample and A_{blank} is the absorbance of a blank.

2.12.3. Hydroxyl Free Radical Scavenging Activity

Then, 1.0 mL of LPB8-0 and LPB8-1 samples at different concentrations were blended with FeSO$_4$ (2 mmol/L) and salicylic acid—ethanol solution (6 mmol/L), and then, H$_2$O$_2$

solution (0.5 mL, 9 mmol/L) was added to the mixed solutions before incubation at 37 °C for 30 min. Subsequently, the absorbance of the mixtures was immediately tested at 536 nm. The scavenging activity of LPB8-0 and LPB8-1 samples was calculated as follows:

$$\text{scavenging ability } (\%) = 1 - (A_2 - A_1)/A_0 \times 100$$

where A_0, A_1, and A_2 represent the absorbance of the reagent blank, control, and samples, respectively.

2.13. Statistical Analysis

Statistical analysis and graph plotting were performed using the SPSS 20.0 (IBM Co., Armonk, NY, USA) and Origin 9.0.0 software (Origin Lab Co., Northampton, MA, USA), and values are expressed as means $\pm$ standard deviations.

3. Results and Discussion

3.1. Extraction, Purification, and Chemical Composition of the EPSs

The crude EPSs of *L. pentosus* B8 were harvested through a series of processing steps, including ethanol precipitation, deproteinization, dialysis, and lyophilization. The EPS production yield was 1401.52 $\pm$ 9.54 mg/L. It was first isolated using a DEAE-52 anion-exchange column, and two fractions (designated as LPB8-0 and LPB8-1, respectively) displayed on the elution profile were obtained (Figure 1A). Then, LPB8-0 and LPB8-1 were purified through Sephacryl S-300 HR and S-400 HR gel-filtration columns, respectively. As depicted in Figure 1B,C, the elution profiles of the two fractions appeared with only one single peak each, showing that LPB8-0 and LPB8-1 were homogeneous. Further dialysis and lyophilization produced LPB8-0 and LPB8-1 fractions with a purity of 96.2% $\pm$ 1.0% and 99.1% $\pm$ 0.5%, but sulfate was not detected in the samples (Table 1). Compared with the crude EPS, no obvious absorption appeared at 260 or 280 nm in the UV–vis spectra (Figure 2A), indicating no nucleic acids or proteins in LPB8-0 and LPB8-1.

Table 1. Chemical compositions and basic properties of LPB8-0 and LPB8-1.

Factions	Carbohydrate Content (%)	Protein Content (%)	Sulfates (%)	Mw (Da)
LPB8-0	96.2 $\pm$ 1.0%	Nd *	Nd *	1.12×10^4
LPB8-1	99.1 $\pm$ 0.5%	Nd *	Nd *	1.78×10^5

* Not detected.

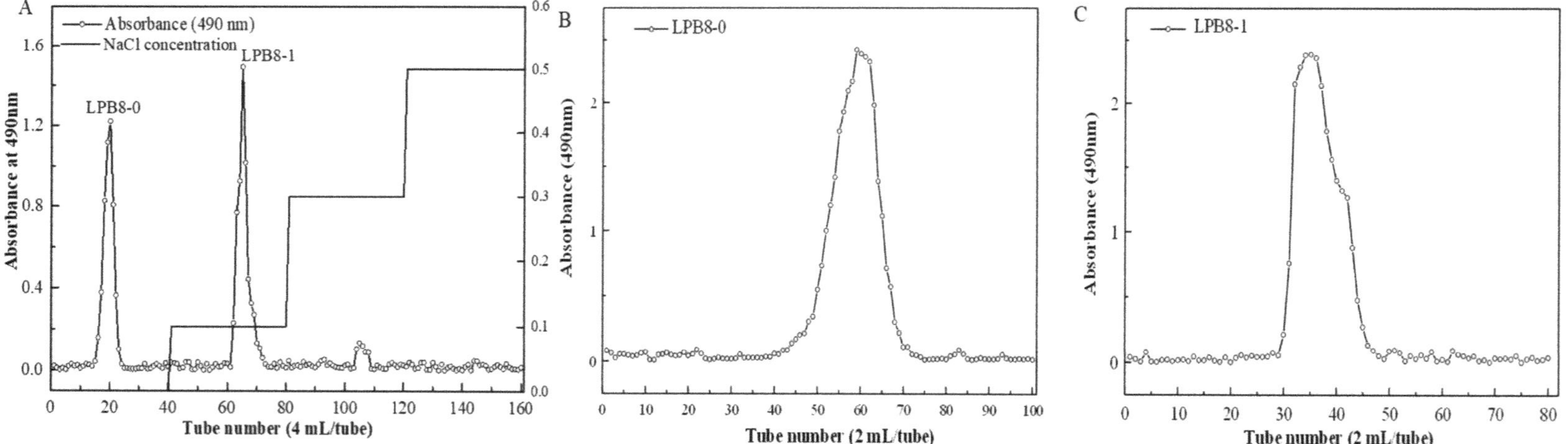

Figure 1. (A) DEAE-52 anion-exchange chromatogram; (B) Sephacryl S-300 HR chromatographic profile of LPB8-0; (C) Sephacryl S-400 HR chromatographic profile of LPB8-1.

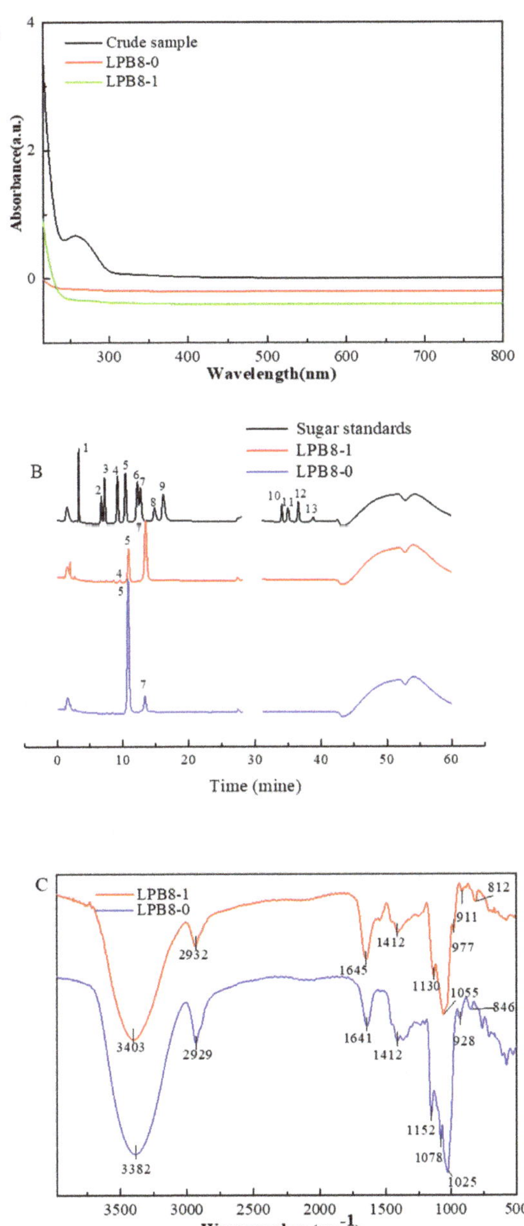

Figure 2. (**A**) UV−vis absorption spectra of crude EPS (black curve), LPB8-0 (red curve), and LPB8-1 (green curve); (**B**) HPAEC-PAD profiles of monosaccharide standards (black curve, peak identities: 1, fucose; 2, rhamnose; 3, arabinose; 4, galactose; 5, glucose; 6, xylose; 7, mannose; 8, fructose; 9, ribose; 10, galacturonic acid; 11, guluronic acid; 12, glucuronic acid; 13, mannuronic acid), LPB8-0 (blue curve) and LPB8-1 (red curve); (**C**) FTIR spectra of LPB8-0 (blue curve) and LPB8-1 (red curve).

3.2. MW Distribution and Monosaccharide Composition Analysis of the EPSs

The MWs of LPB8-0 and LPB8-1 were tested using an HPSEC–RI–MALLS system. As tabulated in Table 1, the MWs of LPB8-0 and LPB8-1 were 1.12×10^4 and 1.78×10^5 Da, respectively. This phenomenon indicated that one strain could produce EPSs of different MWs. These results were consistent with those of a previous study reporting LAB-EPSs in the MW range of 10^4–10^6 g/mol [14].

The monosaccharide composition of LPB8-0 and LPB8-1 was detected using the HPAEC-PAD system (Figure 2B). LPB8-0 was primarily composed of approximately 15.76% mannose and 84.24% glucose, while LPB8-1 was composed of mannose, glucose, and galactose at the molar ratios of 77.74%, 21.08%, and 1.18%, respectively. Notably, no uronic acid was found in LPB8-0 and LPB8-1. The yield, total carbohydrate content, and monosaccharide composition of LPB8-1 were higher than those of LPB8-0, which was used in the subsequent experiments.

3.3. Functional Groups and Glycosidic Linkages

As they represent the most important and reliable analytical method, FTIR absorption spectra were employed to analyze the major characteristic peaks and linkage bonds in the samples. As depicted in Figure 2C, the obvious peaks at 3382 (LPB8-1) and 3403 (LPB8-0) cm^{-1} represented the O–H stretching vibration. The peaks at 2932 (LPB8-1) and 2929 (LPB8-0) cm^{-1} were caused by the stretching vibration of C–H. The absorption peaks at 1645 (LPB8-1) and 1641 (LPB8-0) cm^{-1} might have been due to the associated water, and the peaks at 1412 cm^{-1} could be assigned to the bending vibration of C–OH in this sample [15]. Notably, the characteristic absorption bands in the range of 900–1200 cm^{-1} were attributed to the stretching vibrations of C–O–C and C–O–H [16]. Weak peaks were observed from 800 to 900 cm^{-1}, indicating the presence of α- and β-configurations.

LPB8-1 was subjected to methylation treatment followed by GC–MS analysis to confirm the types of glycosidic linkages of monosaccharide residues. The results are summarized in Table 2 and Figure S1. The reduced LPB8-1 exhibited ten types of linkages, including T-Man*p*-(1→, →3)-Man*p*-(1→, →2)-Man*p*-(1→, →6)-Man*p*-(1→, →6)-Glc*p*-(1→, →4)-Glc*p*-(1→, →3,6)-Man*p*-(1→, →2,6)-Man*p*-(1→, →2,6)-Glc*p*-(1→, and →2,3,6)-Man*p*-(1→ with molar ratios of 36.00:11.94:17.31:2.58:2.93:11.11:0.97:10.10:6.74:0.32, respectively.

Table 2. Glycosidic linkage composition of methylated LPB8-1 by GC–MS analysis.

Time (min)	Methylated Sugars	Deduced Linkages	Molar Ratios
8.9	2,3,4,6-Me$_4$-Man*p*	T-Man*p*-(1→	36.00
12.3	2,4,6-Me$_3$-Man*p*	→3)-Man*p*-(1→	11.94
12.4	3,4,6-Me$_3$-Man*p*	→2)-Man*p*-(1→	17.31
13.6	2,3,4-Me$_3$-Man*p*	→6)-Man*p*-(1→	2.58
13.7	2,3,4-Me$_3$-Glc*p*	→6)-Glc*p*-(1→	2.93
14.1	2,3,6-Me$_3$-Glc*p*	→4)-Glc*p*-(1→	11.11
17.9	2,4-Me$_2$-Man*p*	→3,6)-Man*p*-(1→	0.97
18.2	3,4-Me$_2$-Man*p*	→2,6)-Man*p*-(1→	10.10
18.2	3,4-Me$_2$-Glc*p*	→2,6)-Glc*p*-(1→	6.74
21.4	4-Me-Man*p*	→2,3,6)-Man*p*-(1→	0.32

3.4. NMR Analysis

A deep structural characterization of LPB8-1 was investigated with 1D (1D-1H, 1D-13C) and 2D (H-^{1}H COSY, TOCSY, and NOESY; ^{1}H-^{13}C HSQC and HMBC) NMR data to obtain more comprehensive information about the linkages between the glycosyl residues and signal assignments. As depicted in the ^{1}H NMR spectrum (Figure 3A), a group of the predominant signals (H-1) ranging from δ 4.52 to 5.52 ppm was observed. Intensive anomeric carbon signals (C-1) were located at δ 99.30–104.69 ppm in the ^{13}C NMR spectrum (Figure 3B). The broad signals between δ 3.30 and 4.33 ppm were related to H-2–H-6 signals, and the corresponding carbon signals at δ 61.42–80.99 ppm were the characteristic peaks of

C-2–C-6. A total of eleven anomeric protons (5.43, 5.33, 5.19, 5.18, 5.16, 5.13, 5.09, 5.08, 5.05, 4.93, and 4.56 ppm) appeared in the ^{1}H spectrum, which are labelled with A–K, respectively. The chemical shifts of anomeric proton and carbon signals were confirmed through the 2D NMR (COSY, HSQC, and HMBC) spectrum analysis.

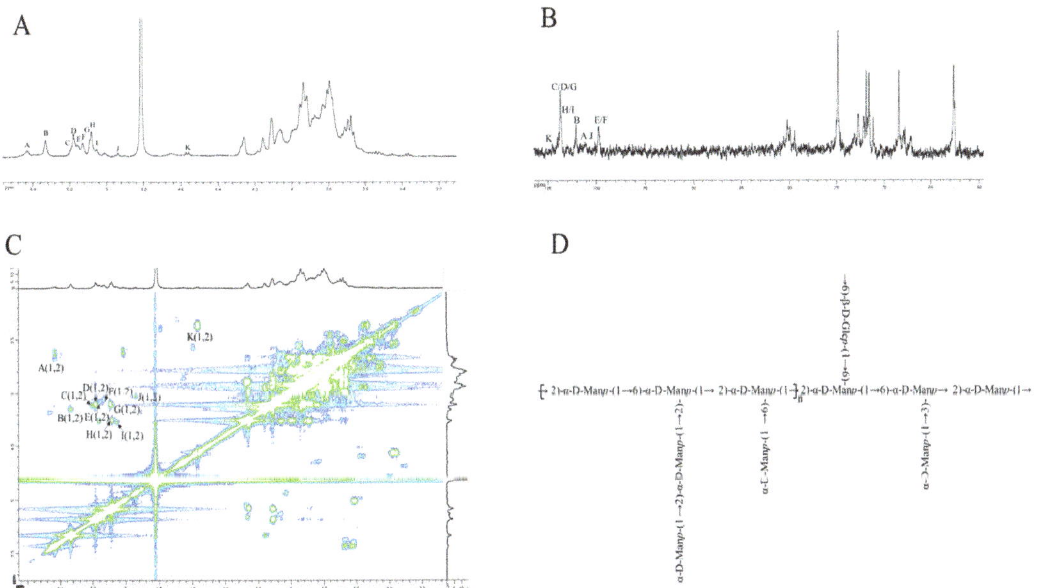

Figure 3. NMR spectra of LPB8-1 recorded in D$_2$O at 298 K: (**A**) 1D-1H spectrum; (**B**) 1D-13C spectrum; (**C**) COSY spectrum; (**D**) one putative structure of LPB8-1.

Based on the eleven sugar signals, a representative signal of residue B was extensively investigated. Residue B had a relatively intensive anomeric proton signal at δ 5.33 ppm, and its corresponding anomeric carbon signal at δ 102.29 ppm was successfully obtained through HSQC spectrum analysis (Figure S2A), showing the presence of an α-configuration. As shown in Figures 3C and S2B, further combining of the COSY and TOCSY spectral data revealed that other proton signals at δ 5.33/4.16 ppm, 4.16/3.96 ppm, 3.96/3.79 ppm, 3.79/3.92 ppm, and 3.92/3.80 (3.66) ppm were mainly caused by the H-1/H-2, H-2/H-3, H-3/H-4, H-4/H-5, and H-5/H-6, respectively, of residue E. Hence, the proton signals of residue B occurred at δ 4.16 ppm, 3.96 ppm, 3.79 ppm, 3.92 ppm, and 3.80 (3.66) ppm for H-2–H-6, respectively. According to the HSQC spectrum of LPB8-1, the strong cross-peaks of 5.33/102.29, 4.16/79.80, 3.96/71.68, 3.79/68.01, 3.92/74.49, and 3.80 (3.66)/62.38 were attributed to the H-1/C-1, H-2/C-2, H-3/C-3, H-4/C-4, H-5/C-5, and H-6/C-6, respectively. Compared with a previous study, the C-2 (δ 79.80 ppm) moved downward, showing that it was substituted at the position of C-2 [17,18]. Hence, residue B was identified as →2)-α-D-Man*p*-(1→. Subsequently, the signal changes of other residues were identified and deduced using the same method, and the detailed signal assignments of protons and carbon are summarized in Table 3 [19–22].

Afterward, the correlations of the linkage sites/types and sequences between different sugar residues in LPB8-1 were evaluated, and the NOESY and HMBC spectra were analyzed (Figure S2C,D). In the HMBC spectrum, the cross-peak between the H-1 of residue B (δ 5.33 ppm) and the C-5 of residue F (δ 74.57 ppm) indicated that H-1 was related to O-6 and O-2, suggesting the existence of →2)-α-D-Man*p*-(1→2,6)-α-D-Man*p*-(1→. Similarly, the H-1 of residue D (δ 5.18 ppm) was found to be linked to the O-6 of residue F (δ 74.57 ppm), suggesting the presence of α-D-Man*p*-(1→2,6)-α-D-Man*p*-(1→. The H-1 signals of α-D-

Man$p$-(1→ (G) at δ 5.09 ppm corresponded to the C-2 signal of →2)-α-D-Man$p$-(1→(B) at δ 79.80 ppm, demonstrating the linkage α-D-Man$p$-(1→2)-α-D-Man$p$-(1→. In the NOESY spectrum, the correlations at 5.09/4.16, 5.08/4.16, and 4.56/3.78 were assigned to G (H-1)/H (H-2), H (H-1)/B (H-2), and K (H-1)/E (H-6), respectively. These related signals confirmed the existence of α-D-Man$p$-(1→2)-α-D-Man$p$-(1→, →2,6)-α-D-Man$p$-(1→3,6)-α-D-Man$p$-(1→, and →6)-β-D-Glc$p$-(1→2,6)-α-D-Man$p$-(1→, respectively. According to the NMR spectrum analysis, a possible structural unit of LPB8-1 was predicted and is depicted in Figure 3D.

Table 3. ^{1}H and ^{13}C NMR chemical shift data for LPB8-1.

Sugar Residue	Chemical Shifts δ (ppm)						
	H-1/C-1	H-2/C-2	H-3/C-3	H-4/C-4	H-5/C-5	H-6/C-6	H-6′
A: →4)-α-D-Glc$p$-(1→	5.43/101.28	3.66/72.05	4.01/72.64	3.67/73.08	3.83/74.25	3.93/61.84	3.69
B: →2)-α-D-Man$p$-(1→	5.33/102.29	4.16/79.80	3.96/71.68	3.79/68.01	3.92/74.49	3.80/62.38	3.66
C: →3)-α-D-Man$p$-(1→	5.19/103.62	4.11/71.36	3.92/74.74	3.68/68.24	3.85/74.64	3.94/62.46	3.70
D: α-D-Man$p$-(1→	5.18/103.54	4.10/71.29	3.91/72.07	3.67/68.24	3.82/74.72	3.93/62.23	3.69
E: →2,6)-α-D-Man$p$-(1→	5.16/99.72	4.07/80.27	3.80/71.91	3.70/68.55	3.83/74.57	3.92/68.08	3.78
F: →2,6)-α-D-Man$p$-(1→	5.13/99.64	4.06/80.19	3.80/71.76	3.71/68.63	3.82/74.62	3.93/68.32	3.78
G: α-D-Man$p$-(1→	5.09/103.39	4.10/71.44	3.91/71.68	3.68/68.32	3.81/74.57	3.92/62.46	3.69
H: →3,6)-α-D-Man$p$-(1→	5.08/103.62	4.28/71.05	3.89/75.05	3.78/67.85	3.95/74.72	3.84/68.01	3.79
I: →6)-α-D-Manp-(1→	5.05/103.54	4.28/71.21	3.97/71.76	3.79/67.85	3.95/74.41	3.82/68.16	3.77
J: →2,6)-β-D-Glc$p$-(1→	4.93/101.04	4.03/79.72	3.65/71.32	3.91/70.82	3.80/74.18	3.91/68.24	3.71
K: →6)-β-D-Glc$p$-(1→	4.56/104.56	3.37/74.64	3.54/77.14	3.67/69.26	3.78/74.80	3.97/67.77	3.62

3.5. Crystalline Features and Thermal Stability

XRD, a technique providing crystallinity information, has been widely used to examine the amorphous and crystalline structure of polysaccharides. As depicted in Figure 4A, LPB8-0 and LPB8-1 exhibited major crystalline reflections at 18.69° and 19.17°, respectively, confirming that they were semicrystalline polymers with low crystallinity. This was consistent with previous study results showing similar crystalline structures for the ESPs from *Bacillus licheniformis* PASS26 [23] and *B. cereus* KMS3-1 [24]. This special semicrystalline structure could be attributed to the order degree within the polysaccharides, which might directly affect the physical and functional properties of the polysaccharides, such as viscosity, solubility, water holding capacity, tensile strength, and swelling power.

Figure 4B depicts the thermal stability analysis results for LPB8-0 and LPB8-1, displaying two stages. In the first phase, an initial mass loss of approximately 5% was observed between 35 and 115 °C, which might have been ascribed to the evaporation and desorption of water. However, the mass remained unchanged from 115 to 240 °C, indicating that LPB8-0 and LPB8-1 were relatively stable below 240 °C. With the temperature increasing, the maximum mass loss was observed from 240 to 550 °C with a mass loss of 75%. This might have been due to the depolymerization of LPB8-0 and LPB8-1. Sharp peaks were observed on the DTG curves at 308.3 and 311.7 °C highlighting the high degradation temperature (Td) of LPB8-0 and LPB8-1, respectively, and especially of LPB8-1. The masses gradually decreased to 24.94% and 14.98% in LPB8-0 and LPB8-1, respectively, with an increase in temperature to 800 °C. The Td values of LPB8-0 and LPB8-1 were higher than those of the EPSs from *Leuconostoc pseudomesenteroides* DRP-5 (298.81 °C) [25] and *Lactobacillus sakei* L3 (272 °C) [26]. These results led to the conclusion that LPB8-0 and LPB8-1 had significant thermal stability, which is vital for food industries requiring a high level of thermal processing.

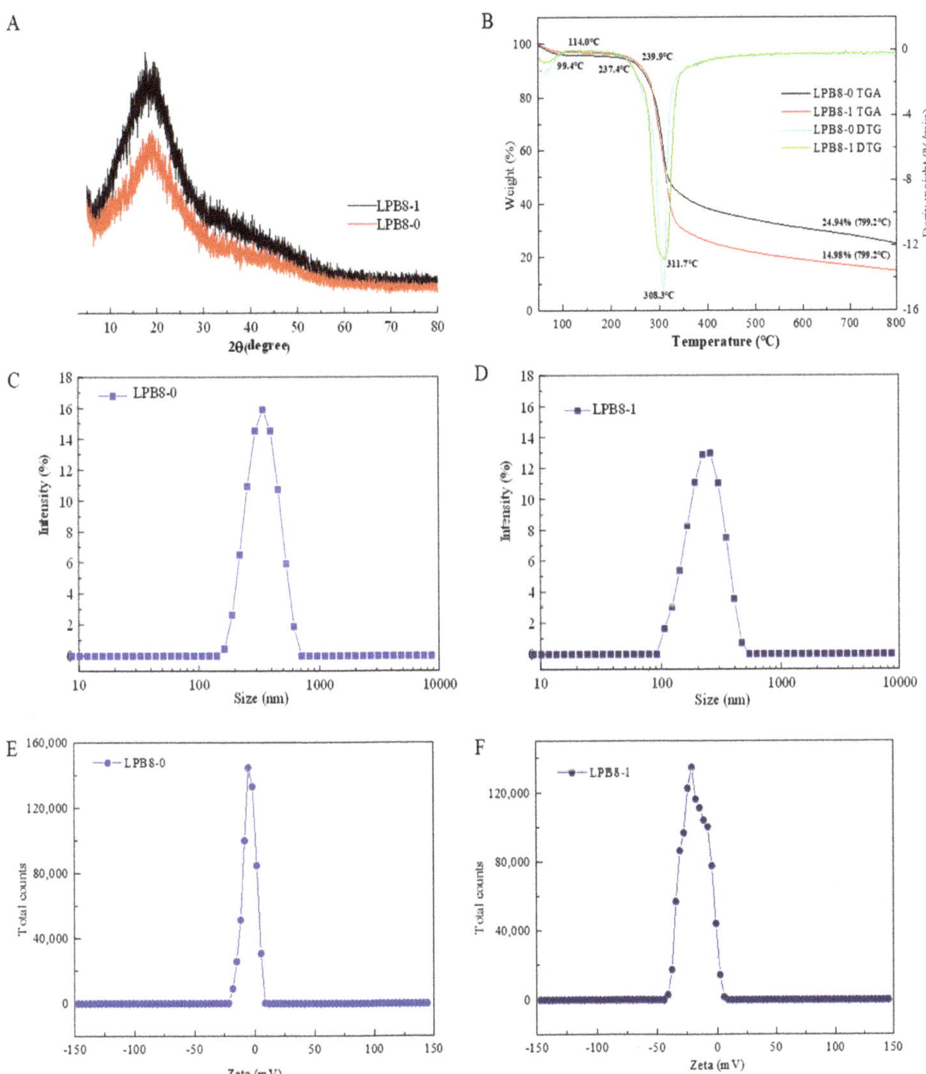

Figure 4. The XRD (**A**) and TGA (**B**) spectra of LPB8-0 and LPB8-1; the zeta potential (**C,D**) and particle size (**E,F**) at various concentrations of LPB8-0 and LPB8-1.

3.6. Particle Size and Zeta Potential Examination

Figure 4C,D presents the particle sizes (nm) of LPB8-0 and LPB8-1 (0.5% *w/v*) in aqueous solution. LPB8-0 and LPB8-1 possessed uniform particles, and the size of LPB8-0 particles (Figure 3C) ranged from 150 to 550 nm, with an average particle size of 286.3 ± 4.1 nm, while the size of LPB8-1 particles (Figure 3D) ranged from 100 to 450 nm, with an average particle size of 254.5 ± 2.3 nm. These particle sizes were smaller than those of the EPSs from *Lactobacillus plantarum* C70 (525.5 nm) [19] and *Pediococcus pentosaceus* M41 (446.8 nm) [27] but larger than that of the EPS produced by *Lactococcus garvieae* C47 (166.6 nm) [28]. The differences in particle size might be ascribed to the MW, linkages, and chemical compositions of the EPSs in question [19].

The zeta potentials (mV) of LPB8-0 and LPB8-1 solutions were detected, and the charges of LPB8-0 and LPB8-1 were -4.31 ± 0.75 mV and -18.3 ± 1.46 mV, respectively (Figure 4E,F). Compared with neutral EPSs, EPSs with negative charges exhibited stronger bioactivity and could alter the process of gel formation, thereby helping to improve the elasticity of the protein network [27]. Additionally, the physicochemical environment (pH, temperature), the range of MW, and charge density might affect the zeta potential [29].

3.7. Morphology Characteristics

As shown in Figure 5A,B, LPB8-0 and LPB8-1 appeared as a soft cotton-like structure. SEM revealed different surface morphologies of LPB8-0 and LPB8-1 (Figure 5C–F) at 1000× and 5000× magnifications. LPB8-0 exhibited an uneven surface primarily composed of an irregular, thin-sliced filiform structure, while the surface morphology of LPB8-1 was a branched, irregular, rod-like spherical body structure. The EPS produced by *Lactobacillus plantarum* HY [16] had a three-dimensional network structure combining sheets and tubes. The EPS obtained from *B. licheniformis* AG-06 [30] exhibited uneven smooth surfaces with network-like structures composed of irregular chains. The different shapes and structures might be attributable to the purification, preparation, and monosaccharide composition of the EPSs [31].

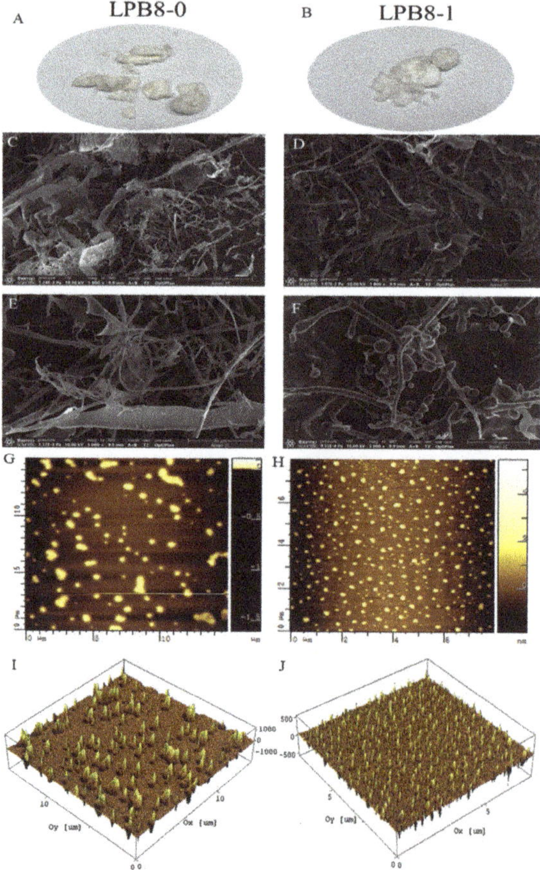

Figure 5. The appearances of LPB8-0 (**A**) and LPB8-1 (**B**); SEM images of LPB8-0 and LPB8-1 under 1000× (**C,D**) and 5000× (**E,F**) magnification; AFM showing the topographic features of LPB8-0 and LPB8-1, planar view (**G,H**) and cubic view (**I,J**).

AFM images of LPB8-0 and LPB8-1 are depicted in Figure 5G–J. LPB8-0 was irregularly agglomerated in the dilute solution, suggesting molecular aggregation, while LPB8-1 was uneven in size and exhibited an irregularly spherical molecular conformation, which might have been ascribed to its branched structure and molecular aggregation through hydrogen bonds. This phenomenon was consistent with the SEM results. A similar spherical molecular conformation was found in the EPSs produced by other probiotic bacteria, such as *Lactobacillus plantarum* YW11 [32] and *B. megaterium* PFY-147 [33].

3.8. Emulsifying Activity and Emulsion Stability of the EPSs

3.8.1. Emulsifying Activities with Several Edible Oils

The emulsifying activity and stability of the prepared emulsions were evaluated against several edible oils. As depicted in Figure 6A,B, after 1 h, LPB8-0 and LPB8-1 showed good EAs (100% $\pm$ 0.0%) against palm oil and olive oil, respectively. The LPB8-0 and LPB8-1 produced by *L. pentosus* B8 were found to possess good emulsion-stabilizing ability against several edible oils, as shown by the EA24 value, retaining at least 50% of the initial volume. An effective emulsifier maintains its capacity to the total emulsion volume 24 h ($\geq$50%) after its formation [34]. Notably, the EA values (70.9% $\pm$ 1.2% and 79.9% $\pm$ 0.8%) of LPB8-0 and LPB8-1 against palm oil and olive oil remained high after 168 h. Compared with other EPS biopolymers, LPB8-0 and LPB8-1 from *L. pentosus* B8 had higher EA values. In previous reports, the EPS produced by *Virgibacillus salarius* BM02 had an EA24 value of 47.06% against olive oil [35]. By contrast, the EPS of *Bacillus coagulans* RK-02 had the highest emulsifying efficiency against sunflower seed oil (70%) than against soybean oil (62%) and rice oil (41%), at a similar concentration [36]. This phenomenon might be attributable to the specificity of the EPSs and certain hydrophobic compounds [35].

3.8.2. Effects of EPS Concentration, pH, and Temperature on Emulsifying Activity

As an important criterion for many fields, an emulsifier should retain its stability when affected by various complicated conditions (concentration, pH, and temperature). As depicted in Figure 6C,D, the EA values significantly increased ($p < 0.05$) with increasing LPB8-0 and LPB8-1 concentrations (0.1–1 mg/mL). However, the emulsion showed good stability ($p > 0.05$) when it was exploited with LPB8-0 and LPB8-1 concentrations ranging from 1 to 2 mg/mL. Hence, 1 mg/mL was considered as the optimum concentration for LPB8-0 and LPB8-1 when used as an emulsifier. A similar concentration was observed for the EPS from *Bacillus amyloliquefaciens* LPL061 [37]. The emulsions of LPB8-0 and LPB8-1 against palm oil and olive oil exhibited excellent stability under different pH environments (Figure 6E). Nevertheless, a contradictory result, that is, a decrease in the EA24 value, was discovered under alkaline environments of pH 10 and 12. This result indicates that LPB8-0 and LPB8-1 could be potential emulsifiers under different pH environments. In a previous study, emulsions prepared using EPS22 and olive oil exhibited lower emulsion activity in the storage pH range of 10 to 12 (alkaline conditions) [34]. As depicted in Figure 6F, two emulsions retained good emulsifying activity in the temperature range of 25–100 °C. Another study reported that the activity of emulsions prepared with the EPS from *Virgibacillus salarius* BM02 and sunflower oil decreased to 47.06% at 100 °C [35]. These differences might be attributable to the excellent thermal stability of LPB8-0 and LPB8-1 at 308.3 and 311.7 °C, respectively, at pH 10 and 12.

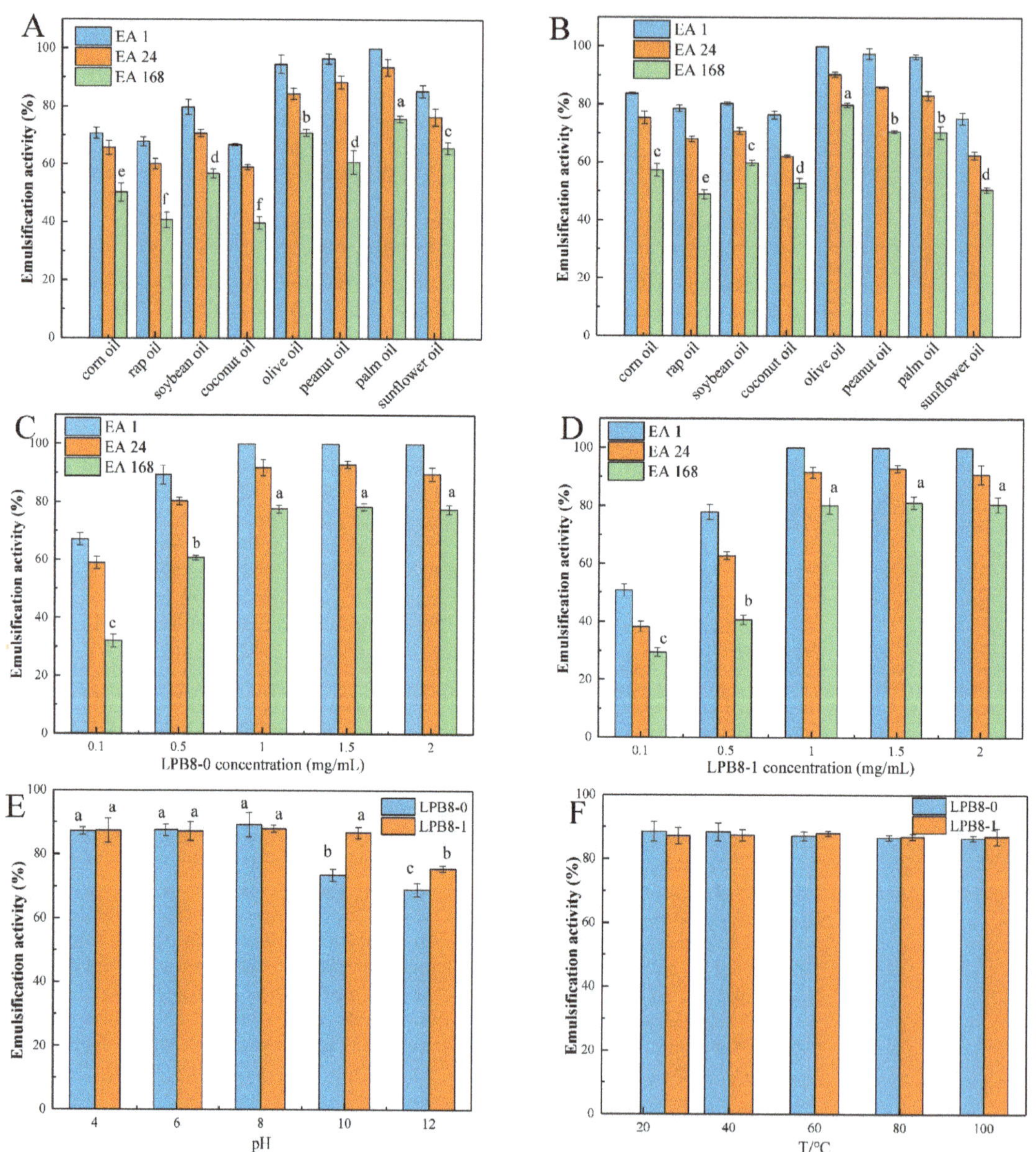

Figure 6. Emulsification activity of emulsions prepared with different oils (**A**,**B**) and different LPB8-0 and LPB8-1 concentrations (**C**,**D**); influence of pH (**E**) and temperature (**F**) on the emulsification activity of the emulsions. Statistical differences are indicated with different lowercase letters ($p < 0.05$).

3.9. Antioxidant Activities of LPB8-0 and LPB8-1

3.9.1. DPPH Radical Scavenging Activities of LPB8-0 and LPB8-1

The scavenging abilities of LPB8-0 and LPB8-1 on DPPH free radicals are shown in Figure 7A. These two fractions exhibited satisfactory DPPH scavenging ability, and a positive correlation was observed between DPPH scavenging ability and the concentrations of LPB8-0 and LPB8-1. LPB8-1 exhibited higher scavenging ability than LPB8-0 at the concentrations of 0–10 mg/mL. When the concentrations reached 10 mg/mL, the DPPH radical scavenging rates of LPB8-0 and LPB8-1 were 50.62% ± 1.5% and 62.82% ± 2.0%, respectively. However, the scavenging abilities of these two fractions were significantly weaker than that of the control vitamin C (VC) ($p < 0.05$). Compared with LPB8-1, an EPS isolated from *Lactobacillus plantarum* CNPC003 exhibited a lower DPPH scavenging activity of 51.52 ± 1.10% at the concentration of 8 mg/mL [7].

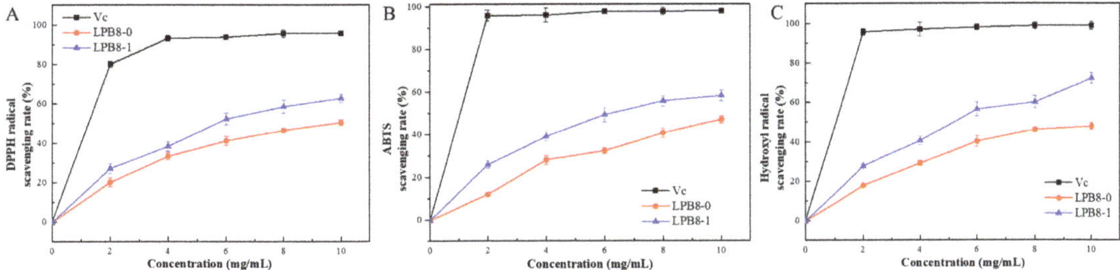

Figure 7. The antioxidant activities of LPB8-0 and LPB8-1 using VC as a control: scavenging activity towards DPPH (**A**); ABTS radical (**B**); hydroxyl radical (**C**).

3.9.2. ABTS Radical Scavenging Activities of LPB8-0 and LPB8-1

Figure 7B depicts the ABTS free radical scavenging activities of LPB8-0 and LPB8-1. The ABTS free radical scavenging activities of these two EPS fractions were considerably lower than that of the VC, exhibiting a concentration-dependent activity. Compared with LPB8-1, LPB8-0 exhibited stronger scavenging activities against ABTS in the range of 0–10.0 mg/mL. This trend was similar to that of the DPPH free radical scavenging ability. At a concentration of 10 mg/mL, the scavenging activities of LPB8-0 and LPB8-1 against ABTS increased to 47.17% ± 1.7% and 58.36% ± 2.4%, respectively. In a previous study, the EPS from *Lactobacillus plantarum* JLAU103 showed an ABTS radical scavenging rate of 65.5% at a concentration of 10 mg/mL [38].

3.9.3. Hydroxyl Free Radical Scavenging Abilities of LPB8-0 and LPB8-1

As depicted in Figure 7C, LPB8-0 and LPB8-1 exhibited a significant, concentration-dependent hydroxyl free radical scavenging activity in the concentration range of 0–10 mg/mL. Their scavenging activities were in the following order: VC > LPB8-1 > LPB8-0. The hydroxyl free radical scavenging capacities of 10 mg/mL LPB8-0 and LPB8-1 were 47.91% ± 1.7% and 72.52% ± 2.7%, respectively. Their hydroxyl free radical scavenging activities were stronger than those of the EPSs from *Lactobacillus rhamnosus* ZFM231 (49.7%) [6] and *Leuconostoc mesenteroides* DRP105 (30.48% ± 0.78%) [39]. Overall, these results indicated that these two polymers (LPB8-0 and LPB8-1) could be potential alternatives to synthetic antioxidants, especially LPB8-1.

3.9.4. Correlation between Structure and Antioxidant Activity

Research fronts have proven that polysaccharides possess excellent antioxidant efficacy, but their antioxidant mechanism is still obscure. Zhang et al. [40] discovered that the antioxidant activity pathway of polysaccharides might be ascribed to the presence of OH groups that could improve hydrogen to stabilize the free radicals or directly react with the free radicals to terminate the radical chain reaction. The antioxidant abilities of polysaccharides largely depend on their purity, monosaccharide compositions, MWs, amounts and positions of functional groups, glycosidic linkages, and chain conformations. The purities of LPB8-0 and LPB8-1 were 96.2% ± 1.0% and 99.1% ± 0.5%, respectively. However, LPB8-1 exhibited better antioxidant activity than LPB8-0. Jiang et al. [41] reported that the polysaccharide (MBP-2) extracted from mung bean skin with higher carbohydrate content (89.26%) exhibited stronger antioxidant activity than that with lower carbohydrate content (MBP-1;

86.69%). These results were in agreement with those reported by Xie et al. [42]. Moreover, the antioxidant properties of the polysaccharides studied herein were closely related to their MWs. In this study, LPB8-1, with a high MW, exhibited better antioxidant properties, indicating that EPSs with higher MWs possess stronger antioxidant activities, which would be consistent with the results of a previous study [43]. However, polysaccharides produced by *Sagittaria sagittifolia* L. with lower MWs (16.62 kDa) had better antioxidant abilities, which might have been due to the fact that low-MW polysaccharides could accept more free radicals [44]. The inconsistency in the aforementioned results indicates that MW does not affect the antioxidant properties of polysaccharides. An et al. [45] reported the significant effects of monosaccharide composition on antioxidant activity. In previous studies, some monosaccharides (mannose, galactose, xylose, rhamnose, arabinose, etc.), the major constituents of polysaccharides, exhibited satisfying antioxidant activity [46,47]. Thus, the high proportions of mannose and galactose in LPB8-1 might have contributed to its high antioxidant efficacy. Furthermore, Feng et al. [48] found that the antioxidant activity of polysaccharides might be governed by their functional groups (hydroxyl and carboxyl) and chain conformation. In this study, we inferred that the antioxidant activity of LPB8-0 and LPB8-1 might have depended not on one factor but rather on a combination of multiple factors (high purity, high MW, and high mannose and galactose contents). However, further exploration is highly recommended to explore the detailed antioxidant mechanism of EPSs produced by LAB.

4. Conclusions

In this study, two novel EPSs (LPB8-0 and LPB8-1) were isolated and purified from *L. pentosus* B8. The MWs, monosaccharide compositions, and surface morphologies of LPB8-0 and LPB8-1 were slightly different, with typical absorption peaks of polysaccharides. The backbone of LPB8-1 was composed of →2)-α-D-Man$p$-(1→ and →6)-α-D-Man$p$-(1→. The results demonstrated that LPB8-0 and LPB8-1 were semicrystalline polymers with outstanding thermal stability. LPB8-0 and LPB8-1 in aqueous solution had average particle diameters of 286.3 ± 4.1 nm and 254.5 ± 2.3 nm and negative charges of −4.31 ± 0.75 mV and −18.3 ± 1.46 mV, respectively. Both LPB8-0 and LPB8-1 showed excellent emulsifying activities against several edible oils, especially palm oil and olive oil. Additionally, DPPH, ABTS, and hydroxyl free radical assays showed that LPB8-0 and LPB8-1 exhibited potential antioxidant activities in vitro in a concentration-dependent manner. LPB8-1, with high purity and a high MW, had better antioxidant ability. Overall, the study results provide a theoretical basis for the potential applications of LBP8-1 as an emulsifier and antioxidant in foods and pharmaceuticals.

Supplementary Materials: The following supporting information can be downloaded at: https://www.mdpi.com/article/10.3390/foods11152327/s1, Figure S1: Methylation analysis results of LPB8-1; Figure S2: HSQC spectrum (A), TOCSY spectrum (B), NOESY spectrum (C), HMBC spectrum (D) of LPB8-1.

Author Contributions: G.J.: conceptualization, formal analysis, writing—original draft preparation; R.L.: revision; J.H.: validation; L.Y. and J.C.: formal analysis, software. Z.X. (Zhe Xu): formal analysis; B.Z. and Y.Y.: software; Z.X. (Zhongmei Xia): funding acquisition, data curation; Y.T.: revision, project administration, supervision. All authors have read and agreed to the published version of the manuscript.

Funding: This work was supported by National Key Research and Development Program of China (2018YFC1802201) and the Si-chuan Key Research and Development Program (2020YFS0287).

Institutional Review Board Statement: Not applicable.

Informed Consent Statement: Not applicable.

Data Availability Statement: Data is contained within the article or supplementary material.

Conflicts of Interest: The authors declare no conflict of interest.

References

1. Riaz Rajoka, M.S.; Wu, Y.; Mehwish, H.M.; Bansal, M.; Zhao, L. Lactobacillus exopolysaccharides: New perspectives on engineering strategies, physiochemical functions, and immunomodulatory effects on host health. *Trends Food Sci. Technol.* **2020**, *103*, 36–48. [CrossRef]
2. Rahbar Saadat, Y.; Yari Khosroushahi, A.; Pourghassem Gargari, B. A comprehensive review of anticancer, immunomodulatory and health beneficial effects of the lactic acid bacteria exopolysaccharides. *Carbohyd. Polym.* **2019**, *217*, 79–89. [CrossRef] [PubMed]
3. Rajoka, M.S.R.; Mehwish, H.M.; Kitazawa, H.; Barba, F.J.; Berthelot, L.; Umair, M.; Zhao, L. Techno-functional properties and immunomodulatory potential of exopolysaccharide from *Lactiplantibacillus plantarum* MM89 isolated from human breast milk. *Food Chem.* **2022**, *377*, 131954. [CrossRef] [PubMed]
4. Riaz Rajoka, M.S.; Mehwish, H.M.; Siddiq, M.; Haobin, Z.; Zhu, J.; Yan, L.; Shi, J. Identification, characterization, and probiotic potential of *Lactobacillus rhamnosus* isolated from human milk. *LWT Food Sci. Technol.* **2017**, *84*, 271–280. [CrossRef]
5. Yang, X.; Ren, Y.; Zhang, L.; Wang, Z.; Li, L. Structural characteristics and antioxidant properties of exopolysaccharides isolated from soybean protein gel induced by lactic acid bacteria. *LWT Food Sci. Technol.* **2021**, *150*, 111811. [CrossRef]
6. Hu, S.M.; Zhou, J.M.; Zhou, Q.Q.; Li, P.; Xie, Y.Y.; Zhou, T.; Gu, Q. Purification, characterization and biological activities of exopolysaccharides from *Lactobacillus rhamnosus* ZFM231 isolated from milk. *LWT Food Sci. Technol.* **2021**, *147*, 111561. [CrossRef]
7. Bomfim, V.B.; Pereira Lopes Neto, J.H.; Leite, K.S.; de Andrade Vieira, É.; Iacomini, M.; Silva, C.M.; Cardarelli, H.R. Partial characterization and antioxidant activity of exopolysaccharides produced by *Lactobacillus plantarum* CNPC003. *LWT Food Sci. Technol.* **2020**, *127*, 109349. [CrossRef]
8. Cao, F.; Liang, M.; Liu, J.; Liu, Y.; Renye, J.A., Jr.; Qi, P.X.; Ren, D. Characterization of an exopolysaccharide (EPS-3A) produced by *Streptococcus thermophilus* ZJUIDS-2-01 isolated from traditional yak yogurt. *Int. J. Biol. Macromol.* **2021**, *192*, 1331–1343. [CrossRef]
9. Gan, L.; Jiang, G.; Li, X.; Zhang, S.; Tian, Y.; Peng, B. Structural elucidation and physicochemical characteristics of a novel high-molecular-weight fructan from halotolerant *Bacillus* sp. SCU-E108. *Food Chem.* **2021**, *365*, 130496. [CrossRef]
10. Yan, J.K.; Li, L.; Wang, Z.M.; Leung, P.H.; Wang, W.Q.; Wu, J.Y. Acidic degradation and enhanced antioxidant activities of exopolysaccharides from *Cordyceps sinensis* mycelial culture. *Food Chem.* **2009**, *117*, 641–646. [CrossRef]
11. Nasir, A.; Sattar, F.; Ashfaq, I.; Lindemann, S.R.; Chen, M.H.; Van den Ende, W.; Anwar, M.A. Production and characterization of a high molecular weight levan and fructooligosaccharides from a rhizospheric isolate of *Bacillus aryabhattai*. *LWT Food Sci. Technol.* **2020**, *123*, 109093. [CrossRef]
12. Jiang, G.; Gan, L.; Li, X.; He, J.; Zhang, S.; Chen, J.; Tian, Y. Characterization of Structural and Physicochemical Properties of an Exopolysaccharide Produced by *Enterococcus* sp. F2 From Fermented Soya Beans. *Front. Microbiol.* **2021**, *12*, 744007. [CrossRef] [PubMed]
13. Jia, Y.N.; Wang, Y.J.; Li, R.I.; Li, S.Q.; Zhang, M.; He, C.W.; Chen, H.X. The structural characteristic of acidic-hydrolyzed corn silk polysaccharides and its protection on the H_2O_2-injured intestinal epithelial cells. *Food Chem.* **2021**, *356*, 129691. [CrossRef] [PubMed]
14. Lynch, K.M.; Zannini, E.; Coffey, A.; Arendt, E.K. Lactic Acid Bacteria Exopolysaccharides in Foods and Beverages: Isolation, Properties, Characterization, and Health Benefits. *Annu. Rev. Food Sci. T* **2018**, *9*, 155–176. [CrossRef]
15. Park, S.; Saravanakumar, K.; Sathiyaseelan, A.; Park, S.; Hu, X.; Wang, M.H. Cellular antioxidant properties of nontoxic exopolysaccharide extracted from Lactobacillales (*Weissella cibaria*) isolated from Korean kimchi. *LWT Food Sci. Technol.* **2022**, *154*, 112727. [CrossRef]
16. Liu, T.; Zhou, K.; Yin, S.; Liu, S.; Zhu, Y.; Yang, Y.; Wang, C. Purification and characterization of an exopolysaccharide produced by *Lactobacillus plantarum* HY isolated from home-made Sichuan Pickle. *Int. J. Biol. Macromol.* **2019**, *134*, 516–526. [CrossRef]
17. Wang, K.; Li, W.; Rui, X.; Chen, X.; Jiang, M.; Dong, M. Structural characterization and bioactivity of released exopolysaccharides from *Lactobacillus plantarum* 70810. *Int. J. Biol. Macromol.* **2014**, *67*, 71–78. [CrossRef] [PubMed]
18. Wang, Q.; Wei, M.; Zhang, J.; Yue, Y.; Wu, N.; Geng, L.; Wang, J. Structural characteristics and immune-enhancing activity of an extracellular polysaccharide produced by marine *Halomonas* sp. 2E1. *Int. J. Biol. Macromol.* **2021**, *183*, 1660–1668. [CrossRef] [PubMed]
19. Ayyash, M.; Abu-Jdayil, B.; Itsaranuwat, P.; Galiwango, E.; Tamiello-Rosa, C.; Abdullah, H.; Hamed, F. Characterization, bioactivities, and rheological properties of exopolysaccharide produced by novel probiotic *Lactobacillus plantarum* C70 isolated from camel milk. *Int. J. Biol. Macromol.* **2020**, *144*, 938–946. [CrossRef] [PubMed]
20. Huo, J.; Lei, M.; Zhou, Y.; Zhong, X.; Liu, Y.; Hou, J.; Wu, W. Structural characterization of two novel polysaccharides from Gastrodia elata and their effects on Akkermansia muciniphila. *Int. J. Biol. Macromol.* **2021**, *186*, 501–509. [CrossRef]
21. Li, M.; Li, W.; Li, D.; Tian, J.; Xiao, L.; Kwok, L.Y.; Sun, Z. Structure characterization, antioxidant capacity, rheological characteristics and expression of biosynthetic genes of exopolysaccharides produced by *Lactococcus lactis* subsp. lactis IMAU11823. *Food Chem.* **2022**, *384*, 132566. [CrossRef]
22. Shi, W.; Zhong, J.; Zhang, Q.; Yan, C. Structural characterization and antineuroinflammatory activity of a novel heteropolysaccharide obtained from the fruits of *Alpinia oxyphylla*. *Carbohyd. Polym.* **2020**, *229*, 115405. [CrossRef] [PubMed]
23. Insulkar, P.; Kerkar, S.; Lele, S.S. Purification and structural-functional characterization of an exopolysaccharide from *Bacillus licheniformis* PASS26 with in-vitro antitumor and wound healing activities. *Int. J. Biol. Macromol.* **2018**, *120*, 1441–1450. [CrossRef] [PubMed]

24. Krishnamurthy, M.; Jayaraman Uthaya, C.; Thangavel, M.; Annadurai, V.; Rajendran, R.; Gurusamy, A. Optimization, compositional analysis, and characterization of exopolysaccharides produced by multi-metal resistant *Bacillus cereus* KMS3-1. *Carbohyd. Polym.* **2020**, *227*, 115369. [CrossRef] [PubMed]
25. Du, R.; Qiao, X.; Zhao, F.; Song, Q.; Zhou, Q.; Wang, Y.; Zhou, Z. Purification, characterization and antioxidant activity of dextran produced by *Leuconostoc pseudomesenteroides* from homemade wine. *Carbohyd. Polym.* **2018**, *198*, 529–536. [CrossRef] [PubMed]
26. Wang, B.; Song, Q.; Zhao, F.; Zhang, L.; Han, Y.; Zhou, Z. Isolation and characterization of dextran produced by *Lactobacillus sakei* L3 from Hubei sausage. *Carbohyd. Polym.* **2019**, *223*, 115111. [CrossRef]
27. Ayyash, M.; Abu-Jdayil, B.; Olaimat, A.; Esposito, G.; Itsaranuwat, P.; Osaili, T.; Liu, S.Q. Physicochemical, bioactive and rheological properties of an exopolysaccharide produced by a probiotic *Pediococcus pentosaceus* M41. *Carbohyd. Polym.* **2020**, *229*, 115462. [CrossRef]
28. Ayyash, M.; Abu-Jdayil, B.; Itsaranuwat, P.; Almazrouei, N.; Galiwango, F.; Esposito, G.; Najjar, Z. Exopolysaccharide produced by the potential probiotic *Lactococcus garvieae* C47: Structural characteristics, rheological properties, bioactivities and impact on fermented camel milk. *Food Chem.* **2020**, *333*, 127418. [CrossRef]
29. Wang, B.H.; Cao, J.J.; Zhang, B.; Chen, H.Q. Structural characterization, physicochemical properties and alpha-glucosidase inhibitory activity of polysaccharide from the fruits of wax apple. *Carbohyd. Polym.* **2019**, *211*, 227–236. [CrossRef]
30. Vinothkanna, A.; Sathiyanarayanan, G.; Balaji, P.; Mathivanan, K.; Pugazhendhi, A.; Ma, Y.; Thirumurugan, R. Structural characterization, functional and biological activities of an exopolysaccharide produced by probiotic *Bacillus licheniformis* AG-06 from Indian polyherbal fermented traditional medicine. *Int. J. Biol. Macromol.* **2021**, *174*, 144–152. [CrossRef]
31. Ji, X.; Hou, C.; Yan, Y.; Shi, M.; Liu, Y. Comparison of structural characterization and antioxidant activity of polysaccharides from jujube (*Ziziphus jujuba* Mill.) fruit. *Int. J. Biol. Macromol.* **2020**, *149*, 1008–1018. [CrossRef] [PubMed]
32. Wang, J.; Zhao, X.; Tian, Z.; Yang, Y.; Yang, Z. Characterization of an exopolysaccharide produced by *Lactobacillus plantarum* YW11 isolated from Tibet Kefir. *Carbohyd. Polym.* **2015**, *125*, 16–25. [CrossRef]
33. Pei, F.; Ma, Y.; Chen, X.; Liu, H. Purification and structural characterization and antioxidant activity of levan from *Bacillus megaterium* PFY-147. *Int. J. Biol. Macromol.* **2020**, *161*, 1181–1188. [CrossRef] [PubMed]
34. Maalej, H.; Hmidet, N.; Boisset, C.; Bayma, E.; Heyraud, A.; Nasri, M. Rheological and emulsifying properties of a gel-like exopolysaccharide produced by *Pseudomonas stutzeri* AS22. *Food Hydrocolloid.* **2016**, *52*, 634–647. [CrossRef]
35. Gomaa, M.; Yousef, N. Optimization of production and intrinsic viscosity of an exopolysaccharide from a high yielding *Virgibacillus salarius* BM02: Study of its potential antioxidant, emulsifying properties and application in the mixotrophic cultivation of Spirulina platensis. *Int. J. Biol. Macromol.* **2020**, *149*, 552–561. [CrossRef] [PubMed]
36. Kodali, V.P.; Das, S.; Sen, R. An exopolysaccharide from a probiotic: Biosynthesis dynamics, composition and emulsifying activity. *Food Res. Int.* **2009**, *42*, 695–699. [CrossRef]
37. Han, Y.; Liu, E.; Liu, L.; Zhang, B.; Wang, Y.; Gui, M.; Li, P. Rheological, emulsifying and thermostability properties of two exopolysaccharides produced by *Bacillus amyloliquefaciens* LPL061. *Carbohyd. Polym.* **2015**, *115*, 230–237. [CrossRef]
38. Min, W.H.; Fang, X.B.; Wu, T.; Fang, L.; Liu, C.L.; Wang, J. Characterization and antioxidant activity of an acidic exopolysaccharide from *Lactobacillus plantarum* JLAU103. *J. Biosci. Bioeng.* **2019**, *127*, 758–766. [CrossRef]
39. Xing, H.; Du, R.; Zhao, F.; Han, Y.; Xiao, H.; Zhou, Z. Optimization, chain conformation and characterization of exopolysaccharide isolated from *Leuconostoc mesenteroides* DRP105. *Int. J. Biol Macromol.* **2018**, *112*, 1208–1216. [CrossRef]
40. Zhang, H.; Cui, S.W.; Nie, S.P.; Chen, Y.; Wang, Y.X.; Xie, M.Y. Identification of pivotal components on the antioxidant activity of polysaccharide extract from *Ganoderma atrum*. *Bioact. Carbohyd. Diet. Fibre* **2016**, *7*, 9–18. [CrossRef]
41. Jiang, L.; Wang, W.; Wen, P.; Shen, M.; Li, H.; Ren, Y.; Xie, J. Two water-soluble polysaccharides from mung bean skin: Physicochemical characterization, antioxidant and antibacterial activities. *Food Hydrocolloid.* **2020**, *100*, 105412. [CrossRef]
42. Xie, J.H.; Jin, M.L.; Morris, G.A.; Zha, X.Q.; Chen, H.Q.; Yi, Y.; Xie, M.Y. Advances on Bioactive Polysaccharides from Medicinal Plants. *Crit. Rev. Food Sci.* **2016**, *56*, 60–84. [CrossRef] [PubMed]
43. Zhang, L.; Hu, Y.; Duan, X.; Tang, T.; Shen, Y.; Hu, B.; Liu, Y. Characterization and antioxidant activities of polysaccharides from thirteen boletus mushrooms. *Int. J. Biol. Macromol.* **2018**, *113*, 1–7. [CrossRef] [PubMed]
44. Gu, J.; Zhang, H.; Wen, C.; Zhang, J.; He, Y.; Ma, H.; Duan, Y. Purification, characterization, antioxidant and immunological activity of polysaccharide from *Sagittaria sagittifolia* L. *Food Res. Int.* **2020**, *136*, 109345. [CrossRef] [PubMed]
45. An, Q.; Ye, X.; Han, Y.; Zhao, M.; Chen, S.; Liu, X.; Wang, W. Structure analysis of polysaccharides purified from *Cyclocarya paliurus* with DEAE-Cellulose and its antioxidant activity in RAW264.7 cells. *Int. J. Biol. Macromol.* **2020**, *157*, 604–615. [CrossRef] [PubMed]
46. Shen, S.; Zhou, C.; Zeng, Y.; Zhang, H.; Hossen, M.A.; Dai, J.; Liu, Y. Structures, physicochemical and bioactive properties of polysaccharides extracted from *Panax notoginseng* using ultrasonic/microwave-assisted extraction. *LWT Food Sci. Technol.* **2022**, *154*, 112446. [CrossRef]
47. Wang, C.; Li, W.; Chen, Z.; Gao, X.; Yuan, G.; Pan, Y.; Chen, H. Effects of simulated gastrointestinal digestion in vitro on the chemical properties, antioxidant activity, alpha-amylase and alpha-glucosidase inhibitory activity of polysaccharides from *Inonotus obliquus*. *Food Res. Int.* **2018**, *103*, 280–288. [CrossRef] [PubMed]
48. Feng, S.; Cheng, H.; Xu, Z.; Yuan, M.; Huang, Y.; Liao, J.; Ding, C. Panax notoginseng polysaccharide increases stress resistance and extends lifespan in Caenorhabditis elegans. *J. Funct. Foods* **2018**, *45*, 15–23. [CrossRef]

Article

Isolation and Purification, Structural Characterization and Antioxidant Activities of a Novel Hetero-Polysaccharide from Steam Exploded Wheat Germ

Lei Hu [1], Xiaodan Zhou [1], Xue Tian [1], Ranran Li [1], Wenjie Sui [1,*], Rui Liu [1], Tao Wu [1] and Min Zhang [2,*]

[1] State Key Laboratory of Food Nutrition and Safety, Tianjin Key Laboratory of Food Quality and Health, College of Food Science and Engineering, Tianjin University of Science & Technology, Tianjin 300457, China; huleistar@126.com (L.H.); dawnzhou5@163.com (X.Z.); xuetian9701@163.com (X.T.); li17862686224@126.com (R.L.); lr@tust.edu.cn (R.L.); wutaoxx@gmail.com (T.W.)

[2] China-Russia Agricultural Processing Joint Laboratory, Tianjin Agricultural University, Tianjin 300392, China

* Correspondence: wjsui@tust.edu.cn (W.S.); zm0102@sina.com (M.Z.)

Abstract: A purified polysaccharide, designated as SE-WGP$_I$, was isolated from wheat germ modified by steam explosion. The primary structure characteristics were determined by HPGPC, GC, periodate oxidation-Smith degradation, methylation analysis, FT-IR, NMR and Congo red test. The results showed that SE-WGP$_I$ was a homogeneous hetero-polysaccharide with the average molecular weight of 5.6×10^3 Da. The monosaccharide composition mainly consisted of glucose, arabinose and xylose with a molar ratio of 59.51: 20.71: 19.77. The main backbone of SE-WGP$_I$ consisted of →4,6)-α-D-Glc$p$(1→6)-α-D-Glc$p$(1→3)-β-D-Xyl$p$(1→5)-α-L-Ara$f$(1→ and the side chain was α-D-Glc$p$(1→ linked at the C4-position of →4,6)-α-D-Glc$p$(1→. SE-WGP$_I$ likely has a complex netted structure with triple helix conformation and good thermal stability. In addition, SE-WGP$_I$ had valid in vitro radical scavenging activities on DPPH and hydroxyl radicals. This study may provide structural information of SE-WGP$_I$ for its promising application in the fields of functional foods or medicines.

Keywords: wheat germ polysaccharide; steam explosion; isolation and purification; modification; structural characterization; antioxidant activities

Citation: Hu, L.; Zhou, X.; Tian, X.; Li, R.; Sui, W.; Liu, R.; Wu, T.; Zhang, M. Isolation and Purification, Structural Characterization and Antioxidant Activities of a Novel Hetero-Polysaccharide from Steam Exploded Wheat Germ. *Foods* **2022**, *11*, 1245. https://doi.org/10.3390/foods11091245

Academic Editor: David Bongiorno

Received: 22 March 2022
Accepted: 20 April 2022
Published: 26 April 2022

1. Introduction

Wheat germ is a highly nutritive milling by-product of wheat processing. About 2.5 million tons of it are produced per year around the world [1]. Wheat germ is considered to be the most nutritious part of wheat grain, providing 409 kcal per 100 g, with 54% provided by carbohydrates, 20% by lipids and 26% by proteins. Many studies have demonstrated that wheat germ has an important effect on human health (such as antihyperlipidemic, anticancer, antioxidant and hypocholesterolemic effects) [2]. Meanwhile, studies on wheat germ are mainly focused on the thermal and non-thermal modification of wheat germ, functional properties of polysaccharide and protein hydrolysates, encapsulation of wheat germ oil, and relevant applications in flour products. Although an abundant amount of wheat germ is removed during processing and widely used in animal feed, its commercial use still remains minimal and results in resource waste [3].

Since carbohydrates are the major component (56.3 g/100 g) in wheat germ, a significant amount of polysaccharides (18.4 g/100 g) has been found [4]. Recently, plant-derived polysaccharides have drawn wide attention owing to their abundant pharmacological activities and biodegradable, naturally non-toxic and biocompatible properties. In our previous study [5], the primary structure of polysaccharides isolated from wheat germ were identified, involving molecular weight, monosaccharide composition, functional groups, glycosidic linkage and branching characteristics. However, the structural and functional activities of polysaccharides could be strongly influenced by various modification treatments of materials. Recent studies have attempted to prepare polysaccharides using extrusion cooking, high-pressure homogenization, enzymatic hydrolysis and fermentation, etc. Several structural features modified by the above methods, including monosaccharide composition, functional groups, molecular weight, conformation and branching, etc., have been identified as responsible for various activity promotions of specific polysaccharides. Furthermore, not only the individual, but also combinations of these structural characteristics seem to influence the direct correlation between polysaccharides and cells or other components of the immune system.

Among many industrial technologies, steam explosion (SE) is a typical hydrothermal treatment in which the material is exposed to high-pressure saturated steam for a period of time and then suddenly decompressed to achieve the desired effect [6]. It has certain economic advantages and eco-friendly effects on the processing of a large number of raw materials, such as wheat bran, olive leaves, sumac fruits, orange peels, etc. [7]. Application of SE in high-valued utilization of by-products from wheat processing have been proven to have certain benefits, including disrupting porous structures to prompt the release of intracellular active substances, converting structural carbohydrates into soluble dietary fibers, inactivating some enzymes and reducing some anti-nutritional factors, etc. [8]. Some studies have shown that polysaccharides with different molecular weights could be obtained, and soluble dietary fiber could be dissolved more easily compared with the untreated group after SE treatment [9]. Hence, the application of SE in wheat germ modification has been promising, and it is necessary to purify and identify the major polysaccharides of wheat germ modified by SE.

In this paper, the polysaccharide from steam exploded wheat germ (SE-WGP$_I$) was purified with DEAE-52 cellulose and Sephadex G-50 gel chromatography. High-performance gel permeation chromatography (HPGPC), infrared spectroscopy (FT-IR), periodate oxidation-Smith degradation, methylation, gas chromatography-mass spectrometry (GC-MS), nuclear magnetic resonance (NMR), Congo red test and thermal analysis were utilized to identify the structural features of purified polysaccharide. The antioxidant activities of purified polysaccharide were further evaluated, including the scavenging performance on DPPH radical and hydroxyl radicals.

2. Materials and Methods

2.1. Materials and Reagents

Wheat germ was provided by FADA Flour Group in Shandong Province, China. All the chemical reagents and solvents used in this study were analytical grade.

2.2. Steam Explosion Process

SE treatment was carried out according to our previous study [10]. Whole wheat germ with a moisture content of 30% (w/w) was exploded by a laboratory scale SE test bench (Hebi Zhengdao Bioenergy Company, China), in which the pressure was 0.8 MPa and the maintenance time was 5 min. The exploded sample was dried at 60 °C for 24 h and then ground and sieved through a 60 mesh screen. The steam exploded wheat germ (SEWG) was refluxed with petroleum ether for 8 h at 55 °C to decolorize and remove grease and petroleum ether soluble constituents. The samples were dried naturally for subsequent experiments.

2.3. Extraction and Purification of Polysaccharide

The extraction procedure of polysaccharide from SEWG was optimized by single factor experiment and response surface experiment with extraction time, temperature, times and ratio of material to liquid as research parameters and polysaccharide yield as detection index. According to the optimum extraction procedure, the defatted SEWG was extracted thrice at 70 °C with distilled water (1: 5, w/v) under gentle stirring for 30 min each time, and all of the supernatants after centrifugation (4000 rpm/min, 20 min) were concentrated by vacuum rotary evaporator. Then, they were deproteinized 15 times by

Sevag reagent [11]. Afterwards, 4 volumes of anhydrous ethanol were added in and stood overnight at 4 °C, and the precipitate was freeze-dried after centrifugation to obtain crude SEWG polysaccharide (SE-WGP). Each SE-WGP sample was weighed with an analytical balance, according to the following equation to calculate the extraction yield (Y):

$$Y\% = \frac{\text{weight of SE} - \text{WGP}}{\text{weight of SEWG}} \times 100 \tag{1}$$

An amount of 20 mg crude SE-WGP was dissolved in appropriate deionized water to place on a DEAE-Cellulose column (1.6 × 45 cm). The elution was carried out with deionized water and 0.1, 0.3, 0.5, 0.7 mol/L NaCl at a flow rate of 0.8 mL/min [12]. The eluent fractions were measured by phenol-sulfuric acid method, and two fractions were obtained and named as SE-WGP$_I$ and SE-WGP$_{II}$. Then, SE-WGP$_I$ was further purified with a Sephadex G-50 (1.6 × 45 cm) column and eluted with distilled water at a flow rate of 0.8 mL/min. The main collected parts were dialyzed and lyophilized to obtain purified SE-WGP$_I$ for further study.

2.4. Chemical Analysis

Phenol sulfuric acid method was used to determine the total sugar content [13]. BCA protein assay method was used to measure the protein content [14], and m-hydroxybiphenyl method was applied to determine the content of uronic acid [15].

2.5. Homogeneity and Molecular Weight Determination

The average molecular weight and homogeneity of SE-WGP$_I$ were measured by using a Waters HPLC instrument (RID-10A detector) equipped with OHpak SB-804 HQ (8.0 mm × 300 mm) column, taking ultra-pure water as mobile phase. The dissolved sample was filtered by 0.22 μm filter membrane and then injected (20 μL), and analyzed at the elution rate of 0.8 mL/min. The standards were dextrans with different molecular weights (41, 15, 5.0, 2.5, 1.2 and 1 kDa).

2.6. Monosaccharide Composition Determination

The determination of monosaccharide composition was conducted based on the published paper with slight adjustments [16]. Briefly, 10 mg SE-WGP$_I$ was hydrolyzed in 2 mol/L TFA in a sealed glass tube at 110 °C for 5 h. Then, the sample was reduced with NaBH$_4$ after removing excessive TFA with repeated methanol, followed by acidification with acetic acid. The alditol acetate was prepared to be acetylated with pyridine-acetic anhydride (1: 1) at 105 °C for 1 h, and then subjected to Agilent 7890A gas chromatography (GC) system equipped with an OV-1701 capillary column (0.32 mm × 0.5 μm × 30 m). The column temperature was maintained at 150 °C for 1 min, then rose to 200 °C for 10 min with a heating rate 10 °C/min, then increased to 220 °C for 5 min with a heating rate 5 °C/min, and finally increased to 240 °C for 20 min with a heating rate 1.5 °C/min. The temperatures of the injector and the FID detector were set at 250 °C. Arabinose, xylose, mannose, rhamnose, glucose, galactose and fucose were used as monosaccharide standards.

2.7. Periodate Oxidation-Smith Degradation

SE-WGP$_I$ (20 mg) and 0.03 M NaIO$_4$ solution were placed in darkness at 4 °C for reaction. At different time points (0, 4, 8, 12, 24, 48 h), the absorbance at 233 nm of mixture was monitored [17]. When the absorbance ceased descending and remained stable, the reaction was completed and the excess NaIO$_4$ was then removed with appropriate amount of ethylene glycol. The formic acid was titrated with 0.099 mol/L NaOH solution to determine its production. The remaining periodate products obtained by the above reactions were dialyzed for 48 h and reduced with 50 mg NaBH$_4$ for 24 h. Then, the solution was concentrated and subjected to complete hydrolysis at 105 °C with 2 mol/L TFA for about 5 h. After removing residual TFA and acetylating process, the products were analyzed by GC system.

2.8. Methylation Analysis

Methylation analysis of SE-WGP$_I$ was performed according to the reported method with some modifications [18]. An amount of 10 mg SE-WGP$_I$ was mixed with 50 mg NaOH and dissolved in 2 mL anhydrous DMSO under nitrogen atmosphere and sonicated for 1 h. Methyl iodide (1 mL) was added and performed with sonicating in the dark for 1 h, and then the reaction was stopped by 2 mL distilled water to obtain the methylated polysaccharide. It was extracted four times with dichloromethane, and dried with sodium sulfate and then evaporated to dryness. Afterwards, the methylated SE-WGP$_I$ was hydrolyzed and acetylated as mentioned before. The result was analyzed by a GC-MS (VARIAN 4000GC-MS, Palo Alto, CA, USA) system and the chromatographic column was a HP-5 capillary column (30 m $\times$ 0.25 mm $\times$ 0.25 μm). The column temperature was raised from 100 $°$C to 280 $°$C at a heating rate of 10 $°$C/min, and remained at 280 $°$C for 12 min.

2.9. Infrared Spectrum Analysis

About 1 mg SE-WGP$_I$ and 150 mg KBr were mixed and ground quickly to a fine powder. FTIR spectroscopy (IS50, Thermo Fisher, Massa, WA, USA) was measured by the potassium bromide pellet method in a vibrations region of 400–4000 cm^{-1} [19].

2.10. NMR Spectrum Analysis

About 20 mg SE-WGP$_I$ was dissolved with 0.6 mL D$_2$O in an NMR tube and the 1D-NMR (^{1}H NMR, ^{13}C NMR) and 2D-NMR (^{1}H-^{1}H COSY, HSQC and HMBC) spectra were obtained using a Bruker Avance III (400 MHz) (Cambridge Isotope Laboratories, Inc., Tewksbury, MA, USA).

2.11. Congo Red Test

The solution of sample (2.0 mL, 2.5 mg/mL) was mixed with 2 mL of Congo red solution (80 μmol/L). Then, 1.5 mol/L NaOH was continually added to final NaOH concentration of 0, 0.1, 0.2, 0.3, 0.4 and 0.5 mol/L. Full wavelength scanning in the scanning range of 400 to 700 nm was analyzed by a TU-1810 (Ultraviolet spectrophotometer, China), and the maximum absorption wavelengths were recorded [20].

2.12. Thermal Analysis

TGA analysis was determined on a TA instrument (model TGA Q50). About 3 mg of SE-WGP$_I$ sample was placed in an aluminum crucible and heated from 30 $°$C to 600 $°$C at a speed of 10 $°$C/min [20].

2.13. Antioxidant Activities Evaluation

The DPPH radical scavenging activity of SE-WGP$_I$ was studied according to previous literature [21] with slight modifications. SE-WGP$_I$ was prepared into aqueous solutions with different concentrations (0.2–1.2 mg/mL, 0.5 mL) and mixed with 2.5 mL of DPPH anhydrous ethanol solution (0.4 mM), respectively. The mixtures were shaken thoroughly at 25 $°$C for 30 min in the darkness, and then were detected at 517 nm by a TU-1810 (Ultraviolet spectrophotometer, China) with ascorbic acid as positive control. The DPPH radical scavenging activity of SE-WGP$_I$ was expressed as:

$$\text{DPPH radical scavenging activity } (\%) = \left[1 - \frac{A_2 - A_1}{A_0}\right] \times 100\% \tag{2}$$

where A_0 represents the absorbance of the control substance (water used instead of sample), A_1 represents the absorbance of the sample without DPPH solution under identical conditions and A_2 represents the absorbance of the sample mixed with DPPH solution.

The ability of SE-WGP$_I$ to scavenge hydroxyl radicals was measured as by a previous study [22] with slight modification. An amount of 1.0 mL SE-WGP$_I$ solutions with different concentrations (0.2–1.2 mg/mL) were blended with salicylic acid-ethanol solution

(6 mmol/L, 1 mL) and $FeSO_4$ (2 mmol/L, 1.0 mL), respectively. Then, H_2O_2 (6 mmol/L, 1.0 mL) was added and the mixed solution was maintained at 37 °C for 30 min. The absorbance of the mixture was immediately measured at 510 nm by a TU-1810 (Ultraviolet spectrophotometer, China) with the positive control of Vc. Hydroxyl radical scavenging activity was calculated in the same way:

$$\text{Hydroxyl scavenging activity (\%)} = \left[1 - \frac{A_1 - A_2}{A_0}\right] \times 100\% \tag{3}$$

where A_0 represents the absorbance of the reagent blank, A_1 represents the absorbance of the sample and A_2 represents the absorbance of the control.

2.14. Statistical Analysis

The statistical analysis was performed by one-way ANOVA followed by the Tukey post hoc test using Origin 9.0 statistical software program. Three independent trials were performed, and all the data were expressed as mean ± standard deviation (SD).

3. Results and Discussion

3.1. Isolation and Purification

The extraction yield of crude polysaccharide prepared from the SEWG was 18.72% (dry weight). In the previous study [5], crude polysaccharides were isolated from defatted wheat germ without steam explosion, and the yield was only 1.16%, indicating that steam explosion technology effectively improved the extraction yield of crude polysaccharides. As we know, the separation mechanism of anion-exchange column chromatography is not only ion exchange, but also adsorption–desorption. Therefore, anion exchange resins are mostly used for the separation of neutral and acidic polysaccharides. Neutral polysaccharides can be obtained by water elution and acidic polysaccharides by salt solution elution [23]. Therefore, SE-WGP was separated and purified by a DEAE-52 column with gradient elution of deionized water, 0.1, 0.3, 0.5 and 0.7 mol/L NaCl. Two independent fractions, SE-WGP$_I$ and SE-WGP$_{II}$, were obtained (1.61% losses) with yields of 50.21% and 48.18% (Figure 1). SE-WGP$_I$ was eluted by distilled water and further purified (0.98% loss) with a high-resolution Sephadex G-50 size-exclusion column taking distilled water as the eluent. The yield of SE-WGP$_I$ was 4.17% from the defatted wheat germ after the dialysis and lyophilization of the first fraction. As shown in Figure 1, SE-WGP$_I$ was a homogeneous polysaccharide with a single and tailed peak. A major part of SE-WGP$_I$ was collected and detected by the phenol-sulfuric acid method for the determination of 98.8% content. This study mainly focused on the structural characterization of SE-WGP$_I$ and its in vitro antioxidant activities. The relevant property of SE-WGP$_{II}$ has been previously reported [5].

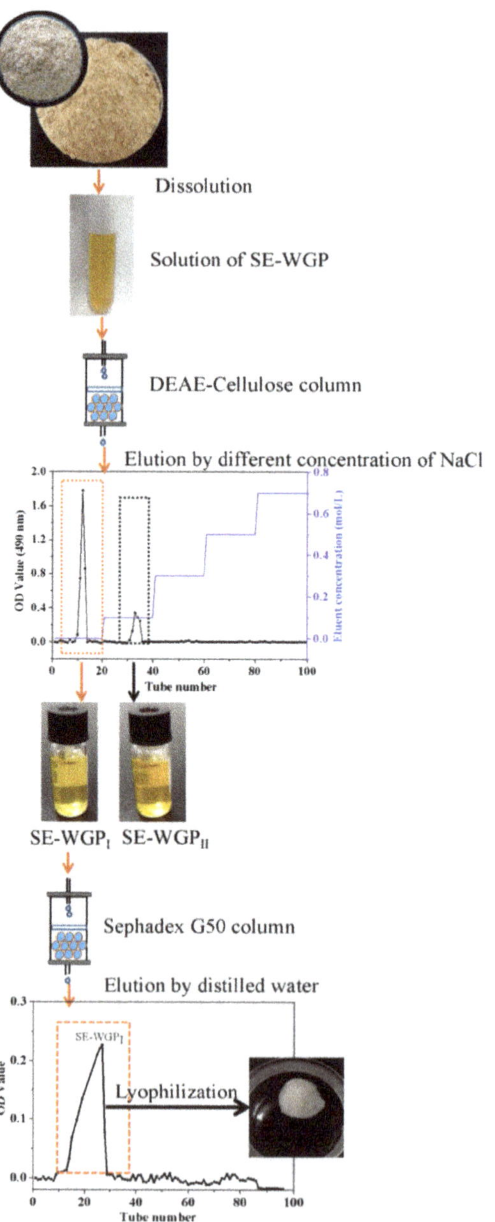

Figure 1. The separation flow diagram of SE-WGP$_I$.

3.2. Chemical Composition

The content of total sugar in SE-WGP$_I$ was determined to be 91.21 ± 3.22% by the phenol-sulfuric acid assay. The content of protein in SE-WGP$_I$ was measured to be 1.20 ± 0.09% according to the BCA protein assay method. The content of uronic acid was 1.01 ± 0.02%. Owing to the low content of protein and uronic acid in SE-WGP$_I$, it was considered to be a neutral hetero-polysaccharide.

3.3. Homogeneity and Average Molecular Weight

The average molecular weight and homogeneity of SE-WGP$_I$ was evaluated by HPGPC, as shown in Figure 2a,b. A single and symmetrical narrow peak was detected at about 11.67 min of elution time, and SE-WGP$_I$ was possibly homogeneous because of its low polydispersity index (Mw/Mn = 1.15). The average molecular weight (Mw) of SE-WGP$_I$ was 5.6×10^3 Da, which was a kind of polysaccharide with lower molecular weight compared to most polysaccharides with hundreds of thousands of Da. For example, the molecular weight of wheat germ polysaccharides reached 4×10^3 kDa when studied by Yun et al. [5].

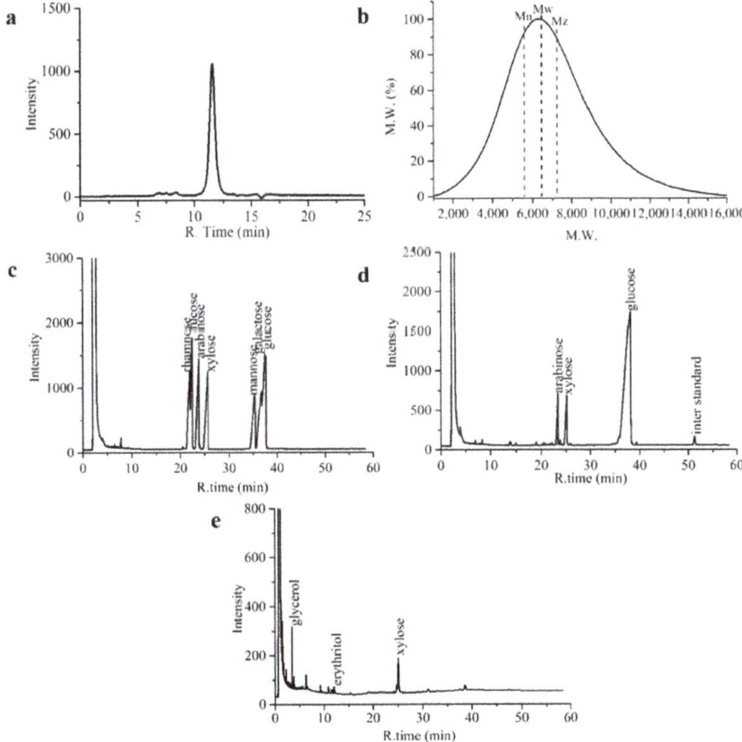

Figure 2. Molecular weight distribution of SE-WGP$_I$ (**a**). Differential mass distribution curve of SE-WGP$_I$ (**b**). GC chromatography of monosaccharide standards (**c**). SE-WGP$_I$ (**d**) and Smith degradation of SE-WGP$_I$ (**e**).

3.4. Monosaccharide Composition

Quantitative monosaccharide analysis of SE-WGP$_I$ was performed using the GC method, and the spectrums are shown in Figure 2c (monosaccharide standards) and Figure 2d. The monosaccharides in SE-WGP$_I$ were mainly detected to be glucose (Glc), arabinose (Ara) and xylose (Xyl), and the molar ratio was 59.51: 20.71: 19.77. The peak associated with uronic acid was not detected, indicating that the content of uronic acid in SE-WGPI was low. The above results confirmed that SE-WGP$_I$ was a homogeneous polysaccharide, in which Glc accounted for the greatest proportion of total monosaccharides. It is easy to find that Ara, Xyl and Glc were found in both wheat germ polysaccharides and bran polysaccharides, but the specific contents were different [5].

3.5. Periodate Oxidation-Smith Degradation

According to the periodate oxidation analysis, 1 mol sugar residue consumed 34.5 mmol periodate and produced 0.08 mmol formic acid, which implied that 1-linked or 1,6-linked sugar residue existed in SE-WGP$_I$ [24]. However, the consumption of periodate was more than twice the production of formic acid, inferring the possible presence of a large number of 1,2-linked, 1,4-linked or 1,4,6-linked glycosidic bonds. Then, Smith degradation was further performed on SE-WGP$_I$, and Xyl, erythritol and glycerol were detected in it (Figure 2e). The existence of Xyl suggested that certain residues of Xyl cannot be oxidized by periodate owing to the existence of 1,3-linked, 1,3,4-linked or 1,3, 6-linked glycosidic bonds. Ara and Glc were not observed, which meant that all the residues of Ara and Glc were oxidized, such as 1-linked or 1,6-linked. Furthermore, the appearance of erythritol suggested the existence of 1,4-linked and 1,4,6-linked glycosidic bonds, and glycerol correlated to the presence of 1-linked, 1,2-linked or 1,6-linked glycosidic bonds in the backbone of SE-WGP$_I$ [25].

3.6. Methylation Analysis

As shown in Table 1, the methylation analysis demonstrated that SE-WGP$_I$ had five partially O-methylated alditol acetate (PMAAs) peaks. By comparing with standard data and other papers [17], it could be discovered that SE-WGP$_I$ contained 1,4,5-tri-O-acetyl-2,3-di-O-methyl-Arabitol (A, 18.56 mol%), 1,3,5-tri-O-acetyl-2,4-di-O-methyl-xylitol (E, 19.28 mol%), 1,5-di-O-acetyl-2,3,4,6-tera-O-methyl-D- glucitol (C, 20.05 mol%), 1,5,6-tri-O-acetyl-2,3,4-tri-O-methyl-D-glucitol (D, 20.34 mol%) and 1,4,5,6-tera-O- acetyl-2,3-di-O-methyl-D-glucitol (B, 21.77 mol%), which were identified as 1,5-linked L-Araf, 1,3-linked D-Xylp, T-D-Glcp, 1,6-linked D-Glcp and 1,4,6-linked D-Glcp, respectively. In all PMAA mass spectra, there was a base peak of m/z 43 formed by the loss of acetyl ions (CH$_3$CO$^+$). The C2 and C3 in 1,5-linked L-Araf were all methylated carbon. The methylated C-C was easy to break, mainly producing m/z 117, m/z 189, m/z 161 and m/z 145 ion fragments. The ion fragments of m/z 161 were degraded into m/z 101 secondary ion fragments, which was eventually degraded into m/z 71, m/z 43 and other smaller ion fragments. In addition, m/z 189 was further broken into m/z 129. In 1,4,6-linked D-Glcp, C2 and C3 were methylated carbon, C4 was acetylated carbon and C2-C3 was easier to break than C3-C4. When the break occurred in C2-C3, ion fragments of m/z 117 and m/z 261 were produced, and m/z 261 was further broken into secondary ion fragments of m/z 201. When the fracture occurred in C3-C4, the ion fragment of m/z 161 was produced and further split into the secondary ion fragment of m/z 101. In T-Glcp, m/z 117 and m/z 205 were ion fragments produced by the fracture between C2-C3, and m/z 205 split into ion fragments m/z 145. When the fracture occurred between C3-C4, the ion fragments of m/z 161 and m/z 162 were produced, which were broken into the secondary ion fragments of m/z 101 and m/z 129. These secondary ion fragments were further broken into ion fragments of m/z 71 and m/z 43. In 1,6-linked D-Glcp, C2, C3 and C4 were all methylated carbon, so the fracture occurred between C2-C3 and C3-C4. When the fracture occurred in C2-C3, the main ion fragments were m/z 117 and m/z 233, and m/z 233 was further broken into m/z 145. When the fracture occurred in C3-C4, the main ion fragments were m/z 161 and m/z 189, which were broken into m/z 101 and m/z 129, respectively. In 1,3-linked D-Xylp, C2 and C4 were methylated carbon. When the break occurred in C1-C2, ion fragments of m/z 233 were produced. When the break occurred in C2-C3, m/z 118 was produced. When the fracture occurred in C3-C4 and C4-C5, the ion fragment of m/z 117 and m/z 234 was produced and further split into the secondary ion fragment. This agrees with the results of periodate oxidation-Smith degradation analysis. It can be seen from Table 1 that glycosidic bonds related to Glc were 62.16% of all monosaccharides, being consistent with the analysis in Section 3.4. The above results revealed that SE-WGP$_I$ contained the linear polysaccharides with branching chain, and the main backbone of SE-WGP$_I$ contained 2,3-Me2-Araf, 2,4-Me2-Xylp, 2,3-Me2-Glcp, 2,3,4,6-Me4-Glcp and 2,3,4-Me3-Glcp.

Table 1. Methylation analysis of SE-WGP$_I$.

Residues	Retention Time (min)	PMAAs	Type of Linkage	Molar Ratio (mol %)	Major Mass Fragments (*m/z*)
A	29.25	1,4,5-tri-O-acetyl-2,3-di-O-methyl-Arabitol	1,5-linked L-Ara*f*	18.56	43, 59, 71, 87, 101, 117, 129, 173, 189
E	29.60	1,3,5-tri-O-acetyl-2,4-di-O-methyl-xylitol	1,3-linked D-Xyl*p*	19.28	43, 59, 71, 87, 101, 117, 129, 189, 201, 233
C	30.87	1,5-di-O-acetyl-2,3,4,6-tera-O-methyl-D-glucitol	T-D-Glc*p*	20.05	43, 59, 71, 87, 101, 113, 117, 129, 145, 161, 162, 205
D	32.54	1,5,6-tri-O-acetyl-2,3,4-tri-O-methyl-D-glucitol	1,6-linked D-Glc*p*	20.34	43, 59, 71, 87, 101, 117, 129, 145, 162, 173, 189,233
B	33.24	1,4,5,6-tera-O-acetyl-2,3-di-O-methyl-D-glucitol	1,4,6-linked D-Glc*p*	21.77	43, 59, 71, 87, 101, 117, 129, 161, 173, 201, 217, 233, 261

3.7. FT-IR Spectrum

As shown in Figure 3, the strong and broad band at around 3412 cm^{-1} was the stretching vibration of OH due to intra- and inter-molecular hydrogen bands. The absorption at 2932 cm^{-1} was due to the C-H stretching vibrations of CH$_2$ groups of free sugars [26]. The strong band at 1653 cm^{-1} indicated the C=O bonds of a carbonyl group. These bands at 3412, 2932 and 1653 cm^{-1} were characteristics of a carbohydrate ring, and C-H led to a weak absorption at approximately 1412 cm^{-1} because of the deformation vibration. The weak band at 1248 cm^{-1} was assigned to the absorption of CH$_3$ from the acetyl group. Three peaks at 1150, 1045 and 1000 cm^{-1} implied the stretching vibrations of a pyranose ring [27]. The weak absorption at 888 cm^{-1} and 838 cm^{-1} represented the presence of β-pyranoside and α-pyranoside, respectively [28]. As a result, the spectra indicated that the major groups of SE-WGP$_I$ were not affected by the purification process, and SE-WGP$_I$ contained α-type and β-type glycosidic linkages with a pyranose ring.

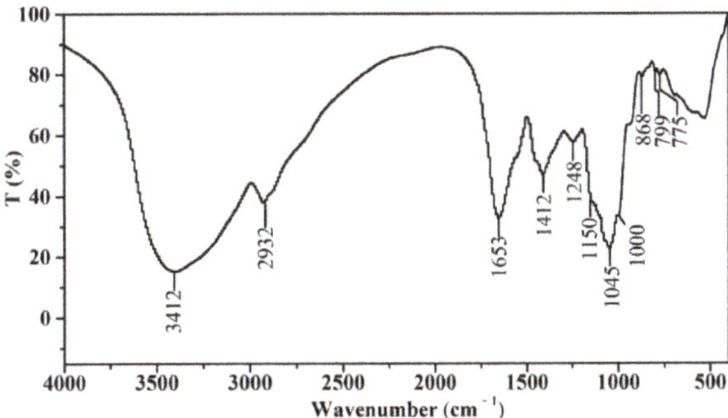

Figure 3. The Fourier transforms infrared spectrogram of SE-WGP$_I$.

3.8. NMR Spectroscopy

The structural characteristics of SE-WGP$_I$ were further elucidated by 1D- (^{1}H, ^{13}C NMR) and 2D-NMR (COSY, HSQC, HMBC spectra). The ^{1}H NMR spectrum (Figure 4a) showed that the chemical shifts of anomeric proton in this study ranged from δ 4.90 ppm to 5.32 ppm, including weak signals at δ 5.02 and 5.16 ppm and strong signals at δ 4.90, 5.28 and 5.32 ppm. The presence of these five signals proved that SE-WGP$_I$ contained both α-configuration and β-configuration, and it was confirmed by the conclusion of FT-IR [29]. In Figure 4b, the distribution of the anomeric carbon ranged from δ 91.90 to 107.44 ppm in the ^{13}C NMR spectrum. The signal at δ 91.90, 98.06, 98.53, 99.96 and 107.44 ppm indicated that five types of linkage patterns existed in SE-WGP$_I$, and this was in good accordance with the conclusions of methylation.

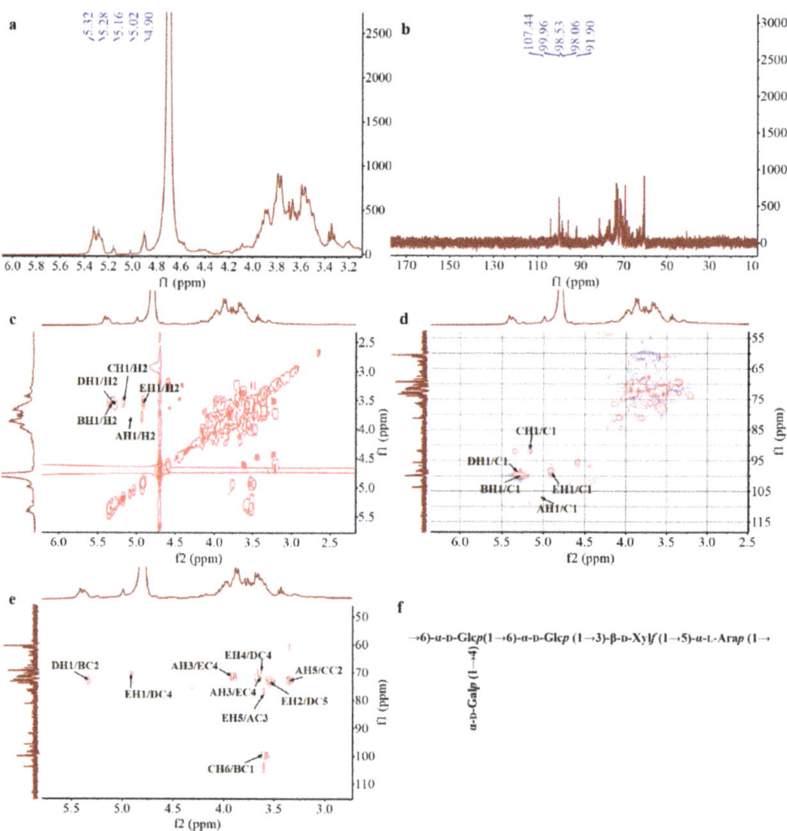

Figure 4. ^{1}H NMR (**a**), ^{13}C NMR (**b**), COSY (**c**), HSQC (**d**), HMBC (**e**) spectra and the proposed structure of SE-WGP$_I$ (**f**).

In the HSQC spectrum of SE-WGP$_I$ (Figure 4d), δ 5.02/107.44, 5.28/99.96, 5.16/91.90, 5.32/98.06 and 4.90/98.53 ppm were detected in the anomeric region and sequentially marked as sugar residues A, B, C, D and E. According to the results of the NMR analysis, methylation analysis and literature data, it could be deduced that A, B, C, D and E were identified as →5)-α-L-Ara*f* (1→, →4,6)-α-D-Glc*p* (1→, α-D-Glc*p* (1→, →6)-α-D-Glc*p* (1→ and →3)- β-D-Xyl*p* (1→, respectively. The chemical shifts of H and C for all the residues are summarized in Table 2.

Table 2. Chemical shifts of resonances in the ^{1}H and ^{13}C NMR spectra of SE-WGP$_I$.

Sugar Residue		Chemical Shift (ppm)					
		H1/C1	**H2/C2**	**H3/C3**	**H4/C4**	**H5/C5**	**H6/C6**
A	→5)-α-L-Ara*f* (1→	5.02/107.44	3.84/81.22	3.90/77.17	3.73/84.07	3.38/67.24	-
B	→4,6)-α-D-Glc*p* (1→	5.28/99.96	3.64/72.73	3.23/73.71	3.61/78.04	3.79/70.98	3.62/66.90
C	α-D-Glc*p* (1→	5.16/91.90	3.54/72.99	3.97/70.28	3.79/71.06	3.80/70.98	3.62/69.90
D	→6)-α-D-Glc*p* (1→	5.32/98.06	3.63/72.30	3.98/73.09	3.65/70.40	3.37/72.30	3.61/66.10
E	→3)-β-D-Xyl*p* (1→	4.90/98.53	3.54/72.50	3.80/81.07	3.62/70.70	3.61/66.01	-

Figure 4e shows the spectrum of HMBC from which the glycosidic linkages between sugar residues can be deduced in more detail. The cross-peaks at δ 3.62/99.96 (CH6/BC1),

5.32/72.73 (DH1/BC2) and 4.90/70.40 (EH1/DC4) indicated the presence of →4,6)-α-D-Glc*p*-(1→6)-α-D-Glc*p*(1→3)-β-D-Xyl*p*(1→ in SE-WGP$_\mathrm{I}$. Based on the methylation analysis, the cross-peak δ 3.62/99.96 (CH6/BC1) and 3.62/71.06 (BH6/CC4) should also be attributed to the correlation between residue B and C, suggesting the residue B was terminated with residue C, showing that there was a sequence of α-D-Glc*p* (1→4,6)-α-D-Glc*p* (1→ in SE-WGP$_\mathrm{I}$. Therefore, the backbone of SE-WGP$_\mathrm{I}$ was mainly made up of →4,6)-α-D-Glc*p*(1→6)-α-D-Glc*p* (1→3)β-D-Xyl*p* (1→5)-α-L-Ara*f* (1→ and the side chain was α-D-Glc*p* (1→ linked at the C4-position of →4,6)-α-D-Glc*p* (1→. A possible predicted repetitive unit structure of SE-WGP$_\mathrm{I}$ is proposed in Figure 4f.

3.9. Conformational Structure

The Congo red assay has been increasingly used due to its simplicity of operation and instrument independence, for example, to identify the triple helix conformation of polysaccharides [30]. As an acid dye, Congo red formed complexes with polysaccharides containing the triple helix configuration. In a certain concentration range, the maximum absorption wavelength (λ_{max}) of the complexes showed a bathochromic shift, thus forming a metastable region in comparison with Congo red (control group) [31]. Figure 5a demonstrates the changes in λ_{max} of Congo red and polysaccharide complex with 0~0.5 mol/L NaOH concentrations. It can be seen from the figure that the λ_{max} of the sample had a red shift phenomenon at all alkaline concentrations. Therefore, we speculated that SE-WGP$_\mathrm{I}$ might have a triple helix conformation in solution. At the same time, the view that polysaccharides with three-helix structures usually had higher biological activities has been confirmed by many studies [32].

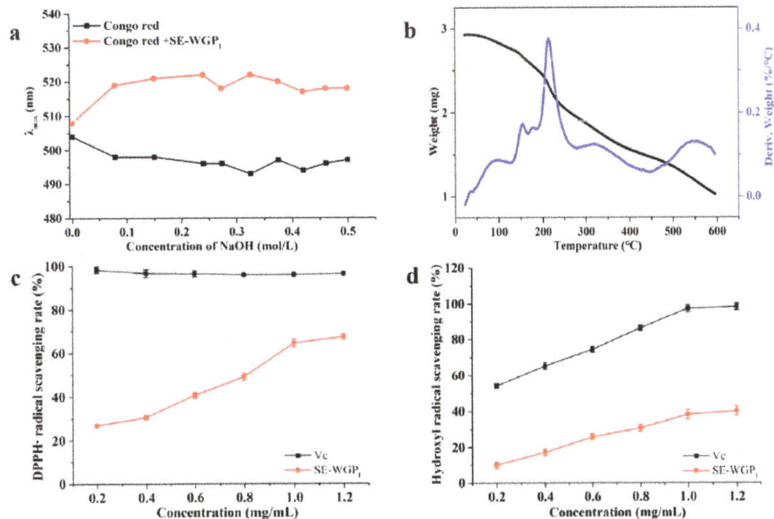

Figure 5. Maximum absorption (λmax) of Congo red and Congo red + polysaccharide at different concentrations of NaOH (**a**). The TGA curves of SE-WGP$_\mathrm{I}$ of different moisture contents (**b**). DPPH radical scavenging activity of SE-WGP$_\mathrm{I}$ (**c**). Hydroxyl radical scavenging activity of SE-WGP$_\mathrm{I}$ (**d**).

3.10. Thermal Analysis

It is very important to analyze the thermodynamic properties of polysaccharides. TGA results showed the thermal degradation characteristics and weight loss of degraded products during the temperature changing process. In Figure 5b, DTG curves showed a slight mass loss ranging from 30 °C to 200 °C, resulting from the loss of free and bound water in SE-WGP$_\mathrm{I}$. A large weight loss from around 200 °C to 600 °C was observed due to the degradation of the polysaccharide structure by dehydroxylation or deoxylation [33].

The measured chemical structure of the polysaccharide determined its thermodynamic properties, both of which were influenced by the modification method. According to our previous study, the degradation temperature of the polysaccharide extracted from wheat germ was 290.03 °C, and the weight loss was 88.42% at 600 °C [5]. In this study, the degradation temperature of SE-WGP$_I$ was determined to be 234.05 °C, and the weight loss was 70.25% at 600 °C. This indicated that the thermodynamic properties of polysaccharides were obviously different before and after steam explosion treatment. The low degradation temperature of SE-WGP$_I$ could be attributed to the hydrothermal degradation of polysaccharides caused by steam explosion, resulting in the generation of some substances that were easy to be degraded. The low weight loss of SE-WGP$_I$ could be ascribed to the polymerization or aggregation of some macromolecular substances, which had strong heat resistance and were not easy to decompose at 600 °C. The results indicated that SE-WGP$_I$ possessed superior thermal stability. From the perspective of food industry production, good thermal stability is conducive to maintaining the nutritional quality of materials during thermal processing such as sterilization and baking [34].

3.11. Antioxidant Activities

In many studies on structure characterization and functional activities of polysaccharides from wheat bran, researchers have proposed and verified that wheat bran polysaccharides had good in vivo and in vitro antioxidant activities [35,36]. Inspired by these articles, we tentatively evaluated the in vitro antioxidant activities of wheat germ polysaccharides.

DPPH radical scavenging activity of SE-WGP$_I$ with different concentrations was investigated as depicted in Figure 5c. SE-WGP$_I$ showed a dose-dependent radical scavenging capacity at 0.2~1.2 mg/mL. It presented the highest DPPH radical scavenging activity of 67.41 ± 1.16% at 1.2 mg/mL. In a previous report [37], two polysaccharides extracted from wheat bran showed weak DPPH radical scavenging activity with EC$_{50}$ estimated at 9.6 and 4.2 mg/mL, respectively. In contrast, SE-WGP$_I$ had an obvious scavenging result on DPPH radicals with the EC$_{50}$ of 0.81 mg/mL. It was also found in the literature that the DPPH radical scavenging ability of SE-WGP$_I$ was higher than that of the wheat germ polysaccharide with an average molecular weight of 2.99 × 10^5 Da, and its scavenging rate at 4 mg/mL was 50.21% [35]. SE-WGP$_I$ contained a low amount of protein, which has been reported to play a significant effect on the overall radical scavenging activity of polysaccharides [38].

In Figure 5d, when the concentration of SE-WGP$_I$ was the highest, the scavenging rate of this polysaccharide on hydroxyl radicals reached 40.21%, which was also dose-dependent. Meanwhile, the hydroxyl radical scavenging ability of SE-WGP$_I$ was similar to that of the wheat germ polysaccharide, with an average molecular weight of 2.99 × 10^5 Da [35]. The most likely pathway of antioxidant activity of the polysaccharide on hydroxyl radicals was that the polysaccharide could provide hydrogen to stabilize free radicals and/or directly react with them to terminate the free radical chain reaction [36]. The above results indicated that SE-WGP$_I$ in this study had relatively good antioxidant activity.

4. Conclusions

This is the first report on the purification and structural characterization of the heteropolysaccharide from wheat germ modified by SE treatment. SE-WGP$_I$, as a novel water-soluble polysaccharide, was composed of arabinose, xylose and glucose with a molar ratio of 20.71: 19.77: 59.51. Based on the results of FT-IR, periodate oxidation-Smith degradation, methylation and NMR spectroscopy, the backbone of SE-WGP$_I$ was →4,6)-α-D-Glc*p* (1→6)-α-D-Glc*p* (1→3)-β-D-Xyl*p* (1→5)-α-L-Ara*f* (1→ and the side chain was α-D-Glc*p* (1→ linked at the C4-position of →4,6)-α-D-Glc*p* (1→. The result of the Congo red test suggested that SE-WGP$_I$ possibly has a triple helix conformation, and it had relatively good thermal stability according to the TGA analysis. In addition, SE-WGP$_I$ showed valid in vitro antioxidant activities in a dose-dependent manner. This study provides a theoretical basis for further research into the structure–activity relationship of SE-WGP$_I$. The bioactivity of SE-WGP$_I$ is currently being studied in our laboratory.

Author Contributions: Methodology, writing—original draft preparation, L.H.; writing—review and editing, X.Z.; investigation, X.T.; investigation, R.L. (Ranran Li); writing—review and editing, supervision, W.S.; formal analysis, R.L. (Rui Liu); formal analysis, T.W.; funding acquisition, validation, M.Z. All authors have read and agreed to the published version of the manuscript.

Funding: The work was supported by National Natural Science Foundation of China (32172169), Tianjin "131" Innovative Talent Team Project (201926).

Institutional Review Board Statement: Not applicable.

Informed Consent Statement: Not applicable.

Data Availability Statement: Data is contained within the article.

Conflicts of Interest: The authors declare no conflict of interest.

References

1. Rizzello, C.G.; Nionelli, L.; Coda, R.; De Angelis, M.; Gobbetti, M. Effect of sourdough fermentation on stabilisation, and chemical and nutritional characteristics of wheat germ. *Food Chem.* **2010**, *119*, 1079–1089. [CrossRef]
2. Mueller, T.; Voigt, W. Fermented wheat germ extract-nutritional supplement or anticancer drug? *Nutr. J.* **2011**, *10*, 89. [CrossRef] [PubMed]
3. Ozdogan, A.; Gunes, R.; Palabiyik, I. Investigating release kinetics of phenolics from defatted wheat germ incorporated chewing gums. *J. Sci. Food Agric.* **2019**, *99*, 6333–6341. [CrossRef] [PubMed]
4. Sudha, M.L.; Srivastava, A.K.; Leelavathi, K. Studies on pasting and structural characteristics of thermally treated wheat germ. *Eur. Food Res. Technol.* **2006**, *225*, 351–357. [CrossRef]
5. Yun, L.; Wu, T.; Liu, R.; Li, K.; Zhang, M. Structural Variation and Microrheological Properties of a Homogeneous Polysaccharide from Wheat Germ. *J. Agric. Food Chem.* **2018**, *66*, 2977–2987. [CrossRef]
6. Liu, C.Y.; Sun, Y.Y.; Jia, Y.Q.; Geng, X.Q.; Pan, L.C.; Jiang, W.; Xie, B.Y.; Zhu, Z.Y. Effect of steam explosion pretreatment on the structure and bioactivity of *Ampelopsis grossedentata* polysaccharides. *Int. J. Biol. Macromol.* **2021**, *185*, 194–205. [CrossRef] [PubMed]
7. Alvira, P.; Tomas-Pejo, E.; Ballesteros, M.; Negro, M.J. Pretreatment technologies for an efficient bioethanol production process based on enzymatic hydrolysis: A review. *Bioresour. Technol.* **2010**, *101*, 4851–4861. [CrossRef]
8. Chadni, M.; Grimi, N.; Bals, O.; Ziegler-Devin, I.; Brosse, N. Steam explosion process for the selective extraction of hemicelluloses polymers from spruce sawdust. *Ind. Crop. Prod.* **2019**, *141*, 111757. [CrossRef]
9. Wang, L.; Xu, H.; Yuan, F.; Fan, R.; Gao, Y. Preparation and physicochemical properties of soluble dietary fiber from orange peel assisted by steam explosion and dilute acid soaking. *Food Chem.* **2015**, *185*, 90–98. [CrossRef]
10. Sui, W.; Xie, X.; Liu, R.; Wu, T.; Zhang, M. Effect of wheat bran modification by steam explosion on structural characteristics and rheological properties of wheat flour dough. *Food Hydrocoll.* **2018**, *84*, 571–580. [CrossRef]
11. Sevag, M.; Maiweg, L. The respiration mechanism of pneumococcus. III. *J. Exp. Med.* **1934**, *60*, 95–105. [CrossRef]
12. Dong, Q.; Yao, J.; Fang, J.N.; Ding, K. Structural characterization and immunological activity of two cold-water extractable polysaccharides from *Cistanche deserticola* Y. C. Ma. *Carbohydr. Res.* **2007**, *342*, 1343–1349. [CrossRef]
13. Cuesta, G.; Suarez, N.; Bessio, M.I.; Ferreira, F.; Massaldi, H. Quantitative determination of pneumococcal capsular polysaccharide serotype 14 using a modification of phenol–sulfuric acid method. *J. Microbiol. Methods* **2003**, *52*, 69–73. [CrossRef]
14. Shibata, N.; Okawa, Y. Chemical structure of beta-galactofuranose-containing polysaccharide and O-linked oligosaccharides obtained from the cell wall of pathogenic dematiaceous fungus *Fonsecaea pedrosoi*. *Glycobiology* **2011**, *21*, 69–81. [CrossRef]
15. Blumenkrantz, N.; Asboe-Hansen, G. New method for quantitative determination of uronic acids. *Anal. Biochem.* **1973**, *54*, 484–489. [CrossRef]
16. Jin, X.; Wang, Q.; Yang, X.; Guo, M.; Li, W.; Shi, J.; Adu-Frimpong, M.; Xu, X.; Deng, W.; Yu, J. Chemical characterisation and hypolipidaemic effects of two purified *Pleurotus eryngii* polysaccharides. *Int. J. Food Sci. Tech.* **2018**, *53*, 2298–2307. [CrossRef]
17. Shen, C.-Y.; Jiang, J.-G.; Li, M.-Q.; Zheng, C.-Y.; Zhu, W. Structural characterization and immunomodulatory activity of novel polysaccharides from *Citrus aurantium* Linn. variant *amara* Engl. *J. Funct. Foods* **2017**, *35*, 352–362. [CrossRef]
18. Dong, Q. Structural characterization of the water-extractable polysaccharides from *Sophora subprostrata* roots. *Carbohydr. Polym.* **2003**, *54*, 13–19. [CrossRef]
19. Li, F.; Pak, S.; Zhao, J.; Wei, Y.; Zhang, Y.; Li, Q. Structural Characterization of a Neutral Polysaccharide from *Cucurbia moschata* and Its Uptake Behaviors in Caco-2 Cells. *Foods* **2021**, *10*, 2357. [CrossRef]
20. Nie, C.; Zhu, P.; Wang, M.; Ma, S.; Wei, Z. Optimization of water-soluble polysaccharides from stem lettuce by response surface methodology and study on its characterization and bioactivities. *Int. J. Biol. Macromol.* **2017**, *105*, 912–923. [CrossRef]

21. Chen, Q.; Wang, R.; Wang, Y.; An, X.; Liu, N.; Song, M.; Yang, Y.; Yin, N.; Qi, J. Characterization and antioxidant activity of wheat bran polysaccharides modified by *Saccharomyces cerevisiae* and *Bacillus subtilis* fermentation. *J. Cereal Sci.* **2021**, *97*, 103157. [CrossRef]
22. Fan, Y.; Lin, M.; Luo, A. Extraction, characterization and antioxidant activities of an acidic polysaccharide from *Dendrobium devonianum*. *J. Food Meas. Charact.* **2022**, *16*, 867–879. [CrossRef]
23. Shi, L. Bioactivities, isolation and purification methods of polysaccharides from natural products: A review. *Int. J. Biol. Macromol.* **2016**, *92*, 37–48. [CrossRef] [PubMed]
24. Wang, K.-p.; Wang, J.; Li, Q.; Zhang, Q.-l.; You, R.-x.; Cheng, Y.; Luo, L.; Zhang, Y. Structural differences and conformational characterization of five bioactive polysaccharides from *Lentinus edodes*. *Food Res. Int.* **2014**, *62*, 223–232. [CrossRef]
25. Zou, P.; Yang, X.; Yuan, Y.; Jing, C.; Cao, J.; Wang, Y.; Zhang, L.; Zhang, C.; Li, Y. Purification and characterization of a fucoidan from the brown algae *Macrocystis pyrifera* and the activity of enhancing salt-stress tolerance of wheat seedlings. *Int. J. Biol. Macromol.* **2021**, *180*, 547–558. [CrossRef] [PubMed]
26. Chen, L.C.; Jiang, B.K.; Zheng, W.H.; Zhang, S.Y.; Li, J.J.; Fan, Z.Y. Preparation, characterization and anti-diabetic activity of polysaccharides from adlay seed. *Int. J. Biol. Macromol.* **2019**, *139*, 605–613. [CrossRef] [PubMed]
27. Deveci, E.; Cayan, F.; Tel-Cayan, G.; Duru, M.E. Structural characterization and determination of biological activities for different polysaccharides extracted from tree mushroom species. *J. Food Biochem.* **2019**, *43*, e12965. [CrossRef]
28. He, Y.; Peng, H.; Zhang, H.; Liu, Y.; Sun, H. Structural characteristics and immunopotentiation activity of two polysaccharides from the petal of Crocus sativus. *Int. J. Biol. Macromol.* **2021**, *180*, 129–142. [CrossRef]
29. Zhang, X.; Yu, Q.; Jiang, H.; Ma, C.; David Wang, H.M.; Wang, J.; Kang, W.Y. A novel polysaccharide from *Malus halliana* Koehne with coagulant activity. *Carbohydr. Res.* **2019**, *485*, 107813. [CrossRef]
30. Guo, X.; Kang, J.; Xu, Z.; Guo, Q.; Zhang, L.; Ning, H.; Cui, S. Triple-helix polysaccharides: Formation mechanisms and analytical methods. *Carbohyd. Polym.* **2021**, *262*, 117962. [CrossRef]
31. Mao, C.-F.; Hsu, M.-C.; Hwang, W.-H. Physicochemical characterization of grifolan: Thixotropic properties and complex formation with Congo Red. *Carbohydr. Polym.* **2007**, *68*, 502–510. [CrossRef]
32. Falch, B.H.; Espevik, T.; Ryan, L.; Stokke, B.T. The cytokine stimulating activity of (1→3)-β-D-glucans is dependent on the triple helix conformation. *Carbohydr. Res.* **2000**, *329*, 587–596. [CrossRef]
33. Iqbal, M.S.; Massey, S.; Akbar, J.; Ashraf, C.M.; Masih, R. Thermal analysis of some natural polysaccharide materials by isoconversional method. *Food Chem.* **2013**, *140*, 178–182. [CrossRef]
34. Zhang, W.; Zeng, G.; Pan, Y.; Chen, W.; Huang, W.; Chen, H.; Li, Y. Properties of soluble dietary fiber-polysaccharide from papaya peel obtained through alkaline or ultrasound-assisted alkaline extraction. *Carbohydr. Polym.* **2017**, *172*, 102–112. [CrossRef]
35. Shang, X.-L.; Liu, C.-Y.; Dong, H.-Y.; Peng, H.-H.; Zhu, Z.-Y. Extraction, purification, structural characterization, and antioxidant activity of polysaccharides from wheat bran. *J. Mol. Struct.* **2021**, *1233*, 130096. [CrossRef]
36. Wang, J.; Cao, Y.; Wang, C.; Sun, B. Wheat bran xylooligosaccharides improve blood lipid metabolism and antioxidant status in rats fed a high-fat diet. *Carbohydr. Polym.* **2011**, *86*, 1192–1197. [CrossRef]
37. Jiang, L.; Wang, W.; Wen, P.; Shen, M.; Li, H.; Ren, Y.; Xiao, Y.; Song, Q.; Chen, Y.; Yu, Q.; et al. Two water-soluble polysaccharides from mung bean skin: Physicochemical characterization, antioxidant and antibacterial activities. *Food Hydrocoll.* **2020**, *100*, 105412. [CrossRef]
38. Hromadkova, Z.; Paulsen, B.S.; Polovka, M.; Kostalova, Z.; Ebringerova, A. Structural features of two heteroxylan polysaccharide fractions from wheat bran with anti-complementary and antioxidant activities. *Carbohydr. Polym.* **2013**, *93*, 22–30. [CrossRef]

Article

Biochemical and Structural Properties of a High-Temperature-Active Laccase from *Bacillus pumilus* and Its Application in the Decolorization of Food Dyes

Tao Li [1,2], Xiuxiu Chu [1], Zhaoting Yuan [1], Zhiming Yao [1], Jingwen Li [1], Fuping Lu [3,*] and Yihan Liu [1,*]

1   Key Laboratory of Industrial Fermentation Microbiology, Ministry of Education, Tianjin Key Laboratory of Industrial Microbiology, College of Biotechnology, Tianjin University of Science and Technology, Tianjin 300457, China; 201096@xxmu.edu.cn (T.L.); chuxiuxiu@126.com (X.C.); yuanzhaoting@mail.tust.edu.cn (Z.Y.); yaozhiming611@126.com (Z.Y.); lijingwen619@126.com (J.L.)
2   School of Life Science and Technology, Xinxiang Medical University, Xinxiang 453003, China
3   State Key Laboratory of Food Nutrition and Safety, Tianjin 300457, China
*   Correspondence: lfp@tust.edu.cn (F.L.); lyh@tust.edu.cn (Y.L.); Tel.: +86-022-6060-1958 (F.L.); +86-022-6060-2949 (Y.L.)

Abstract: A novel laccase gene isolated from *Bacillus pumilus* TCCC 11568 was expressed, and the recombinant laccase (rLAC) displayed maximal activity at 80 °C and at pH 6.0 against ABTS. rLAC maintained its structural integrity at a high temperature (355 K) compared to its tertiary structure at a low temperature (325 K), except for some minor adjustments of certain loops. However, those adjustments were presumed to be responsible for the formation of a more open access aisle that facilitated the binding of ABTS in the active site, resulting in a shorter distance between the catalytic residue and the elevated binding energy. Additionally, rLAC showed good thermostability ($\leq$70 °C) and pH stability over a wide range (3.0–10.0), and displayed high efficiency in decolorizing azo dyes that are applicable to the food industry. This work will improve our knowledge on the relationship of structure–function for thermophilic laccase, and provide a candidate for dye effluent treatment in the food industry.

Keywords: *Bacillus pumilus*; molecular dynamic simulations; enzymatic characterization; food dye decolorization

Citation: Li, T.; Chu, X.; Yuan, Z.; Yao, Z.; Li, J.; Lu, F.; Liu, Y. Biochemical and Structural Properties of a High-Temperature-Active Laccase from *Bacillus pumilus* and Its Application in the Decolorization of Food Dyes. *Foods* **2022**, *11*, 1387. https://doi.org/10.3390/foods11101387

Academic Editor: Alejandra Acevedo-Fani

Received: 7 April 2022
Accepted: 9 May 2022
Published: 11 May 2022

Publisher's Note: MDPI stays neutral with regard to jurisdictional claims in published maps and institutional affiliations.

1. Introduction

Laccases (EC 1.10.3.2), as multi-copper enzymes can catalyze the mono-electronic oxidation of substrates (aromatic amines and phenolic compounds) to yield water as a by-product with molecular oxygen as an electron acceptor [1]. Thus, laccases are more eco-friendly and potentially superior to other oxidoreductases, which require expensive cofactors (NAD^+ or $NADP^+$) or the harmful hydrogen peroxide to complete the oxidative reaction [2,3]. Additionally, the substrate scope of laccases can be expanded to polyphenol polymers and other substrates with the assistance of redox mediators, such as syringaldazine, 2,6-dimethoxyphenol (2,6-DMP), and 2,2′-azino-bis-(3-ethylbenzothiazoline-6-sulphonic acid) (ABTS) [4,5]. Because of their capacity for oxidizing numerous substrates (such as methoxy-substituted monophenols, diphenols, aliphatic, and aromatic amine), laccases demonstrate a huge potential for numerous biotechnological processes, including food industry wastewater treatment, pulp bio-bleaching, and xenobiotics bioremediation [6].

Laccases can be acquired from different sources, such as bacteria, fungi, actinomycetes, and plants. Fungal laccases show more favorable characteristics for commercial applications in comparison to other laccases from plants and bacteria, because of their high reduction potential [7]. However, fungal laccases also demonstrate some drawbacks, including low fungal growth rate and being operative only under low temperature and pH conditions. They are often deactivated in some industrial sections, such as food and textile effluent decolorization, under the extreme conditions that occur, including high temperature, extreme pH value, or high ionic strength [8].

Thermophilic enzymes are favorable for use in both fundamental research and industrial applications [9]. A lower risk of microbial contamination and higher reaction rates were obtained when using thermophilic enzymes at high temperatures [10]. Moreover, thermophilic enzymes are favorable for some application scenarios that feature high temperatures, such as the treatment of dye effluent that is discharged during food processing. Therefore, over the past decade, more and more attention has been paid toward laccases with bacterial origins that show an excellent degree of catalytic ability at high temperatures, such as *Bacillus, Staphylococcus, Streptomyces, Geobacillus, Aquisalibacillus, Rhodococcus, Lysinibacillus, Pseudomonas, Proteobacterium, Enterobacter, Delfia*, and *Alteromonas* [11].

The most well-known bacterial laccase to date is CotA from *Bacillus subtilis*, which is a constituent of the endospore coat. It demonstrates a strikingly intrinsic thermostability with an optimum temperature of 75 °C, as well as a half-life of 2 h or so at 80 °C [12]. Recently, more and more laccases of several *Bacillus* species, including *B. licheniformis, B. amyloliquefaciens, B. halodurans, B. subtilis, B. thuringiensis*, and *B. vallismortis* have exhibited excellent rates of catalytic performance over a broad pH range and at high temperatures [13]. Referring to the heat-resistance features of laccase, researchers have reported their experiment evidence and presented a different proposal. For instance, salt bridges and glycosylation played a vital role in maintaining the structural integrity of thermophilic laccase from *Trametes versicolor* [14]. A similar stimulative effect on laccase activity via thermal incubation has also been demonstrated for some fungal laccases derived from *Physisporinus rivulosus, Marasmius quercophilus, Melanocarpus albomyces*, and *Fomes sclerodermeus* [15,16]. Nevertheless, studies for interpreting the thermo-stable mechanism of *Bacillus* laccase at the molecular level are still rare. Therefore, more research is still required to elucidate the structure–function relationship of bacterial laccases at high temperature. According to the peculiar characteristics of bacterial laccases, including a high tolerance for temperature, an alkaline environment, and salt, it is still necessary to discover novel bacterial laccases to enlarge our knowledge on the bacterial laccases and to expand the range of obtainable biocatalysts for different industrial processes performed under harsh conditions. Here, a novel laccase coding sequence was amplified from *B. pumilus* TCCC 11568 and heterologously expressed to identify its catalytic and biochemical properties. Molecular dynamics (MD) simulations were employed to explore the molecular mechanisms for the thermophilic features of this laccase. The ability of this novel laccase to decolorize synthetic dyes at high temperatures was then investigated.

2. Materials and Methods

2.1. Strains, Plasmid, and Chemicals

E. coli JM109, *E. coli* BL21 (DE3), and plasmid pET-22b (+) were deposited in our lab. Pyrobest DNA Polymerase, restriction enzymes, pMD18-T vector cloning Kit, Plasmid Mini Kit, T4 DNA ligase, Gel Extraction and Purification Kit, and DNA Extraction Kit were supplied by TAKARA Bio Inc. (Dalian, China). ABTS and synthetic dyes (azophloxine, etythrosine, Sunset Yellow, Ponceau 4R, Amaranth, Indigo Carmine, Acid Orange II, Congo Red) were ordered from Sigma-Aldrich (St. Louis, MO, USA).

2.2. Strain Screening and Cultivation

The forest soil sample was collected in Guangzhou, Guangdong Province, China. The screening procedure was performed according to the reported method for isolating *Klebsiella pneumonia*, which produced a new pH-stable laccase [17]. Briefly, 10 g of soil was transferred to a sterile saline solution (0.9% NaCl, 100 mL). Subsequently, 1 mL of the suspension was dispersed in 5 mL of Luria-Bertani (LB) medium (tryptone 10 g/L, yeast extract 5 g/L, and NaCl 10 g/L) with shaking for 30 min at 200 rpm and 37 °C. The enriched cells were spread on LB plates supplemented with 0.2 mM Cu^{2+} (copper sulfate) and incubated at 37 °C for 24 h. Several drops of syringaldazine (SGZ)/absolute ethanol

(0.1%, w/v) were added to the bacterial colonies to distinguish the laccase-producing strains by their pink color. Then the obtained colonies that showed a pink color were picked up and purified using single-colony separation.

2.3. Phylogenetic Analysis of the Laccase-Producing Strain

The 16S rDNA was amplified with the genomic DNA of the laccase-producing strain according to the method described previously [17]. The homologous sequences of the resulting 16S rDNA sequence were searched with GenBank BLAST (http://www.ncbi.nlm.nih.gov/BLAST/ accessed on 12 March 2020). Subsequently, a bootstrap consensus tree was built using MEGA 6.0 software with the neighbor-joining method for phylogenetic analysis [18]. To further identify this strain, taxonomic analysis was conducted as noted in Bergey's Manual of Determinative Bacteriology.

2.4. Heterologous Expression of Laccase

The gene encoding laccase (*lac*) was amplified with two primers, Lac-F (5'-CGCGG ATCCGATGGCACTGGAAAAATTTG-3'; the underlined bases encode the *Bam*HI site) and Lac-R (5'-ACGCGTCGACCTGCTTATCCGTGACGTCC-3'; the underlined bases encode the *Sal*I site) from the genomic DNA of *B. pumilus*. After double digestion with *Bam*HI and *Sal*I, it was integrated into the *Bam*HI-*Sal*I-linearized expression plasmid pET-22b (+) to construct the plasmid pET-*lac*. It was then transformed into *E. coli* BL21 (DE3). A positive colony was initially selected to grow at 37 °C in LB medium containing ampicillin (100 µg/mL) for 12 h. Subsequently, the preculture (1 mL) was diluted with 50 mL of fresh LB medium with ampicillin (100 µg/mL). When OD_{600} reached 0.6–0.8, 1 mM isopropyl-β-D-1-thiogalactopyranoside (IPTG) was used to induce the recombinant laccase (rLAC) at 16 °C for 20 h. *E. coli* BL21 cells harboring the empty plasmid pET-22b (+) were used as the control.

2.5. Molecular Docking and Molecular Dynamics Simulation Analysis of rLAC

The software AutoDock Vina was used to perform molecular docking between the rLAC and the ligand. The 3D structure of rLAC was BLAST-searched in the UniProt database using SWISS-model (https://swissmodel.expasy.org/ accessed on 15 May 2021). A protein structure (Protein Data Bank entry 1GSK) with more than 70% homology was used as the template for homology modeling. The 3D structure of ABTS was obtained from PubChem (https://pubchem.ncbi.nlm.nih.gov/ accessed on 15 May 2021). A method of semi-flexible docking was utilized to allow the chemical bonds of the ligand to freely rotate, while fixing the coordinates of the atoms in rLAC. The grid parameters were set to build a sphere grid near the active sites of rLAC1 (H419, C492, H497, and M502) with a radius of 5.0 Å in the docking area.

The docking conformation with the lowest protein–ligand binding energy was selected for molecular dynamics simulation by Gromacs 5.1.4, combined with the AMBER99SB forcefield. The parameters for the molecular dynamics simulations were set based on the previous study [19,20], in which the SPC216 model was set as the water molecule. The simulation system was neutralized by a total of 11 Na^+, which were added into a cubic box with a distance of 1.5 nm in all three dimensions. The incorrect geometries and collisions among the atoms in the simulation system were removed by the steepest descent minimization method, using a maximum number of 50,000 steps. Equilibration simulations for the solvent around the protein were performed under constant volume–temperature (NVT) and constant pressure–temperature (NPT) ensembles under harmonic restraints. Short-range van der Waals interactions and short-range electrostatic interactions were truncated smoothly at 1 nm. The leap-frog integrator algorithm with a step size of 2 fs was utilized to integrate the equation of motion for equilibrium dynamics. The LINCS algorithm was used to constrain the hydrogen bonds. The Particle Mesh Ewald (PME) method was used to calculate the long-range electrostatic interactions with a grid size of 0.16 Å. The pressure (1 bar) and temperature (300 K) under the isothermal–isostatic (NPT) ensemble were maintained using Parrinello-Rahman and the V-rescale method, respectively. Both the NVT and NPT simulation times were set as 100 ps and the final MD simulations were carried out for 100 ns. The root-mean-square fluctuation (RMSF) and the root-mean-square deviation (RMSD) were analyzed using the *gmx rmsf* and *gmx rms* commands respectively. The snapshots were performed using Visual Molecular Dynamics (VMD) 1.9.4.

The approach of MM/PBSA (Molecular Mechanics Poisson Boltzmann Surface Area) was utilized for energy decomposition according to the trajectory of the molecular dynamics simulations. The binding energy within a time period of 0~10 ns was calculated using the tool g_mmpbsa.py and the software Numpy (Numerical Python). The binding free energy is calculated using Equation (1):

$$\Delta G_{bind} = \Delta E_{vdw} + \Delta E_{ele} + \Delta G_{solv} + \Delta G_{SASA} \tag{1}$$

Among them, ΔG_{bind} stands for the binding free-energy, ΔE_{ele} stands for the electrostatic interaction, ΔE_{vdw} stands for the van der Waals force, ΔG_{SASA} stands for the non-polar contribution energy calculated using the solvent accessible surface (SASA), and ΔG_{solv} stands for the solvent free-energy.

2.6. Purification of rLAC

Cells expressing rLAC were centrifugally harvested via centrifugation at $8000\times g$ and 4 °C for 15 min, and then resuspended in 20 mM Tris–HCl buffer (pH 7.0) with 500 mM NaCl and 20 mM imidazole. Subsequently, the cells were disrupted via sonication at 320 W with 4 s strokes and 3 s intervals. After centrifugation (12,000 $\times$ g, 30 min), the supernatant was injected into a nickel-nitrilotriacetic acid (Ni-NTA) agarose gel column (Shenggong, Shanghai, China). rLAC was eluted using Tris-buffer containing 500 mM imidazole and 500 mM NaCl (pH 7.0) after removing the impurities. After dialysis, the purity and molecular mass of rLAC were determined using sodium dodecyl sulfate polyacrylamide gel electrophoresis (SDS-PAGE) analysis.

2.7. Enzyme Assay

The activity of rLAC was determined with ABTS as the substrate at 80 °C using the previously described method [21,22]. Concisely, the enhanced absorbance at 420 nm (ε_{420} = 36,000 M^{-1} cm^{-1}) was monitored for ABTS (6 mM) oxidation in 0.1 M citrate–phosphate buffer (pH 6.0). One unit of the activity for laccase was designated as the amount of laccase that oxidized 1 μmol of substrate per minute.

2.8. Characterization of rLAC

The optimal temperature of rLAC was determined with ABTS as the substrate at different temperatures from 30 °C to 90 °C. To identify the optimal pH of rLAC, the enzymatic reaction was conducted over a pH range of 3.0 to 9.0 with citrate–phosphate buffer (0.1 M, pH 3.0 to 8.0) or glycine–sodium hydroxide buffer (0.1 M, pH 9.0). The relative enzymatic activity was determined with the maximal activity as 100%.

To identify the thermostability of rLAC, it was incubated at three temperatures (60 °C, 70 °C, and 80 °C) for 0 h to 2 h without substrate. The pH stability was identified by keeping the purified rLAC at pH 3.0, 7.0, and 10.0 (4 °C) for various periods (0–10 days). After incubation, the residual activities of the samples were measured with the substrate ABTS under optimal conditions (80 °C, pH 6.0). The activity of the initial rLAC was taken as 100%.

The effects of the metal ions (K^+, Na^+, Cu^{2+}, Ca^{2+}, Fe^{2+}, Fe^{3+}, Zn^{2+}, Mn^{2+}, Co^{2+}, Mg^{2+}, and Ba^{2+}), as well as the inhibitors (EDTA, SDS, L-cysteine, dithiothreitol, and β-mercaptoethanol) on rLAC activity, were assayed with ABTS. rLAC and metal ions or inhibitors were mixed before adding ABTS to determine the relative activity. The laccase activity measured without adding metal ion or inhibitor was marked as 100%.

2.9. Dye Decolorization

Decolorization was conducted using azo dyes, such as etythrosine (λ_{max} = 520 nm), azophloxine (λ_{max} = 530 nm), Sunset Yellow (λ_{max} = 480 nm), Ponceau 4R (λmax = 510 nm), Indigo Carmine (λ_{max} = 612 nm), Amaranth (λ_{max} = 520 nm), Acid Orange II (λ_{max} = 480 nm), and Congo Red (λ_{max} = 500 nm). Decolorization was assayed with or without the mediator (acetosyringone, syringaldehyde, and ABTS). The reaction solution (6 mL) for decolorization was composed of individual dye (azophloxine, Indigo Carmine, Ponceau 4R, 50 mg/L; Amaranth, Acid Orange II, 100 mg/L; etythrosine, Congo Red, Sunset Yellow, 200 mg/L), purified rLAC (80 U), 0.1 M buffer with different pH (citrate–phosphate buffer, pH 5.0 and 7.0; 0.1 M glycine–sodium hydroxide buffer, pH 9.0), and 0.1 mM mediator. After a 6 h reaction at 60 °C, the decolorization ability of each dye was identified by recording the reduction of each dye at the maximum absorbance using the equation: decolorization (%) = [((initial absorbance) − (final absorbance))/(initial absorbance)] × 100%, reflecting the reduction in concentration because of the oxidation reaction by rLAC. The control reactions were carried out with no laccase under the same conditions.

3. Results and Discussion

3.1. Identification of the Strain with Laccase Activity

Numerous studies have been focused upon bacterial laccases over the past decade, especially for those derived from the *Bacillus* genus, due to their advantages over fungal laccases, such as a higher pH stability, wider pH adaptation, and thermostability [12,23]. In this work, a bacterial strain demonstrated a high degree of laccase activity towards syringaldazine among 216 bacteria screened from a soil sample gathered from Guangdong Province of China. After aligning the 16S rDNA sequence of this strain with BLAST included in the NCBI database, the result suggested this strain was a member of the *Bacillus* family. Phylogenetic tree analysis suggested that this strain was most closely associated with *B. pumilus* (Figure S1 in Supplementary Materials).

The morphological properties of this strain are as follows: rod-shaped, Gram-positive, rough-surfaced, and with spore-producing colonies (data not shown). Therefore, it was finally characterized and named *B. pumilus* TCCC 11568 on the basis of its biochemical tests and morphological characteristics, along with the 16S rDNA analysis. In summary, we isolated a new *Bacillus* strain showing laccase activity from a soil sample, and our study enriched the family of laccase-producing bacteria.

3.2. Heterologous Expression of Laccase

The obtained laccase gene (*lac*) (GenBank: MT150577) contains 1530 bp nucleotides coding for 510 amino acids (calculated molecular mass: 57 kDa). Multiple protein sequence alignments indicated that its four histidine-rich regions for copper-binding were greatly conserved with other laccases. A high amino acid identity was demonstrated with the following laccases of other *Bacillus* species, such as 97.65% to *B. pumilus* ATCC 7061 [24], 94.71% to *B. pumilus* W3 [25], 67.96% to *B. vallismortis* fmb103 [26], 66.54% to *B. amyloliquefaciens* TCCC 111018 [27], 66.21% to *B. subtilis* X1 [28], 64.98% to *B. velezensis* TCCC 111904 [29], and 61.36% to *B. licheniformis* ATCC 14580 (Figure S2) [30].

The tertiary structure of the laccase from *B. pumilus* TCCC 11568 was built using homology modeling using the crystal structure (PDB ID: 2WSD) of laccase from *B. subtilis* MB24 as the template (Figure 1).

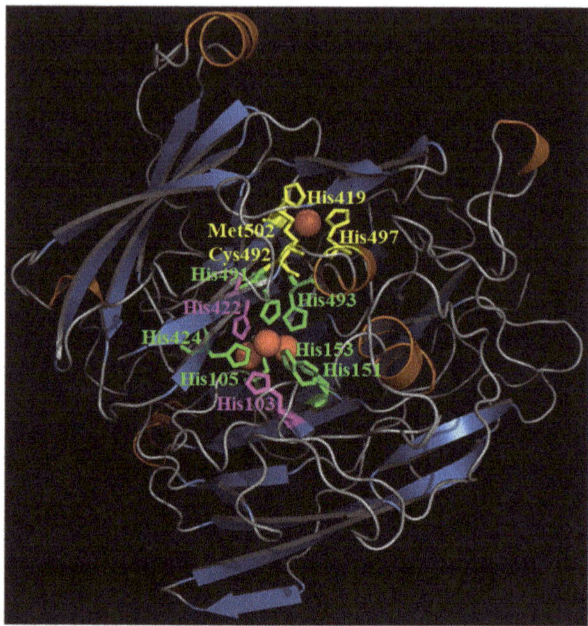

Figure 1. Homology model of the *B. pumilus* TCCC 11568 laccase. Three copper centers, including Type 1, Type 2, and Type 3, are depicted in yellow, purple, and green, respectively. The α-helices and β-sheets are denoted in orange and blue, respectively. Cu^{2+} located at the active site of rLAC is labeled in red.

Bacillus laccases commonly have a highly conserved catalytic site with four copper atoms bound to the T1, T2, and T3 copper centers [31]. Similarly, three Cu oxidase domains, such as T1 (H419, C492, H497, and M502), T2 (H103 and H422), and T3 (H105, H151, and H493/H153, H424, and H491), also exist in the active center of the *B. pumilus* TCCC 11568 laccase.

In this work, a significant band of approximately 57 kDa was observed on the SDS-PAGE gel of the cell extracts, while no band was found in the cell extract of the control (Figure S3a). A target band with a molecular mass of approximately 57 kDa was detected on the SDS-PAGE gel after purification (Figure S3b). The obtained molecular mass was similar to some other reported bacterial laccases, including *B. amyloliquefaciens* [27] and *B. halodurans* C-125 (56 kDa) [32]. Its specific activity towards ABTS was 190 U/mg, which was notably higher than other *Bacillus* laccases to ABTS, such as *B. amyloliquefaciens* LC02 (20.7 ± 1.2 U/mg for ABTS) [33]. According to the sequence alignment, protein modelling, and enzyme activity analysis, it can be concluded that a new laccase with a high level of activity and a conserved catalytic motif that has been observed in other reported laccases was obtained in our study.

3.3. Effect of Temperature and pH on the Activity and Stability of rLAC

The maximum activity of rLAC for ABTS oxidation was detected at 80 °C, and a relatively high activity level was sustained at 60–90 °C (Figure 2a). To investigate its thermostability, the residual activities were determined after incubation at differing temperatures (50–80 °C) for 0 min to 120 min (Figure 2c). It suggested that rLAC was quite stable at 50 °C or 60 °C, and retained 87.3% or 70% of the original activity at these temperatures after a 120-min incubation, respectively. Although rLAC showed an evidently decreased stability after incubation at a high temperature (70 °C or 80 °C), it still retained 50.3% and 24.6% of the original activity after a 120-min incubation, respectively (Figure 2c).

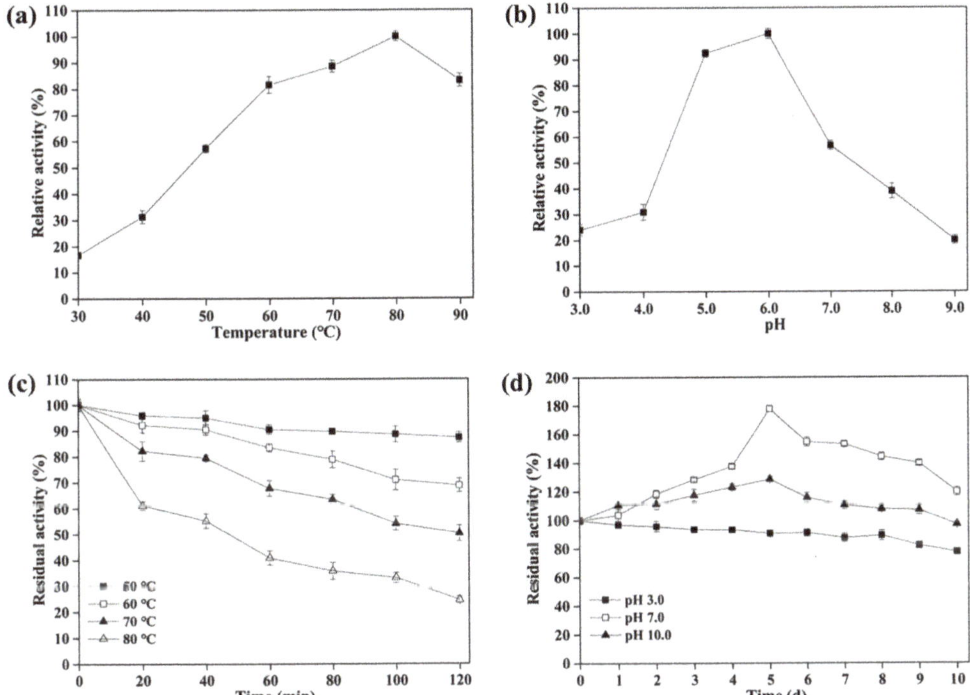

Figure 2. Effects of temperature or pH on the purified rLAC. (**a**) The laccase activity was recorded at different temperatures (30–90 °C) and (**b**) pH values. (**c**) The thermostability of rLAC was evaluated through monitoring the residual activity after incubation at 50 °C, 60 °C, 70 °C, and 80 °C for different lengths of time (0–120 min). (**d**) The pH stability was investigated through measurements of its residual activity after incubation for different periods (0–10 d) at 4 °C and pH 3.0, 7.0, and 10.0. All assays were conducted in triplicate, and the data were exhibited as mean ± SD.

The high optimum temperature was one of the most significant properties demonstrated by the bacterial laccases, especially for those derived from the *Bacillus* genus, such as 80 °C for CotA laccase from *B. amyloliquefaciens* TCCC 111018 [27], *B. velezensis* [29], and *B. licheniformis* [34]. By contrast, a similar phenomenon was not observed for fungal laccases whose optimum temperatures were usually below 60 °C. The optimal temperatures for laccases from *Cerrena unicolorstrain* GSM-01, *Trametes* sp. F1635, and *Aureobasidium melanogenum* strain 11-1 were 45 °C, 50 °C, and 40 °C, respectively [35–37]. Compared to fungal laccases, the high thermostability is another impressive feature of bacterial laccases. The thermostability of rLAC from *B. pumilus* TCCC 11568 was similar to laccases isolated from other *Bacillus* strains, such as *B. amyloliquefaciens* TCCC 111018 [27], and *B. tequilensis* SN4 [38]. Moreover, its thermostability was even higher than *B. licheniformis* DSM 13 laccase, which was deprived of approximately 92% of its initial activity after 1 h of incubation at 80 °C [33]. Thus, rLAC might be directly applied in treating hot effluents from the dyeing process, which is generally performed under high-temperature conditions [39]. In addition, using thermostable laccase to treat hot textile effluents is economically attractive because the recycling of hot water saves a significant amount of energy in the process.

rLAC was capable of oxidizing ABTS over a wide pH range and demonstrated its maximum activity at pH 6.0 (Figure 2b), which was higher than most *Bacillus* laccases, such as *B. velezensis* TCCC 111904 [29], *B. amyloliquefaciens* TCCC 111018 [27], *B. vallismortis* fmb-103 [26], *B. subtilis* X1 [28], and *B. clausii* KSM-K16 [13]. Generally, fungal laccases are merely stable under acidic and neutral pH conditions, but bacterial laccases are stable under alkaline conditions [40]. Additionally, rLAC exhibited a higher degree of stability (Figure 2d) than other *Bacillus* laccases over a wide pH range, such as the *B. velezensis* TCCC 111904 laccase, which retained 67.6% of its initial activity after incubation at 4 °C and pH 9.0 for 24 h [29]. Additionally, rLAC also showed quite a good degree of stability in an acidic environment, retaining 77.1% of its original activity after incubation at pH 3.0 for 10 d. The above results demonstrate that rLAC from *B. pumilus* TCCC 11568 has a high potential for applications in a wide range of pH environments, especially under alkaline conditions.

In summary, the *B. pumilus* TCCC 11568 laccase demonstrated thermophilic features and a higher degree of thermostability than some reported *Bacillus* laccases. Therefore, this laccase has excellent potential for becoming a new industrial enzyme formulation.

3.4. Influence of Metal Ions and Inhibitors on the Activity of rLAC

rLAC activity was seriously affected by 5 mM Co^{2+} and Mn^{2+}, retaining 24% and 9.4% of the control, respectively, but other metal ions did not lead to critical activity loss, indicating the high tolerance of rLAC towards most metal ions (Table 1). A comparable phenomenon was also found with the *B. safensis* sp. S31 laccase, which retained only 13.2% activity when incubated with 1 mM Mn^{2+} [41].

Table 1. Effect of metal ions or inhibitors on the rLAC activity.

Metal Ions/Inhibitors	Concentration (mM)	Relative Activity (%) [a]
None	-	100.0 ± 1.5
KCl	0.5	102.6 ± 2.7
	5	93.2 ± 1.9
$CaCl_2$	0.5	96.2 ± 3.2
	5	90.8 ± 1.6
$CuCl_2$	0.5	95.4 ± 2.1
	5	92.7 ± 2.0
$MgCl_2$	0.5	98.0 ± 2.9
	5	91.1 ± 3.1
$ZnSO_4$	0.5	97.4 ± 2.9
	5	105.5 ± 1.5
$BaCl_2$	0.5	101.7 ± 1.2
	5	87.1 ± 0.5
$NiSO_4$	0.5	90.9 ± 0.6
	5	105.1 ± 3.2
$CoCl_2$	0.5	77.5 ± 0.5
	5	24.0 ± 0.7
$FeSO_4$	0.5	75.8 ± 1.0
	5	42.1 ± 0.7
$FeCl_3$	0.5	94.0 ± 3.5
	5	79.1 ± 1.8
$MnCl_2$	0.5	17.5 ± 2.3
	5	9.4 ± 0.8
NaCl	0.5	105.5 ± 3.0
	5	89.5 ± 2.3
	10	79.9 ± 2.6
	100	68.7 ± 1.7
	500	50.7 ± 0.9
	1000	0

Table 1. *Cont.*

Metal Ions/Inhibitors	Concentration (mM)	Relative Activity (%) [a]
Dithiothreitol	0.5	6.2 ± 1.0
	5	0
L-Cysteine	0.5	35.2 ± 3.5
	5	27.6 ± 1.6
β-Mercaptoethanol	0.5	6.5 ± 2.0
	5	0
SDS	0.5	108.7 ± 0.7
	5	103.9 ± 2.1
EDTA	0.5	90.8 ± 1.1
	5	62.1 ± 2.8

[a] Data represent the means ± SD (n = 3) relative to the untreated control samples.

Electrolytes are required to obtain a high degree of efficiency in the industrial dyeing process. NaCl is the most extensively used electrolyte, and its concentration is approximately 25–30 g/L (around 0.5 M). Most fungal laccases are deactivated in solutions containing over 100 mM NaCl because they are intrinsically sensitive to halides [42]. This is a major barrier for applying fungal laccases to treat wastewater containing chloride ions at a high concentration. By contrast, some *Bacillus* laccases were highly tolerant to chloride, and their activities were even promoted by increased NaCl concentrations [32]. Different tolerances to halide inhibition among fungal and bacterial laccases may result from their different target localizations after protein translation. Laccases from fungal species are primarily secreted for degrading lignin, but *Bacillus* laccases anchor in the spore coat. This protein produced a high degree of stability as an indirect consequence of its evolutionary constraints, since it must adapt to a tightly packed protein coat structure [43]. Here, we found that rLAC from *B. pumilus* TCCC 11568 also exhibited a high tolerance to NaCl, and it even retained 50.7% activity of the control in a solution containing 500 mM NaCl. Similar phenomena were also reported for laccases from *B. pumilus* W3 and *B. vallismortis* fmb-103 [26,44]. Therefore, rLAC with its high salinity tolerance would demonstrate more advantages in dealing with dye effluents that are high in salinity [45]. Additionally, no vital activity loss was found during incubation with 5 mM EDTA (62.1% activity retained, Table 1). On the contrary, a complete loss was found with laccase isoforms of *Aspergillus ochraceus* and *Thermus thermophilus* in 1 mM EDTA [46,47].

The effects of some reported inhibitors, such as L-cysteine, sodium dodecyl sulfate (SDS), β-mercaptoethanol, and dithiothreitol (DTT) on their rLAC activity are also exhibited in Table 1. rLAC activity was greatly decreased by 0.5 mM DTT, L-cysteine, and β-mercaptoethanol; this phenomenon was also noted in other fungal and bacterial laccases [26,44,48]. Additionally, laccases from diverse *Bacillus* strains also demonstrated differing amounts of tolerance against the same inhibitor. SDS (0.1 mM) significantly inhibited *B. safensis* sp. S31 laccase activity, retaining only 1.3% the activity of the control [41]. On the contrary, the activity of rLAC was not even affected by SDS at a high concentration (5 mM). A similar phenomenon was also noted for the *B. vallismortis* fmb-103 laccase [26].

3.5. Investigation of the Thermophilic Features of rLAC via Molecular Docking and MD Simulations

To present a reasonable explanation for the thermo-stable features of rLAC, MD simulations were performed at 325 K, 355 K, and 365 K, respectively. The root mean square deviation (RMSD), displayed as a function of simulation time, is a key indicator for assessing the stability of protein structure upon the binding of a ligand. In general, the lower the RMSD value displayed by the protein complex during MD simulations, the higher the stability shown [49]. Our results suggested that the complex systems reached equilibrium in the 75 ns simulation (Figure 3a). The RMSD for the rLAC-ABTS complex at high temperatures (355 K and 365 K) was lower in comparison to that at a low temperature (325 K), indicating an enhanced degree of stability for the complex upon its increase in temperature.

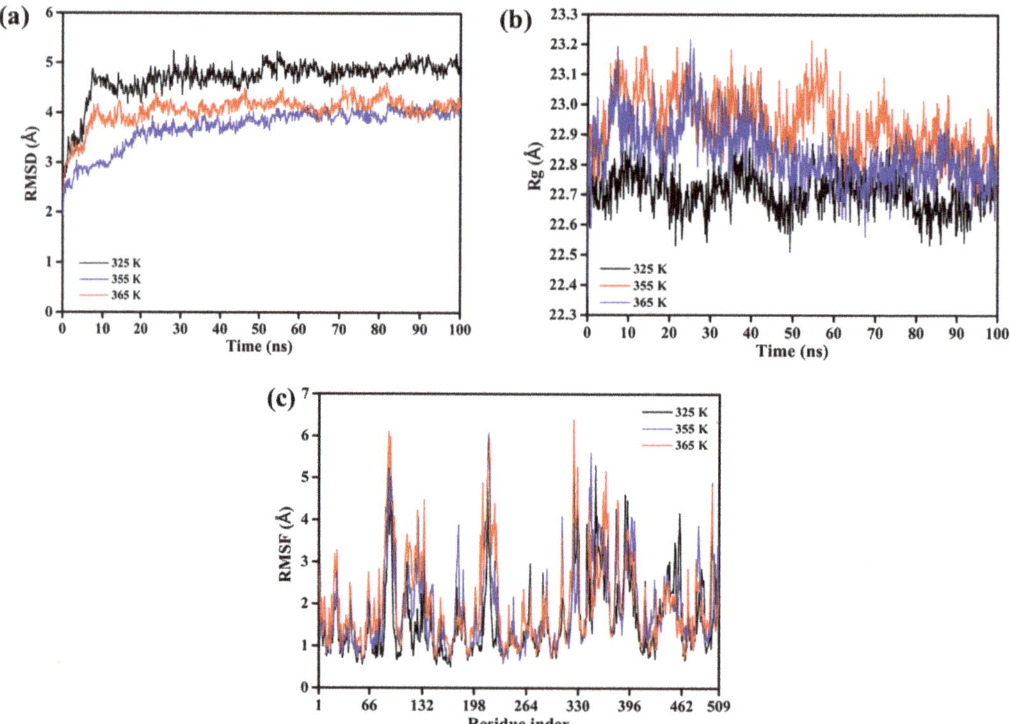

Figure 3. RMSD (**a**) and Rg (**b**) of rLAC binding with ABTS as a function of time (100 ns) at differing temperatures (325 K, 355 K, and 365 K); RMSF (**c**) value of each residue in rLAC-ABTS complex at differing temperatures after MD simulations.

Rg, calculated as the root mean square distance from the center of mass, indicates the level of the secondary structure backbone rigidity and the compactness of the protein complex system. The Rg values of the rLAC-ABTS complex at a high temperature were larger than those at low temperature, suggesting a reduction in the compactness and structural backbone rigidity (Figure 3b). The above results suggested that the stability of the rLAC-ABTS complex is not merely determined by the protein compactness, but by a synergic conformation change. In fact, the subsequent RMSF analysis that is commonly used to characterize the structural fluctuation and protein structural integrity also displayed expansion in some of the loop regions of rLAC, indicating a decrease in protein compactness in some regions, along with an increase in temperature from 325 K to 365 K. Actually, the RMSF values for all rLAC residues under high-temperature conditions (355 K and 365 K) did not demonstrate remarkable differences to those of rLAC under mild temperature conditions (325 K), except for a few residues that were all within the loop region with the largest enhancement of RMSF values, such as Glu90, Leu219, Gly323, and Asp379 (Figures 3c and 4).

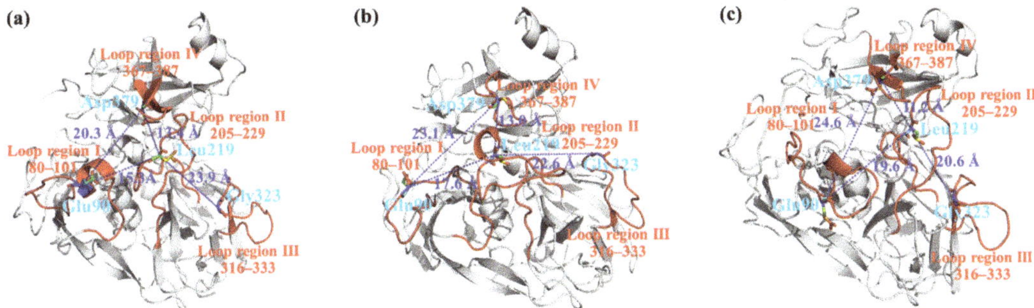

Figure 4. The effect of temperature on the rLAC conformation at the end of MD simulations. (**a**) 325 K; (**b**) 355 K; (**c**) 365 K. The distance among the representative amino acid residues in each loop (Glu90 in loop region I; Leu219 in loop region II; Gly323 in loop region III; Asp379 in loop region (IV) were marked with a green dashed line and specific values.

The distance between residues Glu90 and Asp379, Glu90 and Leu219, Asp379, and Leu219 gradually increased along with the elevated temperature. As a result, the loop regions of I, II, and IV were far away from each other at high temperatures (Figure 4b, 355 K; Figure 4c, 365 K) at the end of the MD simulations. On the contrary, the distance between loop region II and loop region III became shorter. In conclusion, the overall conformation of rLAC did not show significant alterations at high temperatures (355 K, 365 K), but minor location adjustments in some loop regions occurred at high temperatures. As can be supposed, the good flexibility of the loop region may potentially make it simple to expose the active center, and may therefore facilitate the binding of rLAC with the substrate (ABTS) and the subsequent catalytic reaction. In fact, this speculation was also supported by observations of the surface representation and binding modes of ABTS with rLAC (Figure 5).

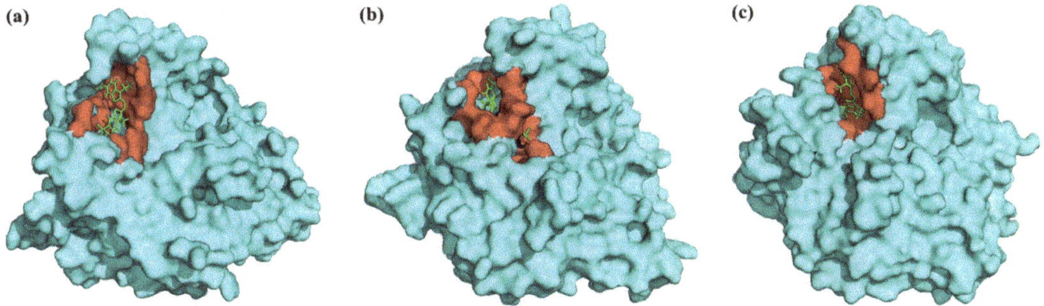

Figure 5. Surface representation and binding modes of the rLAC-ABTS complex. The structures were acquired from cluster analysis of the MD trajectories during 100 ns of simulation at different temperatures, (**a**) 325 K, (**b**) 355 K, and (**c**) 365 K. Surface representation of rLAC (cyan) in complex with ABTS (green) during the initial equilibrated MD trajectory. The substrate-binding pocket is presented as the red region.

As can be seen from the trajectories recorded after 100 ns of MD simulations, a different spatial conformation of the binding pocket in rLAC at different temperatures was observed (Figure 5). A narrow aisle for substrate access to the active site formed under low temperatures, which obviously did not facilitate effective binding of ABTS with the rLAC in the right manner, nor subsequent catalysis (Figure 5a). By contrast, the conformation changed at high temperatures (355 K, 365 K), and a more open space was formed. As a result, ABTS could enter the binding pocket easily and it formed a tighter complex with rLAC in the active site, which was one of the most important prerequisites for an effective catalytic reaction by laccase (Figure 5b,c).

The distance between the catalytic residue and ABTS was also analyzed to explore the influence of temperature on the catalysis performance of rLAC. The distance between the C-atom of H497 located in the active site and the C-atom of ABTS decreases along with an increase in temperature, indicating that a closer interaction between ABTS and rLAC occurred at the active center (Figure 6). In fact, the released binding free-energy analysis for rLAC-ABTS verified this phenomenon further (Table 2).

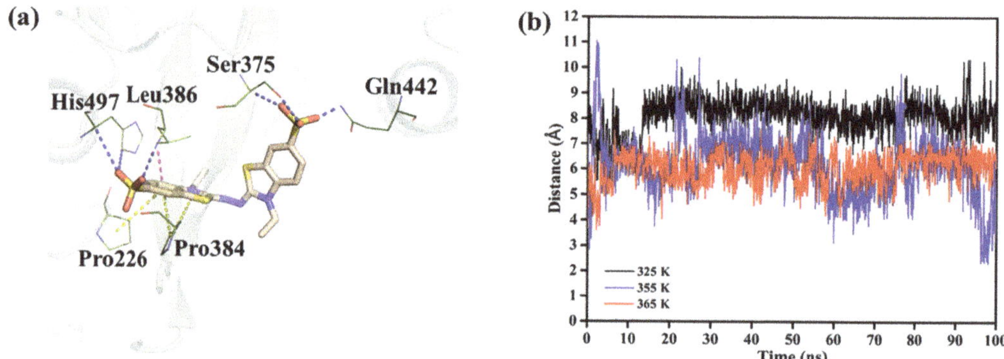

Figure 6. (**a**) In silico binding analysis of ABTS at the active site of rLAC and (**b**) the average distance variation between the C-atom of H497 and ABTS over 100 ns of the simulation at different temperatures.

Table 2. The released binding energy of rLAC to the ligand ABTS.

Temperature (K)	Binding Energy [a] (kJ/mol)	van der Waals Energy (kJ/mol)	Electrostatic Energy (kJ/mol)	Polar Solvation Energy (kJ/mol)	SASA Energy (kJ/mol)
325	-106.77 ± 21.59	-189.47 ± 20.04	-51.29 ± 20.60	155.13 ± 27.40	-21.14 ± 1.61
355	-154.14 ± 27.74	-251.44 ± 30.52	-40.50 ± 13.12	160.48 ± 17.43	-22.67 ± 1.42
365	-135.00 ± 22.68	-234.27 ± 27.52	-28.83 ± 16.34	150.44 ± 27.52	-22.34 ± 1.70

[a] The binding energy was calculated through the summation the values of the van der Waals energy, electrostatic energy, polar solvation energy, and SASA energy.

The electrostatic interaction and the van der Waals interaction made a major contribution to the production of the binding energy during the binding process of ABTS in the active center of rLAC. Moreover, the released binding energy was much higher for rLAC-ABTS at high temperatures, especially at 355 K. These results also matched the RMSD analysis in Figure 3, in which the RMSD value obtained at 355 K was the lowest, indicating a high thermostability of the rLAC-substrate complex and a much higher substrate-binding affinity at high temperature. Mollania et al. (2017) investigated the variation of the protein structure of the laccase obtained from a local *Bacillus* species (HR03) after thermal activation, using gas-phase electrophoretic mobility macromolecule analysis, far-UV CD-spectra, common biochemical methods, fluorescence analysis, etc. [50]. They speculated that the improved activity of *Bacillus* sp. HR03 laccase was not the result of oligomerization, but of the generation of a more active conformation due to the thermal treatment at a high temperature (70 °C). In other words, the incubation of *Bacillus* laccase at high temperatures did not severely affect the overall structure in an irreversible way, but instead led to a more beneficial conformation of the Cu sites and of the active site for substrate binding. According to our literature investigation, the mechanism for the high thermophilic features of *Bacillus* laccase is still not very clear. We herein presented another reasonable and vivid explanation for interpreting the thermophilic features of rLAC via docking and molecular dynamics simulations analyses. We hypothesized that the formation of a correctly matching substrate access aisle to the active site, through the minor adjustment of protein spatial conformation, was responsible for the improved catalytic activity of *Bacillus* laccase when incubated at high temperatures. As a result, the released binding energy greatly increased, due to the tighter binding of ABTS at the active site driving the accelerated catalytic reaction.

3.6. Dye Decolorization

Color is a vital quality attribute of food in stimulating customers' appetites and is, therefore, one of the most important concerns for food manufacturers when considering market acceptance. Synthetic dyes as colorants are widely applied in the food industry to dye foodstuffs such as mustard, sweets, jams, cakes, and juices beverages, and they are also used in the production of drugs and cosmetics, because they have certain fascinating advantages compared to natural pigments, including brighter colors, low production costs, high water solubility, and high fading resistance against exposure to chemicals, light, and water [51].

However, large amounts of effluent with intense colors are produced after the dyeing process, as approximately 10–20% of the dyes are lost during this process [52]. Therefore, without treatment, the discharged effluents are an important source of pollution for water bodies, and bring in severe damage to aquatic ecosystems through impeding light penetration to the water bodies, which consequently reduces the photosynthetic rate of aquatic plants and the dissolved oxygen levels in the water [23,53]. In recent years, more and more bioremediation methods with favorable properties, such as higher efficiencies, lower costs, and eco-friendly properties, have been developed and applied to decolorizing dye effluents [54–56].

Laccases can degrade numerous environmental pollutants, including micropollutants, personal-care products, and textile dyes [6,57]. Since the typical dye effluents are generally identified via their high temperatures and high pH values (e.g., over 40 °C and around pH 9.0) [39], most fungal laccases will be deactivated under these harsh conditions, e.g., laccases from *Sclerotium rolfsii* and *T. hirsuta* can only work best with degrading indigo carmine under an acidic pH environment [45,58]. Thus, for bacterial laccases, conditions of much higher pH stability and thermostability show a higher potential for treating dye effluents.

In this study, a new laccase with a high thermostability and pH stability was obtained from the newly identified bacterium *B. pumilus* TCCC 11568 (Figure 2). Therefore, we assessed its decolorization ability against several azo dyes at 60 °C and over a broad pH range, starting from pH 5.0 to 9.0 (Figure 7). The decolorization rates for each azo dye were all above 70% under optimum conditions (an appropriate redox mediator and pH environment) (Figure 7). Taking the dye Sunset Yellow for example, the decolorization rate was below 20% or 40% when using ABTS or syringaldehyde as a redox mediator, respectively. However, the decolorization rate soared up to 84.4% when incubated with syringaldehyde at pH 9.0 (Figure 7). Additionally, we also found that rLAC displayed a higher decolorizing ability against all of the dyes in a neutral (pH 7.0) or alkaline environment (pH 9.0), except for the dye etythrosine, which is more easily decolorized in an acid environment.

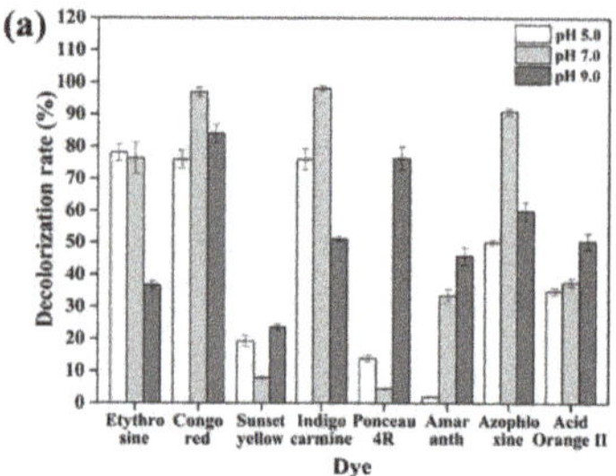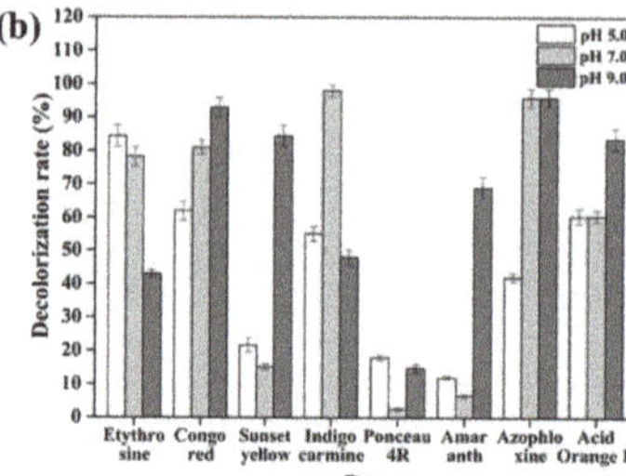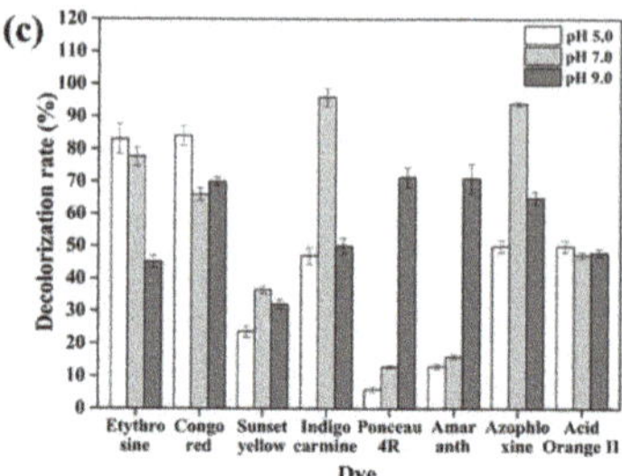

Figure 7. Decolorizing synthetic dyes with purified rLAC using mediator ABTS (**a**), acetosyringone (**b**), and syringaldehyde (**c**) at different pH values (5.0, 7.0, and 9.0). The operation was conducted through incubation of 80 U rLAC at 60 °C. All experiments were conducted in triplicate, and the data represent mean ± SD.

This differs from *B. velezensis* TCCC 111904 laccase, which was more effective for decolorizing the dyes when ABTS was used as the mediator at pH 5.5 and pH 7.0 [25]. Therefore, it was inferred that synthetic dyes with diverse structures could be efficiently decolorized by rLAC isolated from *B. pumilus* TCCC 11568 over a broad pH range (5.0–9.0) at a high temperature (60 °C) with an appropriate mediator. However, depending on the extensive specificity of the CotA laccase, no tendency of laccase could be summarized for the decolorizing activity towards synthetic dyes with different structures [59].

As is known, most dye effluents are released at high temperatures [39]. Thus, it would be helpful to decrease the unnecessary cost of cooling the dye effluents to meet the required reaction conditions due to the alkaline stability and high thermostability of rLAC. Moreover, recycling hot water after decolorization could save a lot of energy. Therefore, the *B. pumilus* TCCC 11568 laccase presented promising applications because of its excellent efficiency in decolorizing dyes under high temperatures and alkaline conditions.

4. Conclusions

The recombinant laccase (rLAC) of *Bacillus pumilus* TCCC 11568 demonstrated a high thermostability and pH stability over a wide pH range. The overall conformation of rLAC did not show significant alterations at a high temperature (355 K), but minor location adjustments in some loop regions occurred at a high temperature. However, those adjustments were presumed to be responsible for the formation of a more open access aisle that facilitated ABTS binding in the active site, resulting in a shorter distance from the catalytic residue and elevated binding energy. Due to its thermophilic feature, rLAC could effectively decolorize azo dyes at high temperatures over an extensive pH range in the presence of the appropriate mediator. These aforementioned features make rLAC a potential candidate for industrial applications in dye decolorization.

Supplementary Materials: The following supporting information can be downloaded at: https://www.mdpi.com/article/10.3390/foods11101387/s1, Figure S1: 16S rDNA-based phylogenetic comparison of *B. pumilus* TCCC 11568 and other *Bacillus* species. The phylogenetic tree was engineered using MEGA 6.0 software with the neighbor-joining method. The numbers at branch points suggested the percentage of bootstrap sampling from 1000 replications; Figure S2: Multiple alignments of rLAC protein sequence from *B. pumilus* TCCC 11568 and other laccases derived from the *Bacillus* genus (*B. pumilus* ATCC 7061, *B. pumilus* W3, *B. vallismortis* fmb103, *B. amyloliquefaciens* TCCC 111018, *B. subtilis* X1, *B. velezensis* TCCC 111904, and *B. licheniformis* ATCC 14580). The identical and similar amino acids are marked in black and solid grey, respectively. These protein sequences were obtained from NCBI database, and the alignment was carried out with DNAMAN software; Figure S3: Detection of rLAC expressed in *E. coli*. (a) Lane M: protein standard ladder; Lane 1: cell extract of *E. coli* BL21/pET-22b; Lane 2: cell extract of *E. coli* BL21/pET-lac. (b) Lane M: protein standard ladder; Lane 1: the purified rLAC. The black arrowhead indicates the rLAC band.

Author Contributions: Conceptualization, T.L.; Methodology, T.L.; Writing—original draft preparation, T.L.; Validation, J.L.; Investigation, J.L., X.C. and Z.Y. (Zhiming Yao); Software, Z.Y. (Zhaoting Yuan); Supervision, F.L.; Funding Acquisition, Y.L. and T.L.; Writing—Review and Editing, Y.L. All authors have read and agreed to the published version of the manuscript.

Funding: This research was funded by the National Key Research and Development Program of China, grant number [2021YFC2100300], the Tianjin Key-Training Program of the "Project and Team" of China [XC202032], and the Key Research Program of the Henan Higher Education Institutions of China [22A350008].

Data Availability Statement: The data presented in this study are available in article and Supplementary Material.

Conflicts of Interest: The authors declare no conflict of interest.

References

1. Ghodake, G.S.; Yang, J.; Shinde, S.S.; Mistry, B.M.; Kim, D.Y.; Sung, J.S.; Kadam, A.A. Paper waste extracted α-cellulose fibers super-magnetized and chitosan functionalized for covalent laccase immobilization. *Bioresour. Technol.* **2018**, *261*, 420–427. [CrossRef] [PubMed]

2. Rodriguez-Couto, S. Laccases for denim bleaching: An eco-friendly alternative. *Open Text. J.* **2012**, *5*, 1–7. [CrossRef]

3. Sharma, B.; Dangi, A.K.; Shukla, P. Contemporary enzyme based technologies for bioremediation: A review. *J. Environ. Manage.* **2018**, *210*, 10–22. [CrossRef] [PubMed]

4. Forootanfar, H.; Faramarzi, M.A. Insights into laccase producing organisms, fermentation states, purification strategies, and biotechnological applications. *Biotechnol. Prog.* **2015**, *31*, 1443–1463. [CrossRef] [PubMed]

5. Giardina, P.; Faraco, V.; Pezzella, C.; Piscitelli, A.; Vanhulle, S.; Sannia, G. Laccases: A never-ending story. *Cell. Mol. Life Sci.* **2010**, *67*, 369–385. [CrossRef] [PubMed]

6. Sondhi, S.; Sharma, P.; Saini, S.; Puri, N.; Gupta, N. Purification and characterization of an extracellular, thermo-alkali-stable, metal tolerant laccase from *Bacillus tequilensis* SN4. *PLoS ONE* **2014**, *9*, e96951. [CrossRef] [PubMed]

7. Augustine, A.J.; Kragh, M.E.; Sarangi, R.; Fujii, S.; Liboiron, B.D.; Stoj, C.S.; Kosman, D.J.; Hodgson, K.O.; Hedman, B.; Solomon, E.I. Spectroscopic studies of perturbed T1 Cu sites in the multicopper oxidases *Saccharmyces cerevisiae* Fet3p and *Rhus vernicifera* laccase: Allosteric coupling between the T1 and trinuclear Cu sites. *Biochemistry* **2008**, *47*, 2036–2045. [CrossRef]

8. Baldrian, P. Fungal laccases occurrence and properties. *FEMS Microbiol. Rev.* **2006**, *30*, 215–242. [CrossRef]

9. Vieille, C.; Zeikus, G.J. Hyperthermophilic enzymes: Sources, uses, and molecular mechanisms for thermostability. *Microbiol. Mol. Biol. Rev.* **2001**, *65*, 1–43. [CrossRef]

10. Hilde'n, K.; Hakala, T.; Lundell, T. Thermotolerant and thermostable laccases. *Biotechnol. Lett.* **2009**, *31*, 1117–1128. [CrossRef]

11. Chauhan, P.S.; Goradia, B.; Saxena, A. Bacterial laccase: Recent update on production and industrial applications. *3 Biotech.* **2017**, *7*, 323. [CrossRef]

12. Martins, L.O.; Soares, C.M.; Pereira, M.M.; Teixeira, M.; Costa, T.; Jones, G.H.; Henriques, A.O. Molecular and biochemical characterization of a highly stable bacterial laccase that occurs as a structural component of the *Bacillus subtilis* endospore coat. *J. Biol. Chem.* **2002**, *277*, 18849–18859. [CrossRef]

13. Brander, S.; Mikkelsen, J.D.; Kepp, K.P. Characterization of an alkali- and halide-resistant laccase expressed in *E. coli*: CotA from *Bacillus clausii*. *PLoS ONE* **2014**, *9*, e99402.

14. Christensen, N.J.; Kepp, K.P. Stability mechanisms of a thermophilic laccase probed by molecular dynamics. *PLoS ONE* **2013**, *8*, e61985. [CrossRef]

15. Farnet, A.M.; Criquet, S.; Cigna, M.; Gil, G.; Ferré, E. Purification of a laccase from *Marasmius quercophilus* induced with ferulic acid: Reactivity towards natural and xenobiotic aromatic compounds. *Enzym. Microb. Technol.* **2004**, *34*, 549–554. [CrossRef]

16. Papinutti, L.; Dimitriu, P.; Forchiassin, F. Stabilization studies of *Fomes sclerodermeus* laccases. *Bioresource Technol.* **2008**, *99*, 419–424. [CrossRef]

17. Liu, Y.H.; Huang, L.; Guo, W.; Jia, L.B.; Fu, Y.; Gui, S.; Lu, F.P. Cloning, expression, and characterization of a thermostable and pH-stable laccase from *Klebsiella pneumoniae* and its application to dye decolorization. *Process. Biochem.* **2017**, *53*, 125–134. [CrossRef]

18. Tamura, K.; Stecher, G.; Peterson, D.; Filipski, A.; Kumar, S. MEGA6: Molecular evolutionary genetics analysis version 6.0. *Mol. Biol. Evol.* **2013**, *30*, 2725–2729. [CrossRef]

19. Liu, Y.H.; Huang, L.; Shan, M.Y.; Sang, J.C.; Li, Y.Z.; Jia, L.G.; Wang, N.; Wang, S.; Shao, S.L.; Liu, F.F.; et al. Enhancing the activity and thermostability of *Streptomyces mobaraensis* transglutaminase by directed evolution and molecular dynamics simulation. *Biochem. Eng. J.* **2019**, *151*, 107333. [CrossRef]

20. Morris, G.M.; Huey, R.; Lindstrom, W.; Sanner, M.F.; Belew, R.K.; Goodsell, D.S.; Olson, A.J. AutoDock4 and AutoDockTools4: Automated Docking with selective receptor flexibility. *J. Comput. Chem.* **2009**, *30*, 2785–2791. [CrossRef]

21. Benkert, P.; Biasini, M.; Schwede, T. Toward the estimation of the absolute quality of individual protein structure models. *Bioinformatics* **2011**, *27*, 343–350. [CrossRef]

22. Kittl, R.; Mueangtoom, K.; Gonaus, C.; Khazaneh, S.T.; Sygmund, C.; Haltrich, D.; Ludwig, R. A chloride tolerant laccase from the plant pathogen ascomycete *Botrytis aclada* expressed at high levels in *Pichia pastoris*. *J. Biotechnol.* **2012**, *157*, 304–314. [CrossRef]

23. Lu, L.; Zhao, M.; Li, G.F.; Li, J.; Wang, T.N.; Li, D.B.; Xu, T.F. Decolorization of synthetic dyes by immobilized spore from *Bacillus amyloliquefaciens*. *Catal. Commun.* **2012**, *26*, 58–62. [CrossRef]

24. Reiss, R.; Ihssen, J.; Thöny-Meyer, L. *Bacillus pumilus* laccase: A heat stable enzyme with a wide substrate spectrum. *BMC Biotechnol.* **2011**, *11*, 9. [CrossRef]

25. Zhou, W.; Guan, Z.B.; Chen, Y.; Zhang, F.; Cai, Y.J.; Xu, C.W.; Chen, X.S.; Liao, X.R. Production of spore laccase from *Bacillus pumilus* W3 and its application in dye decolorization after immobilization. *Water Sci. Technol.* **2017**, *76*, 147–154. [CrossRef]

26. Zhang, C.; Diao, H.W.; Lu, F.X.; Bie, X.M.; Wang, Y.F.; Lu, Z.X. Degradation of triphenylmethane dyes using a temperature and pH stable spore laccase from a novel strain of *Bacillus vallismortis*. *J. Bioresour. Technol.* **2012**, *126*, 80–86. [CrossRef]

27. Wang, H.B.; Huang, L.; Li, Y.Z.; Ma, J.Y.; Wang, S.; Zhang, Y.F.; Ge, X.Q.; Wang, N.; Lu, F.P.; Liu, Y.H. Characterization and application of a novel laccase derived from *Bacillus amyloliquefaciens*. *Int. J. Biol. Macromol.* **2020**, *150*, 982–990. [CrossRef]

28. Guan, Z.B.; Zhang, N.; Song, C.M.; Zhou, W.; Zhou, L.X.; Zhao, H.; Xu, C.W.; Cai, Y.J.; Liao, X.R. Molecular cloning, characterization, and dye-decolorizing ability of a temperature-and pH-stable laccase from *Bacillus subtilis* X1. *Appl. Biochem. Biotech.* **2014**, *172*, 1147–1157. [CrossRef]

29. Li, T.; Huang, L.; Li, Y.Z.; Xu, Z.H.; Ge, X.Q.; Zhang, Y.F.; Wang, N.; Wang, S.; Yang, W.; Lu, F.P.; et al. The heterologous expression, characterization, and application of a novel laccase from *Bacillus velezensis*. *Sci. Total Environ.* **2020**, *713*, 136713. [CrossRef]

30. Koschorreck, K.; Richter, S.M.; Ene, A.B.; Roduner, E.; Schmid, R.D.; Urlacher, V.B. Cloning and characterization of a new laccase from *Bacillus licheniformis* catalyzing dimerization of phenolic acids. *Appl. Microbiol. Biotechnol.* **2008**, *79*, 217–224. [CrossRef]

31. Solomon, E.I.; Sundaram, U.M.; Machonkin, T.E. Multicopper oxidases and oxygenases. *Chem. Rev.* **1996**, *96*, 2563–2606. [CrossRef] [PubMed]

32. Ruijssenaars, H.J.; Hartmans, S. A cloned *Bacillus halodurans* multicopper oxidase exhibiting alkaline laccase activity. *Appl. Microbiol. Biotechnol.* **2004**, *65*, 177–182. [CrossRef] [PubMed]

33. Chen, B.; Xu, W.Q.; Pan, X.R.; Lu, L. A novel non-blue laccase from *Bacillus amyloliquefaciens*: Secretory expression and characterization. *Int. J. Biol. Macromol.* **2015**, *76*, 39–44. [CrossRef] [PubMed]

34. Li, T.; Wang, H.B.; Li, J.W.; Jiang, L.Y.; Kang, H.W.; Guo, Z.H.; Wang, C.; Yang, W.; Liu, F.F.; Lu, F.P.; et al. Enzymatic characterization, molecular dynamics simulation, and application of a novel *Bacillus licheniformis* laccase. *Int. J. Biol. Macromol.* **2021**, *167*, 1393–1405. [CrossRef]

35. Aung, T.; Jiang, H.; Chen, C.C.; Liu, G.L.; Hu, Z.; Chi, Z.M.; Chi, Z. Production, gene cloning, and overexpression of a laccase in the marine-derived yeast *Aureobasidium melanogenum* strain 11-1 and characterization of the recombinant laccase. *Mar. Biotechnol.* **2019**, *21*, 76–87. [CrossRef]

36. Wang, S.S.; Ning, Y.J.; Wang, S.N.; Zhang, J.; Zhang, G.Q.; Chen, Q.J. Purification, characterization, and cloning of an extracellular laccase with potent dye decolorizing ability from white rot fungus *Cerrena unicolor* GSM-01. *Int. J. Biol. Macromol.* **2017**, *95*, 920–927. [CrossRef]

37. Wang, S.N.; Chen, Q.J.; Zhu, M.J.; Xue, F.Y.; Li, W.C.; Zhao, T.J.; Li, G.D.; Zhang, G.Q. An extracellular yellow laccase from white rot fungus *Trametes* sp. F1635 and its mediator systems for dye decolorization. *Biochimie* **2018**, *148*, 46–54. [CrossRef]

38. Shraddha; Shekher, R.; Sehgal, S.; Kamthania, M.; Kumar, A. Laccase: Microbial sources, production, purification, and potential biotechnological applications. *Enzym. Res.* **2011**, *2011*, 217861.

39. Yang, Q.H.; Zhang, M.L.; Zhang, M.M.; Wang, C.Q.; Liu, Y.Y.; Fan, X.J.; Li, H. Characterization of a novel, cold-adapted, and thermostable laccase-like enzyme with high tolerance for organic solvents and salt and potent dye decolorization ability, derived from a marine metagenomic library. *Front. Microbiol.* **2018**, *9*, 2998. [CrossRef]

40. Ossiadacz, J.; Al-Adhami, A.J.H.; Bajraszewska, D.; Fischer, P.; Peczyñska-Czoch, W. On the use of trametes versicolor laccase for the conversion of 4-methyl-3-hydroxyanthranilic acid to actinocin chromophore. *J. Biotechnol.* **1999**, *72*, 141–149. [CrossRef]

41. Siroosi, M.; Amoozegar, M.A.; Khajeh, K.; Dabirmanesh, B. Decolorization of dyes by a novel sodium azide-resistant spore laccase from a halotolerant bacterium, *Bacillus safensis* sp. strain S31. *Water Sci. Technol.* **2018**, *77*, 2867–2875. [CrossRef]

42. Jimenez-Juarez, N.; Roman-Miranda, R.; Baeza, A.; Sánchez-Amat, A.; VazquezDuhalt, R.; Valderrama, B. Alkali and halide-resistant catalysis by the multipotent oxidase from *Marinomonas mediterranea*. *J. Biotechnol.* **2005**, *117*, 73–82. [CrossRef]

43. Enguita, F.J.; Martins, L.O.; Henriques, A.O.; Carrondo, M.A. Crystal structure of a bacterial endospore coat component: A laccase with enhanced thermostability properties. *J. Biol. Chem.* **2003**, *278*, 19416–19425. [CrossRef]

44. Guan, Z.B.; Song, C.M.; Zhang, N.; Zhou, W.; Xu, C.W.; Zhou, L.X.; Zhao, H.; Cai, Y.J.; Liao, X.R. Overexpression, characterization, and dye-decolorizing ability of a thermostable, pH-stable, and organic solvent-tolerant laccase from *Bacillus pumilus* W3. *J. Mol. Catal. B-Enzym.* **2014**, *101*, 1–6. [CrossRef]

45. Rodrigues, C.S.D.; Madeira, L.M.; Boaventura, R.A.R. Treatment of textile effluent by chemical (Fenton's Reagent) and biological (sequencing batch reactor) oxidation. *J. Hazard. Mater.* **2009**, *172*, 1551–1559. [CrossRef]

46. Telke, A.A.; Kadam, A.A.; Jagtap, S.S.; Jadhav, J.P.; Govindwar, S.P. Biochemical characterization and potential for textile dye degradation of blue laccase from *Aspergillus ochraceus* NCIM-1146. *Biotechnol. Bioprocess. Eng.* **2010**, *15*, 696–703. [CrossRef]

47. Miyazaki, K. A hyperthermophilic laccase from *Thermus thermophiles* HB27. *Extremophiles* **2005**, *9*, 415–425. [CrossRef]

48. Forootanfar, H.; Faramarzi, M.A.; Shahverdi, A.R.; Yazdi, M.T. Purification and biochemical characterization of extracellular laccase from the ascomycete *Paraconiothyrium* variabile. *Bioresour. Technol.* **2011**, *102*, 1808–1814. [CrossRef]

49. Wang, J.Y.; Zhang, Y.; Wang, X.J.; Shang, J.Z.; Li, Y.; Zhang, H.T.; Lu, F.P.; Liu, F.F. Biochemical characterization and molecular mechanism of acid denaturation of a novel α-amylase from *Aspergillus niger*. *Biochem. Eng. J.* **2018**, *137*, 222–231. [CrossRef]
50. Mollania, N.; Heidari, M.; Khajeh, K. Catalytic activation of *Bacillus* laccase after temperature treatment: Structural & biochemical characterization. *Int. J. Biol. Macromol.* **2018**, *109*, 49–56.
51. Abe, F.R.; Machado, A.L.; Soares, A.M.V.M.; Oliveira, D.P.; Pestana, J.L.T. Life history and behavior effects of synthetic and natural dyes on *Daphnia magna*. *Chemosphere* **2019**, *236*, 124390. [CrossRef]
52. Gao, J.F.; Zhang, Q.; Wang, J.H.; Wu, X.L.; Wang, S.Y.; Peng, Y.Z. Contributions of functional groups and extracellular polymeric substances on the biosorption of dyes by aerobic granules. *Bioresour. Technol.* **2011**, *102*, 805–813. [CrossRef]
53. Abe, F.R.; Soares, A.M.V.M.; Oliveira, D.P.; Gravato, C. Toxicity of dyes to zebrafish at the biochemical level: Cellular energy allocation and neurotoxicity. *Environ. Pollut.* **2018**, *235*, 255–262. [CrossRef]
54. Saratale, R.G.; Saratale, G.D.; Chang, J.S.; Govindwar, S.P. Bacterial decolorization and degradation of azo dyes: A review. *J. Taiwan Inst. Chem. Eng.* **2011**, *42*, 138–157. [CrossRef]
55. Imran, M.; Ashraf, M.; Hussain, S.; Mustafa, A. Microbial biotechnology for detoxification of azo-dye loaded textile effluents: A critical review. *Int. J. Agric. Bio.* **2019**, *22*, 1138–1154.
56. Wang, F.H.; Xu, Z.H.; Wang, C.; Guo, Z.H.; Yuan, Z.T.; Kang, H.W.; Li, J.W.; Lu, F.P.; Liu, Y.H. Biochemical characterization of a tyrosinase from *Bacillus aryabhattai* and its application. *Int. J. Biol. Macromol.* **2021**, *176*, 37–46. [CrossRef]
57. Couto, S.R.; Herrera, J.L.T. Industrial and biotechnological applications of laccases: A review. *Biotechnol. Adv.* **2006**, *24*, 500–513. [CrossRef]
58. Santhanam, N.; Vivanco, M.; Decker, S.R.; Reardon, K.F. Expression of industrially relevant laccases: Prokaryotic style. *Trends Biotechnol.* **2011**, *29*, 480–489. [CrossRef]
59. Enguita, F.J.; Matias, P.M.; Martins, L.O.; Placido, D.; Henriques, A.O.; Carrondo, M.A. Spore-coat laccase CotA from *Bacillus subtilis*: Crystallization and preliminary X-ray characterization by the MAD method. *Acta Crystallogr. D Biol. Crystallogr.* **2002**, *58*, 1490–1493. [CrossRef]

Article

Production of Corn Protein Hydrolysate with Glutamine-Rich Peptides and Its Antagonistic Function in Ulcerative Colitis In Vivo

Yan Jing [1,2], Xiaolan Liu [2,3,*], Jinyu Wang [1,2], Yongqiang Ma [1,*] and Xiqun Zheng [2,3]

[1] College of Food Engineering, Harbin University of Commerce, Harbin 150076, China
[2] Key Laboratory of Corn Deep Processing Theory and Technology of Heilongjiang Province, College of Food and Bioengineering, Qiqihar University, Qiqihar 161006, China
[3] College of Food, Heilongjiang Bayi Agricultural University, Daqing 163319, China
* Correspondence: 01275@qqhru.edu.cn (X.L.); 101729@hrbcu.edu.cn (Y.M.); Tel./Fax: +86-452-2738341 (X.L.); Tel./Fax: +86-451-84844281 (Y.M.)

Abstract: Ulcerative colitis is a typical chronic inflammatory disease of the gastrointestinal tract, which has become a serious hazard to human health. The purpose of the present study was to evaluate the antagonistic effect of corn protein hydrolysate with glutamine-rich peptides on ulcerative colitis. The sequential hydrolysis of corn gluten meal by Alcalase and Protamex was conducted to prepare the hydrolysate, and then the mouse ulcerative colitis model induced by dextran sulfate sodium was applied to evaluate its biological activities. The results indicated that the hydrolysate significantly improved weight loss ($p < 0.05$), reduced the colonic shortening and the disease activity index, diminished the infiltration of inflammatory cells in the colonic tissue, and reduced the permeability of the colonic mucosa in mice. In addition, the hydrolysate decreased the contents of pro-inflammatory factors IL-1β, IL-6, and TNF-α, increased the anti-inflammatory factor IL-10 and oxidative stress markers GSH-Px and SOD in the animal tests. Moreover, the hydrolysate also regulated the abundance and diversity of the intestinal microbiota, improved the microbiota structure, and increased the content of beneficial bacteria including *Lactobacillus* and *Pediococcus*. These results indicated that the hydrolysate might be used as an alternative natural product for the prevention of ulcerative colitis and could be further developed into a functional food.

Keywords: corn gluten meal; glutamine; ulcerative colitis

Citation: Jing, Y.; Liu, X.; Wang, J.; Ma, Y.; Zheng, X. Production of Corn Protein Hydrolysate with Glutamine-Rich Peptides and Its Antagonistic Function in Ulcerative Colitis In Vivo. *Foods* **2022**, *11*, 3359. https://doi.org/10.3390/foods11213359

Academic Editors: Fuping Lu and Wenjie Sui

Received: 17 September 2022
Accepted: 24 October 2022
Published: 25 October 2022

Publisher's Note: MDPI stays neutral with regard to jurisdictional claims in published maps and institutional affiliations.

1. Introduction

Inflammatory bowel disease (IBD) mainly includes ulcerative colitis (UC) and Crohn's disease (CD), and it is a complex disease of the digestive system characterized by repeated chronic intestinal inflammation, intestinal epithelial dysfunction, and mucosal tissue damage in the gastrointestinal tract [1]. Epidemiology shows that in recent years, the prevalence of UC has been on the rise worldwide [2,3]. Multiple causes of UC have been identified, and its consequences usually include the impaired structure and function of the intestinal mucosa [4]. At present, the main drugs for the treatment of UC include salicylazosulfapyridine, immunosuppressants, and an anti-tumor necrosis factor-α monoclonal antibody. Unfortunately, the long-term use of these drugs can cause serious side effects and are expensive, which severely limits their use [5,6]. Hence, safer and more efficient naturally active substances for the treatment of UC are urgently needed.

Glutamine (Gln) is one of the most abundant amino acids in the human body and has many physiological functions, such as enhancing the immune capacity of the body and protecting the intestinal barrier function [7,8]. However, the application of Gln is limited due to its unstable monomer, easy decomposition, and low solubility. A few studies have found that glutamine peptide not only retains a variety of physiological functions and nutrition of Gln, but also has good water solubility and stability, which is a stable substitute for Gln. Researchers have also found that glutamine peptide has antioxidant activity, promotes protein synthesis, and protects human intestinal health function. Under stress conditions such as fatigue, intense exercise, and disease, glutamine peptides can be rapidly decomposed into glutamine to meet the needs of the body. For example, Juan Tian et al. reported that alanine–glutamine could attenuate soya saponin-induced growth inhibition and intestinal injury in zebrafish [9]. Bie Tan et al. found that alanine–glutamine could replace glutamine as a source of energy and protein in the gastrointestinal tract [10]. Therefore, the development of bioactive peptides rich in Gln can effectively solve the problem of the limited application of Gln.

Corn is one of the three major cereals in the world, and China's yield of corn reached 272.55 million tons in 2021. Nearly one-sixth of corn is used to produce corn starch using the wet milling process. It generates a large amount of corn gluten meal (CGM) as a main co-product, which contains about 67–72% (*w/w*) protein [11,12]. These proteins mainly include zein and gluten; the content of Gln is about one-third of the total amino acids in gluten [13,14] and is an excellent raw material for preparing glutamine active peptides. Therefore, corn protein hydrolysate with glutamine-rich peptides (APH) was first prepared from the hydrolysis of CGM by proteases in the present study, and then its antagonistic effect on DSS-induced UC in vivo was systematically evaluated.

2. Materials and Methods

2.1. Materials

CGM was purchased from Longjiang Fufeng Biotechnology Co., Ltd. (Qiqihar, China). The proteases Alcalase and Protamex were purchased from Novo Nordisk (Bagsvaerd, Denmark). DSS (36–50 kDa) was purchased from Macklin (Shanghai, China). All of the other reagents and chemicals were of analytical grade.

2.2. Preparation of CGM Hydrolysates

The CGM was pretreated using the method described in our previous report [15,16]. Pretreated CGM (20 g) was added to 200 mL of distilled water, and then Alcalase and Protamex were added successively for hydrolysis. The optimum hydrolysis conditions for Alcalase were a temperature of 64 °C, hydrolysis time of 2.5 h, amount of enzyme of 1300 U/g, and a pH of 8.0. The optimum hydrolysis conditions for Protamex were a temperature of 50 °C, hydrolysis time of 2.5 h, amount of enzyme of 200 U/g, and a pH of 7.0. For the sequential hydrolysis of the two proteases, first of all, the CGM was hydrolyzed with one protease (Alcalase or Protamex), and then the second protease was added after adjusting the reaction conditions. After the reaction, the hydrolysates were centrifuged at 4500 rpm for 15 min to collect the supernatant and were lyophilized.

2.3. Determination of Gln Content in Hydrolysates

The Gln content of the corn protein hydrolysate was determined by Katharina's method [17]. Briefly, 500 µL corn protein hydrolysate (100 mg/mL) was transferred into an ampoule (Hawk, Harbin, China), and then 2 mL bis-1,1-trifluoroacetoxy-iodobenz-ene (BTI) acetonitrile-aqueous solution (10 mg/mL) and 500 µL pyridine (50 µmol/mL) were added into the ampoule. The reaction was carried out at 50 °C for 2 h to protect the Gln in the hydrolysate, and the other hydrolysate sample was not subjected to BTI protection reaction. Then, 5 mL HCl (6 mol/L) was added to the two ampoules for acid hydrolysis, and the ampoules were sealed in a vacuum and hydrolyzed at 110 °C for 24 h. After cooling, the pH of the hydrolysate was adjusted to 6–8 with NaOH, and the glutamic acid content was determined using a glutamate biosensor analyzer; the difference in the glutamic acid between the two samples was the content of Gln.

2.4. Animals and Experimental Design

Male Kunming mice (6–8 weeks) were provided by the Laboratory Animal Technology of Changchun Yisi (Changchun, Jilin, China, Permission No. SCXK (JI) 2020-0002). The animal study protocol was approved on 25 March 2021 by the Animal Ethics Committee of College of Food and Bioengineering, Qiqihar University (Approval No. 2021-003). The care and treatment of the mice were in accordance with international guidelines for laboratory animals (Association for Assessment and Accreditation of Laboratory Animal Care, AAALAC). Before the experiment, all animals were fed and watered freely for one week, and the room temperature was maintained at 23 ± 3 °C. Then, the mice were randomly divided into five groups ($n = 10$). All mice were fed a normal diet during the experiment, and dextran sulfate sodium (DSS) was used as an inducer to establish the colitis model. In GroupI (control group), the mice received drinking water; in GroupII (DSS model group), the mice received drinking water for the first 7 days, and drinking water containing 3% DSS for the next 7 days; GroupIII (APH low-dose group) received 100 mg kg^{-1} APH + 3% DSS; GroupIV (APH medium-dose group) received 300 mg kg^{-1} APH + 3% DSS; and GroupV (APH high-dose group) received 500 mg kg^{-1} APH + 3% DSS. During the experiment, the APH group mice were given the corresponding APH dose by gavage daily and treated with drinking water containing 3% DSS from days 8 to 14. The weight and stool changes of the mice were recorded daily, and the disease activity index (DAI) was established (Table 1) [18]. At the end of the experiment, blood was taken from the retro-orbital sinus, and the serum was obtained and centrifuged at 8000 rpm at 4 °C for 20 min, and stored at −80 °C. Then, the mice colonic samples were collected for further analysis.

Table 1. Disease activity index scoring system.

Score	Weight Loss (%)	Stool Consistency	Occult/Gross Bleeding
0	0	Normal	Normal
1	1–5	-	-
2	6–10	Loose	Hemoccult positive
3	11–15	-	-
4	>16	Diarrhea	Gross bleeding

2.5. Histopathology

The colon specimens were fixed in 10% formalin for 24 h and embedded in paraffin. After the paraffin-embedded sections were dewaxed and stained with haematoxylin and eosin, the degree of inflammation, depth of the lesion, crypt, and epithelial damage of the colon tissue were observed under a light microscope.

2.6. Determination of DAO, SOD, GSH-Px, and Inflammatory Cytokine Concentrations

The concentrations of diamine oxidase (DAO), IL-1β, IL-6, IL-10, and TNF-α in serum were determined with mouse ELISA kits (Shanghai Jianglai Biotechnology Co., Ltd., Shanghai, China) following the manufacturer's instructions. The concentrations of superoxide dismutase (SOD) and glutathione peroxidase (GSH-Px) of the colon tissue were assessed with commercial kits (Jianglai, Shanghai, China).

2.7. 16S rRNA High Throughput Sequencing

Genomic DNA was obtained from frozen fecal samples using a DNA isolation kit (Jianglai, Shanghai, China). The structural changes of the intestinal microbiota were conducted by 16S rRNA sequencing analysis. The data were analyzed on the online platform Majorbio Cloud Platform (Majorbio, Shanghai, China).

2.8. Statistical Analysis

The results were represented as mean ± standard deviation (SD). DAI was assessed with the unpaired two-tailed Student's *t*-test. Multiple comparisons were evaluated using one-way ANOVA followed by Tukey's multiple-comparison test. $p < 0.05$ was considered significant.

3. Results

3.1. Preparation of APH

The preparation of bioactive peptides by enzymatic hydrolysis is widely used due to its high safety, mild reaction conditions and easy control of the hydrolysis process. In order to obtain corn protein hydrolysate with glutamine-rich peptides, Alcalase (A) and Protamex (P) were used to hydrolyze the CGM in this study, and the effect of a different order of enzyme addition (A + P and P + A) on the Gln content in the hydrolysate was compared. The content of Gln in the hydrolysate by A + P and P + A were 12.53% and 9.99%, respectively, showing a significant difference ($p < 0.05$). The sequential hydrolysis of A + P was more conducive to the release of glutamine peptides from CGM than P + A. Thus, the sequential hydrolysis of CGM by A + P was selected to prepare the APH.

3.2. APH Alleviates Colitis Symptoms in Mice

It is well known that the rate of change in body weight and DAI score are the main evaluation indexes of UC. The body weight change rate of all groups of mice before DSS treatment was low, and their activity status was normal (Figure 1A,B). DSS treatment was started on the eighth day: the activity of the mice decreased and the DAI score increased gradually with the increase of treatment time in the model group. As soon as the mice were treated with DSS for 4 days, they developed loose, sticky, and bloody stools. According to the above phenomena, the model could be considered successful. Compared with the DSS group, the rate of change in body weight and the degree of diarrhea and bloody stools in the APH groups were significantly improved ($p < 0.05$), especially in the APH (500 mg/kg·d) group.

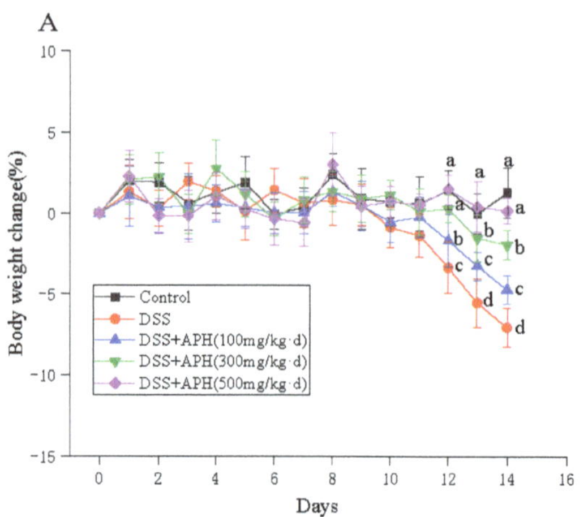

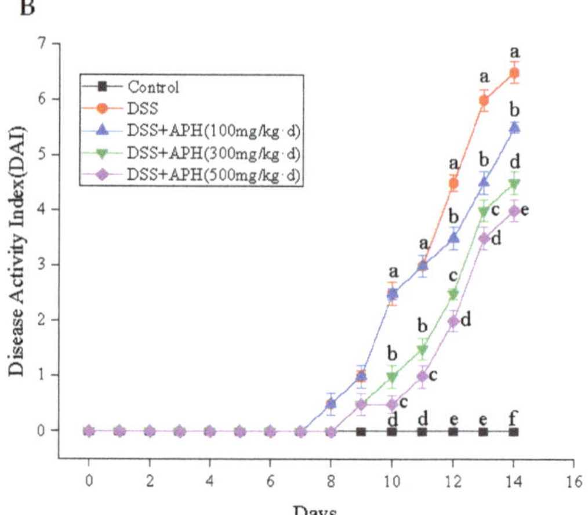

Figure 1. *Cont.*

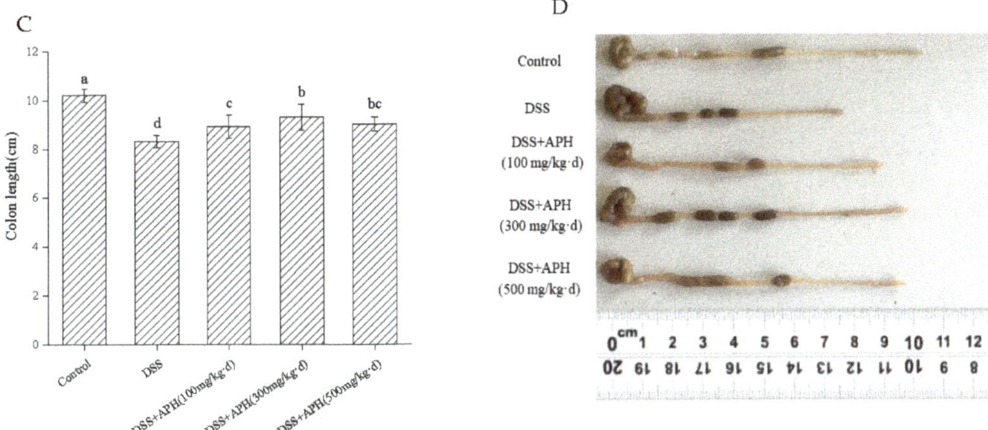

Figure 1. APH ameliorates clinical signs of colitis in mice. (**A**) Changes in body weight; (**B**) DAI scores; (**C,D**) colon length. Data are expressed as mean ± SD (*n* = 8). Different letters indicate significant differences (*p* < 0.05).

The colon length of the animals usually shortens under inflammatory stress. The colon length was measured in each group (Figure 1C,D) and, compared to the control group, the model group's colon length significantly shortened. The length of the colon of mice in the APH groups was considerably longer than those in the model group, especially in the medium-dose APH (300 mg/kg·d) group. This indicated that APH inhibited the intestinal stress response and alleviated colitis symptoms.

3.3. APH Alleviates Histopathological Changes in Mice

A pathologic examination was conducted to observe the extent of colonic injury (Figure 2). In the control group, the colonic mucosal layer was intact, there was a neat arrangement of glands, and no inflammatory cell infiltration was observed. The DSS group showed serious damage to the colonic mucosa, including edema, loss of some glands and crypt, and the infiltration of inflammatory cells. However, in the APH groups, the degree of inflammatory cell infiltration and colonic mucosa injury were reduced, especially in the high-dose APH group, where the structure of the colonic epithelial was more complete, and the colonic tissue injury was significantly improved. The colonic tissue morphology was similar to the control group, indicating that APH can substantially reduce the damage of colon tissue, and has a specific protective effect on the colon.

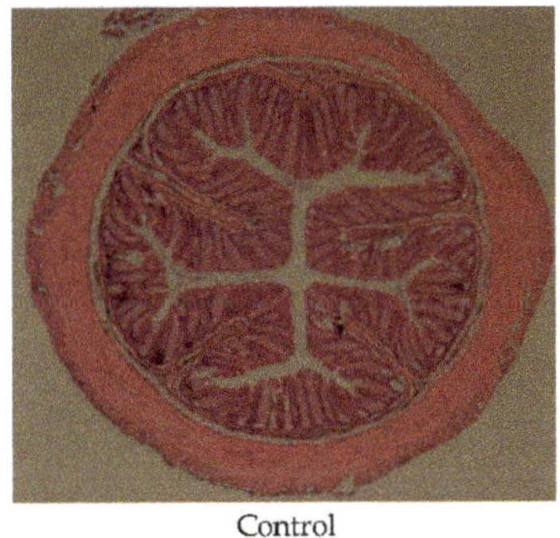

Control

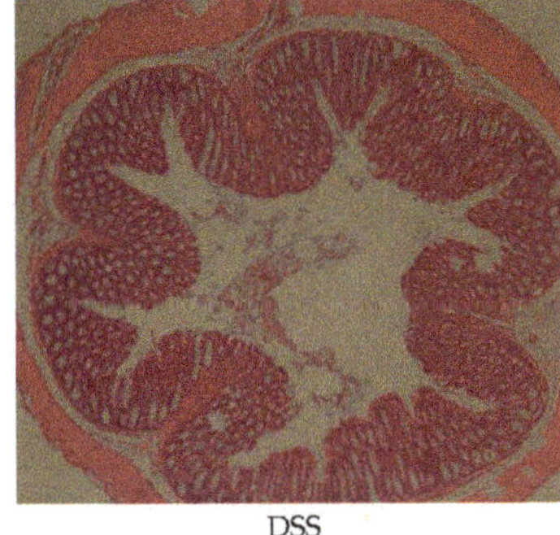

DSS

Figure 2. *Cont.*

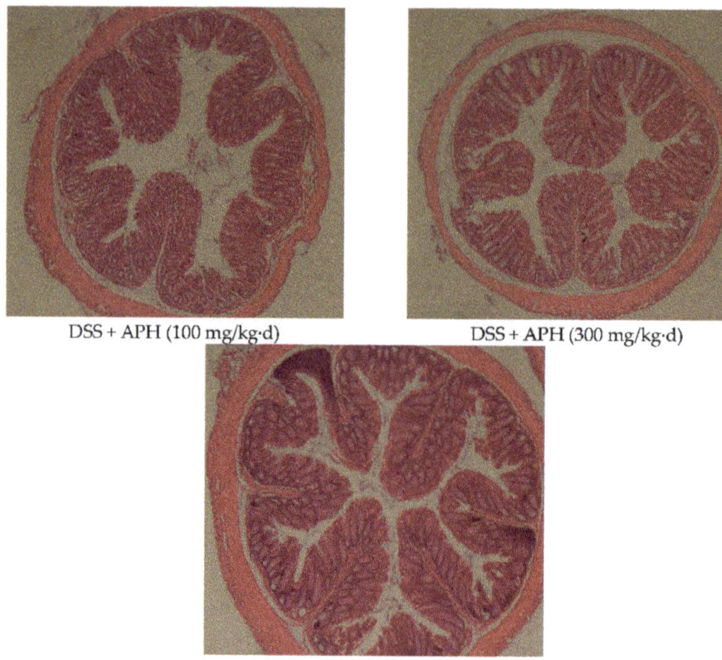

DSS + APH (100 mg/kg·d) DSS + APH (300 mg/kg·d)

DSS + APH (500 mg/kg·d)

Figure 2. APH alleviates DSS-induced histopathological changes.

3.4. Effects of APH on the Inflammatory Cytokines in Mice

Inflammatory cytokines are critical signaling molecules of inflammatory response in vivo, and colitis severity is closely related to their expression levels. The contents of inflammatory cytokines in serum were detected by ELISA (Figure 3). The serum IL-10 levels of mice in the DSS group were significantly lower than the control group ($p < 0.05$) (Figure 3A). In contrast, the levels of TNF-α, IL-6, and IL-1β were increased considerably (Figure 3B–D), indicating that colon inflammation was induced in mice after DSS treatment. After treatment with different doses of APH, the contents of IL-10 in the serum increased, and the contents of TNF-α, IL-6, and IL-1β decreased. These results indicated that APH played a specific role in regulating the levels of inflammatory factors, and it further revealed that APH had a particular anti-inflammatory effect on colitis in vivo.

3.5. Effects of APH on Oxidative Stress-Related Indicators in Mice

To further evaluate the effect of APH on UC from the perspective of oxidative stress, two oxidative stress markers, GSH-Px and SOD in the colon tissue, were examined (Figure 4). The contents of GSH-Px and SOD in the model group were significantly lower than in the control group ($p < 0.05$). The levels of the two oxidative stress markers were significantly increased in a dose-dependent manner after treatment with different doses of APH, indicating that APH could improve the antioxidant capacity of mice with colitis, and had a specific regulatory effect on oxidative stress.

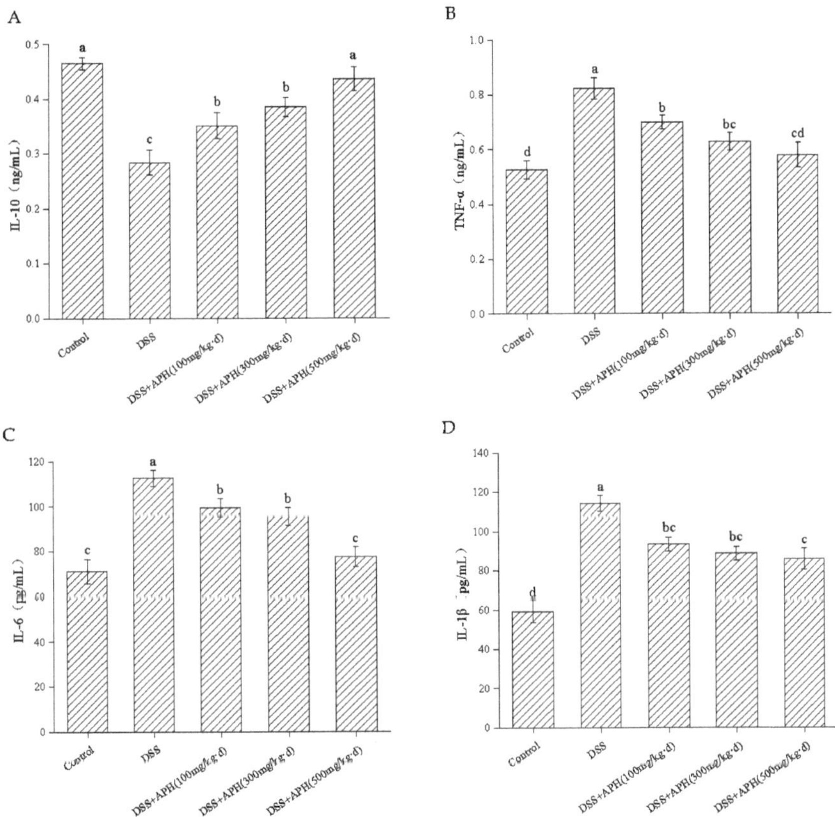

Figure 3. Effects of APH on the inflammatory cytokines, including the serum levels of IL-10 (**A**), TNF-α (**B**), IL-6 (**C**) and IL-1β (**D**). Data are expressed as mean ± SD (n = 8). Different letters indicate significant differences ($p < 0.05$).

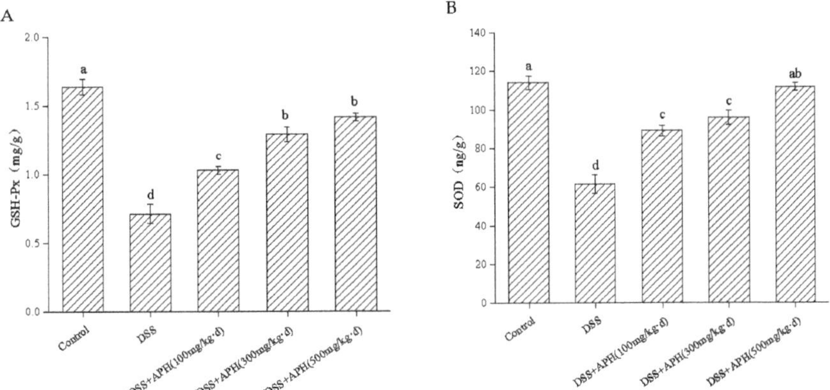

Figure 4. Effects of APH on the oxidative stress-related indicators, including the colonic concentrations of GSH-Px (**A**) and SOD (**B**). Data are expressed as mean ± SD (n = 8). Different letters indicate significant differences ($p < 0.05$).

3.6. Effects of APH on Intestinal Permeability in Mice

Diamine oxidase (DAO) is an intracellular enzyme with strong activity located in the upper layer of the intestinal villi in mammals. It is also a marker enzyme of intestinal mucosal cells. When intestinal mucosal epithelial cells are damaged, the intracellular DAO releases into the intestine and blood, which increases the level of DAO in the blood, indicating the increase of intestinal permeability [19]. Therefore, the change of DAO level is closely related to the integrity and damage degree of the intestinal mucosal mechanical barrier, and can be used as a sensitive indicator to reflect the change of intestinal permeability. The serum DAO levels in the DSS group were significantly higher ($p < 0.05$) than in the control group (Figure 5), indicating that the intestinal mucosa of the mice in the model group was severely damaged and the permeability increased. After APH treatment with different doses, the serum DAO levels of the mice significantly decreased ($p < 0.05$), especially in the high-dose APH group. These results indicated that APH could reduce the damage to the intestinal mucosa, reduce the permeability of the intestinal mucosa, and exhibit a particular protective effect on the mechanical barrier of the intestinal mucosa.

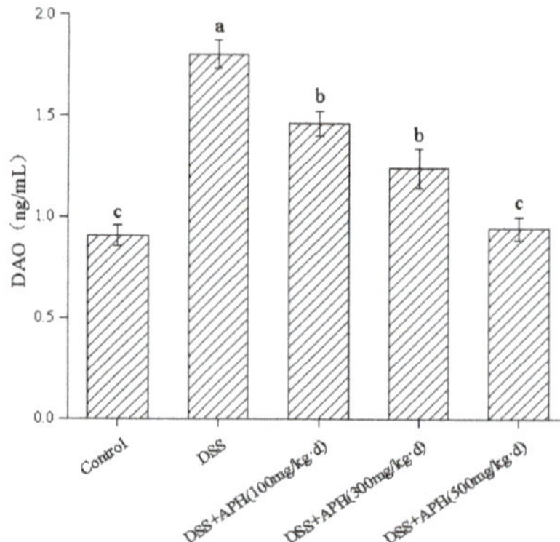

Figure 5. Effects of APH on DAO level in serum of DSS-induced colitis mice. Data are expressed as mean $\pm$ SD ($n = 8$). Different letters indicate significant differences ($p < 0.05$).

3.7. Effects of APH on Intestinal Microbiota in Mice

The intestinal microbiota of mice were measured using 16S rRNA high-throughput sequencing. A total of 203 OTUs were shared by the five groups, and the unique OTUs of each group was 61, 6, 19, 28, and 12, respectively (Figure 6A). This indicated that DSS reduced the diversity of the intestinal microbiota in mice, and the number of OTUs increased after APH treatment with different doses. Principal coordinate analysis (PCoA) can reflect the similarity between the samples: the shorter the distance between the sample and the sample in the figure, the more similar the samples are. The control group was closely clustered (Figure 6B), and there was an obvious dividing line and a long distance with the DSS group, which indicated that DSS had a particular influence on the intestinal microbiota. After APH treatment with different concentrations, the abundance and structure of the intestinal microbiota of the mice improved. The results of alpha diversity analysis demonstrated that DSS reduced the diversity and richness of the intestinal microbiota (Figure 6C), while APH improved it.

The analysis results of intestinal microbiota abundance at the phylum level showed that the overall microbiota composition predominantly changed in the DSS group: the relative abundance of *Firmicutes*, *Actinobacteriota*, *Desulfobacterota*, and *Patescibacteria* significantly decreased, while *Bacteroidota* and *Verrucomicrobiota* increased (Figure 6D). After treatment with different concentrations of APH, the relative abundance of Firmicutes in the high-dose APH (500 mg/kg·d) group and *Actinobacteriota*, *Desulfobacterota*, and *Patescibac-*

teria in the medium-dose APH (300 mg/kg·d) group significantly increased, while the relative abundance of *Bacteroidota* and *Verrucomicrobiota* significantly decreased.

The intestinal microbiota also changed significantly at the genus level (Figure 6E). The abundance of Lactobacillus, a family that regulates intestinal health, decreased in the DSS group. The abundance of other genera including *Desulfovibrio* and *Pediococcus* in the DSS group also decreased. After treatment with different concentrations of APH, the relative abundance of *Lactobacillus* and *Pediococcus* in the high-dose APH (500 mg/kg·d) group and *Desulfovibrio* in the medium-dose APH (300 mg/kg·d) group significantly increased, while the relative abundance of *Aerococcus* significantly decreased.

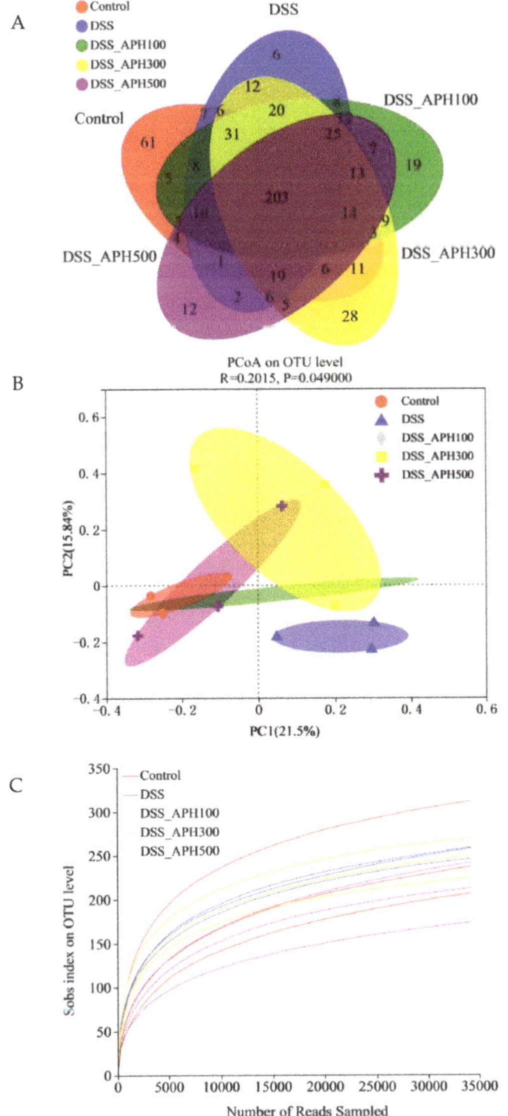

Figure 6. *Cont.*

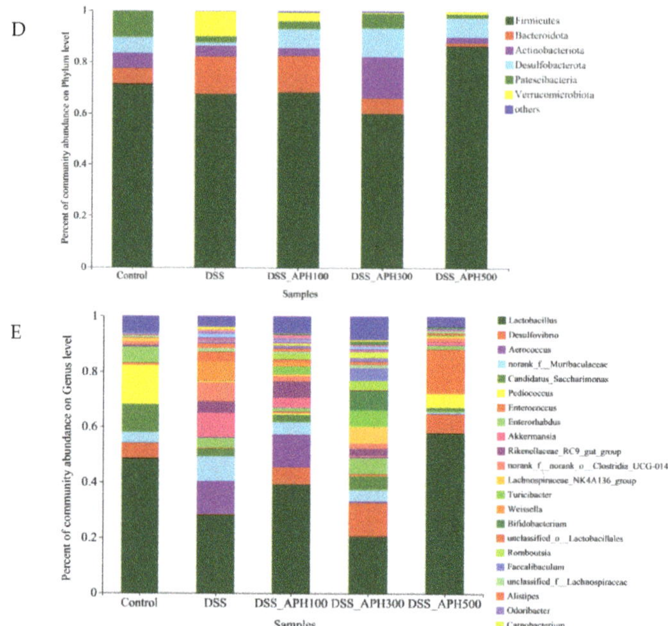

Figure 6. Effects of APH on the intestinal microbiota in mice with DSS-induced colitis. (**A**) Venn diagrams of OTU of each group; (**B**) principal coordinate analysis (PCoA); (**C**) rarefaction curve; (**D**) microbiota analysis at the phylum level; (**E**) microbiota analysis at the genus level.

4. Discussion

As a high-quality protein raw material rich in Gln, CGM was hydrolyzed by two commercial microbial proteases to prepare APH in the present study. Alcalase is produced by the fermentation of *Bacillus licheniformis*, which is an endopeptidase with serine in its catalytic site. Protamex, produced from *Bacillus subtilis*, is also an endopeptidase with broad specificity for peptide bonds. It has been reported that Alcalase and Protamex have little effect on the hydrolysis of the amide group in Gln, and can effectively protect the Gln in the raw material [20,21]. According to the results of this study, the content of Gln in the hydrolysate prepared by the sequential hydrolysis of A + P was significantly higher than that of P + A. It is possible that the cleavage sites of the two proteases are different, and the change of enzyme addition order will lead to different hydrolysates, which will result in the difference of Gln content in the hydrolysates. During the first step of hydrolysis, Alcalase showed a more intense cleavage effect on CGM than Protamex, which resulted in a relatively high Gln content in the hydrolysate. More importantly, more sites were exposed in CGM after the hydrolysis of Alcalase, which was beneficial for Protamex to play a better role in the second step of hydrolysis. On the contrary, the cleavage site of Alcalase was preempted if the Protamex was used as the first hydrolase, which is the reason why Alcalase exhibited lower hydrolysis effects on the residual peptides after the action of Protamex, and the Gln content in the hydrolysate was lower than that of A + P. Thus, the sequential hydrolysis of CGM by A + P was more suitable for the preparation of the APH.

Intestinal mucosal barrier injury is a typical symptom experienced by UC patients, and the pathological manifestation is increased intestinal permeability. With the increase of intestinal permeability, harmful substances such as bacteria and endotoxins can enter the body and affect human health. Preventing mucosal damage or promoting the efficient regeneration of mucosal cells is one of the ways to alleviate UC. As an essential amino acid of the intestinal tract, Gln is the main energy source of the intestinal cells that contribute to enhancing the metabolism and synthesis of mucosal proteins and cells, promote the

Article

Sacha Inchi Oil Press-Cake Protein Hydrolysates Exhibit Anti-Hyperuricemic Activity via Attenuating Renal Damage and Regulating Gut Microbiota

Kun Wang, Shanshan Wu, Pan Li, Nan Xiao, Jiamin Wen, Jinming Lin, Siming Lu, Xin Cai, Yanan Xu and Bing Du *

College of Food Science, South China Agricultural University, Guangzhou 510642, China
* Correspondence: dubing@scau.edu.cn

Abstract: The incidence of hyperuricemia has increased globally due to changes in dietary habits. The sacha inchi oil press-cake is generally discarded, resulting in the waste of resources and adverse environmental impact. For the purpose of developing sacha inchi oil press-cake and identifying natural components with anti-hyperuricemic activities, we systemically investigated the underlying mechanisms of sacha inchi oil press-cake protein hydrolysates (SISH) in the hyperuricemic rat model. SISH was obtained from sacha inchi oil press-cake proteins after trypsin treatment, and 24 peptides with small molecular weight (<1000 Da) were identified. The results of animal experiments showed that SISH significantly decreased the serum uric acid (UA) level by inhibiting the xanthine oxidase (XOD) activity and regulating the gene expression related to UA production and catabolism in hyperuricemia rats, such as Xdh and Hsh. In addition, SISH attenuated the renal damage and reduced the gene expression related to inflammation (Tlr4, Map3k8, Pik3cg, Pik3ap1, Ikbke, and Nlrp3), especially Tlr4, which has been considered a receptor of UA. Notably, SISH reversed high purine-induced gut microbiota dysbiosis, particularly by enhancing the relative abundance of butyric acid-producing bacteria (unidentified_*Ruminococcaceae*, *Oscillibacter*, *Ruminiclostridium*, *Intestinimonas*). This research provided new insights into the treatment of hyperuricemia.

Keywords: sacha inchi oil press-cake protein hydrolyzates; hyperuricemia; renal injury; gut microbiota

Citation: Wang, K.; Wu, S.; Li, P.; Xiao, N.; Wen, J.; Lin, J.; Lu, S.; Cai, X.; Xu, Y.; Du, B. Sacha Inchi Oil Press-Cake Protein Hydrolysates Exhibit Anti-Hyperuricemic Activity via Attenuating Renal Damage and Regulating Gut Microbiota. *Foods* **2022**, *11*, 2534. https://doi.org/10.3390/foods11162534

Academic Editors: Fuping Lu and Wenjie Sui

Received: 22 July 2022
Accepted: 12 August 2022
Published: 22 August 2022

Publisher's Note: MDPI stays neutral with regard to jurisdictional claims in published maps and institutional affil-iations.

1. Introduction

Hyperuricemia has become prevalent due to changes in dietary habits [1,2]. It is a leading cause of gout [3,4] and a risk factor for hypertension [5], obesity [6], cardiovascular disease [7], and renal failure [8]. Hyperuricemia is characterized by an increase in uric acid (UA) levels in serum due to the imbalance between the blood UA generation and excretion [9]. UA is one product in the purine metabolism pathway. Xanthine oxidase (XOD), a critical enzyme in the purine catabolic pathway, catalyzes the oxidation and conversion of hypoxanthine and xanthine to UA [10]. In in vivo settings, kidneys are responsible for most UA excretion, while approximately 25% of UA is excreted into the intestines and further metabolized by gut microbiota [9]. In fact, a recent report indicated that gut microbiota could serve as a candidate intervention target to relieve hyperuricemia pathogenesis [9]. Therefore, both kidney health and gut microbiota integrity are critical in hyperuricemia prevention. At present, three categories of medicines are commonly used to treat hyperuricemia: inhibitors of xanthine oxidase, uricosuric agents, and recombinant uricases [11,12]. However, these drugs have unfavorable side effects, such as serious allergic reactions and rashes resulting from allopurinol [13]. Consequently, it is critical to identify more effective alternatives to manage hyperuricemia with fewer side effects.

Protein hydrolysates are more digestible than protein, releasing more bioactive pep-tides and exhibiting a variety of biological functions, such as anticancer and immunoreg-ulation [14]. Recently, it has been reported that certain protein hydrolysates exhibited anti-hyperuricemic activity. Whey protein hydrolysates lessened renal dysfunction caused by hyperuricemia by restraining the production of UA and promoting the excretion of UA [15]. Bonito protein hydrolysates significantly reduce the serum UA level, inhibit the XOD enzyme activity, and alleviate renal damage in hyperuricemic rats [16]. In addition, walnut protein hydrolysate, tuna protein hydrolysates, and kidney bean protein hydrolysates exhibited anti-hyperuricemic activity [17–19]. Therefore, protein hydrolysates are reasonable resources to treat hyperuricemia. However, despite these previous studies that documented the anti-hyperuricemic activity of protein hydrolysates, a systemic and comprehensive investigation is needed to mechanistically address several essential unknowns. For instance, the major components and compositions of these hydrolysates need to be clarified, and the physiological impacts of protein hydrolysates on the kidney and gut biology and physiology demand specifications.

Sacha inchi (*Plukenetia volubilis* L.) is a plant belonging to the Euphorbiaceae family native to the Amazon Rainforest in Peru [20]. Sacha inchi kernels serve as an enriched source of various nutrients, especially in the forms of oils (35~65%) and proteins (25~30%) [21]. Sacha inchi kernels are mainly used for oil extraction, with the press-cake as a by-product. Although sacha inchi oil press-cake is rich in high-quality proteins (47~59%), it has been generally ignored, causing a massive waste of protein resources [22]. In addition, improper handling of the press-cake can readily result in environmental pollution. As reported, green tea residue contained 18~20% protein, and its hydrolysates showed various biological activity, including antioxidant and anti-hypertensive activity [23]. Sacha inchi oil press-cake is also a good source of protein hydrolysates with various biological activities [24,25]. Consistently, upon hydrolyzation of the sacha inchi oil press-cake by calotropis proteases and crude papain, the obtained protein hydrolysates possess substantial antioxidant properties [25]. In line with these findings, hydrolyzation of the sacha inchi oil press-cake by alcalase, neutrase, and flavourzyme yields protein hydrolysates that have antioxidant activities, as well [24]. Therefore, sacha inchi oil press-cake protein hydrolysates may be an effective way to address resource waste. However, the therapeutic value of the sacha inchi oil press-cake protein hydrolysates on hyperuricemia has not been defined yet.

Given the above considerations, there are multiple advantages to further developing sacha inchi oil press-cake to bring gains in the economy, environment, and the sacha inchi industry. In addition, it is also very important to explore more effective alternatives to treat and prevent hyperuricemia. Therefore, the aim of this study was to find the sacha inchi oil press-cake protein hydrolysates (SISH) with anti-hyperuricemic activity. Additionally, 16S rRNA gene sequencing, UPLC-Q-TOF/MS-based metabolomics, and RNA-Seq transcriptomics were used to determine the anti-hyperuricemic mechanisms underlying SISH. This study provides novel insights into the utilization of SISH and the treatment of hyperuricemia.

2. Materials and Methods

2.1. Preparation and Characterization of SISH

About 5 kg of defatted and squeezed sacha inchi seed flour was provided by the Pure U-Multitude Biological Resources Co., Ltd. (Yunnan Province, China). The sacha inchi seed flour was soaked in water (1:25, w/v) for 3 h. After that, the solution was hydrolyzed with 5% trypsin at 55 °C for 6 h, and the enzyme was inactivated by heating at 90 °C for 15 min [26]. Upon centrifugation ($4000 \times g$, 10 min), the supernatant was passed through an ultrafiltration membrane (3000 Da). The permeated solvent was then concentrated by evaporation and lyophilized to obtain SISH.

The amino acid analysis of SISH was conducted similarly as previously described [27]. The Shimadzu LC-20A HPLC system (Shimadzu, Japan) was used to measure the molecular weight distribution of SISH content according to the method described by Li et al. [28]. The peptide sequence of SISH was similarly identified using high-performance liquid chromatography-mass spectrometry (LC/MS-MS) [29].

2.2. Animals and Experimental Design

All animal experiments were conducted in the Guangdong Medical Laboratory Animal Center (Approval No: SCXY(Yue) 2018-0137) in accordance with Chinese laws and regulations on the use and care of laboratory animals. A total of 36 male Sprague Dawley (SD) rats were provided by the Medical Experimental Animal Center of Guangdong Province and housed in standard laboratory cages (temperature, 22~26 °C; humidity, 40~60%) with 10 h light/14 h dark cycle.

After adaptive feeding for 7 days, the rats were randomly divided into the blank control (CD) group ($n = 9$) and the model groups ($n = 27$). The model groups were orally gavaged with adenine (200 mg/kg BW) and injected intraperitoneally with potassium oxonate (100 mg/kg BW), while the CD group was fed with a normal diet. After feeding for 10 days, eyelid blood was collected to determine the serum UA level. UA levels above 110 μmoL/L were considered to be qualified as hyperuricemic in the rat model [16]. Then, the administration of adenine and potassium oxonate was halted, and the model group rats were randomly divided into three groups ($n = 9$ each), the hyperuricaemic model group (MG), the SISH treatment group (SISH), and the allopurinol treatment group (DG). Rats in the SISH and DG groups were intragastrically administered SISH (0.42 g/kg BW) and allopurinol (10 mg/kg BW), respectively, while the CD and MG groups were given the same volume of 0.9% saline. After SISH treatment for 4 weeks, the blood, liver, kidney, and colon contents were sampled and stored at −80 °C until use.

2.3. Biochemical Determination

XOD activity and the levels of UA, creatinine (Cr), and blood urea nitrogen (BUN) in the rat serum were determined by commercial detection kits (Jiancheng Bioengineering Institute, Nanjing, China) according to instructions of the manufacturer. The liver tissues were homogenized and centrifuged (3000 r/min, 10 min) at 4 °C, and XOD activity in the supernatant was then determined by the kit accordingly.

2.4. Histology

The kidney samples were stained with hematoxylin-eosin (HE) and examined by light microscopy. Scoring the renal tubule and interstitial injury was set up based on the following criteria: 0 points = no significant lesion; 1 point = the range of lesions does not exceed 1/4 area; 2 points = the range of lesions does not exceed 1/2 area; 3 points = the lesion range does not exceed 3/4 area; 4 points = the lesion range exceeds 3/4 area.

2.5. 16S rRNA Gene Amplification and Sequencing

The colon contents samples of each rat ($n = 6$) were collected aseptically for 16S rRNA gene sequencing analysis. According to the manufacturer's recommendation, genomic DNA from the colonic content was extracted through a fecal DNA isolation kit (MoBio Laboratories, Carlsbad, CA, USA). The primers 515F (5′-GTGCCAGCMGCCGCGGTAA-3′) and 806R (5′-GGACTACHVGGGTWTCTAAT-3′) were used to amplify the V4 region of 16S rRNA genes. The resulting PCR products were purified with GeneJET™Gel Extraction Kit (Thermo Scientific, Waltham, MA, USA). According to the manufacturer's recommendation, the Ion Plus Fragment Library Kit 48 rxns (Thermo Scientific) was used to generate the sequencing library. The sequencing test was performed by the TNovogene Bioinformatics Technology Co., Ltd. (Tianjin, China) using the Illumina MiSeq sequencing platform.

In order to obtain high-quality clean reads, the raw reads were trimmed and filtered under standard operations based on Cutadapt (V1.9.1, http://cutadapt.readthedocs.io/en/stable/, accessed on 13 January 2021). The operational taxonomic units (OTUs) with 97% sequence similarity were obtained using the Uparse software (Version 7.0.1001; Edgar, Robert C; Belvedere Tiburon, CA, USA). With the Mothur algorithm, the Silva Database (https://www.arb-silva.de/, accessed on 13 January 2021) was used to annotate the taxonomic information of each representative sequence. The alpha diversity and beta diversity of the gut microbiota were calculated by QIIME (Version 1.7.0; Rob Knight;

Fort Collins, CO, USA). Principal Coordinate Analysis (PCoA) was performed with the WGCNA package, stat packages, and ggplot2 package in the R software (Version 2.15.3; Mathsoft Company; Cambridge, MA, USA). Anosim and multi-response permutation procedure (MRPP) analyses were conducted with QIIME. The Galaxy platform was used for linear discriminant analysis effect size (LEfSe) analysis based on $p < 0.05$ and LDA score > 3 [30].

2.6. Serum Metabolomic Analysis

A total of 100 μL of serum samples were mixed with 400 μL of prechilled methanol. All samples were centrifugated at 15,000 rpm for 5 min, and the supernatant was diluted to a final concentration of 60% methanol. After filtration with a 0.22 μm filter, samples were centrifuged at $15,000 \times g$ for 10 min. Each sample was subjected to ultra-high performance liquid chromatography analysis in combination with quadrupole time-of-flight mass spectrometry (UHPLC-MS/MS) (Thermo Fisher). Samples were injected onto a Hyperil Gold column (100 × 2.1 mm, 1.9 μm) at a flow rate of 0.2 mL/min by a 16 min linear gradient. Eluent A (aqueous solution containing 0.1% fomic acid) and eluent B (methanol) were used for the positive polar mode, while eluent A (5 mM ammonium acetate, pH 9.0) and eluent B (methanol) were used for the negative polar mode. The solvent gradient was set as follows: 2% B, 1.5 min; 2–100% B, 12.0 min; 100% B, 14.0 min; 100–2% B, 14.1 min; 2% B, 16 min. Q ExActive HF-X mass spectrometer was operated in positive/negative polarity mode with the following parameters: spray voltage of 3.2 kV, capillary temperature of 320 °C, sheath gas flow rate of 35 arb, and auxiliary gas flow rate of 10 arb. Compound Discoverer 3.0 (Thermo Fisher) was used to analyze the original data files generated by the UHPLC–MS/MS, and peak alignment, peak calling, and quantitation were performed for each metabolite. In order to obtain accurate qualitative and relative quantitative results, peaks were aligned with mzCloud (https://www.mzcloud.org/, accessed on 21 December 2020) and ChemSpider (http://www.chemspider.com/, accessed on 21 December 2020) databases. Multivariate analyses, including principal component analysis (PCA) and Orthogonal partial least-squares discriminant analysis (OPLS-DA), were performed using the SIMCA-P 11.0 software (Umetrics AB, Umea, Sweden). Metabolites with significant differences between groups were obtained under variable influence on projection (VIP) > 1, and p-value < 0.05 based on the peak areas.

2.7. Renal Transcriptome Analysis

Total RNA was isolated from kidneys in compliance with the protocol of TRIzol reagent (Beyotime, Shanghai, China). The integrity and quality of total RNA were evaluated with the RNA Nano 6000 Assay Kit of the Bioanalyzer 2100 system (Agilent Technologies, Palo Alto, CA, USA). Then, 3 μg of total RNA for each sample was used as the input material for the RNA sample preparations. Based on the manufacturer's recommendations and index codes, TruSeq RNA Sample Preparation Kit was applied to generate the sequencing libraries, which were then sequenced on the Illumina HiSeq 2500 platform by the Novogene Bioinformatics Technology Co., Ltd. (Tianjin, China).

Index of the reference genome was established, followed by alignment using Hisat2 (Version2.0.5; Daehwan Kim; Missouri City, TX, USA) with the paired-end clean reads. The relative expression of each gene was normalized and inferred using fragments per kilobase of transcript per million mapped reads (FPKM). Differential expressions with statistical analysis were calculated using the software DESeq2 R package (Version 1.16.1; Michael I Love; Heidelberg, Germany), and the threshold was set as Fold Change > 1.5 and p-value < 0.05. Finally, the Gene Ontology (GO) program was applied to annotate the unigenes and analyze the pathway enrichment for differential expression genes (DEGs) [31].

2.8. Correlation Analysis and Statistical Analysis

The statistical analysis of quantitative data was conducted with the one-way variance analysis (one-way ANOVA) followed by Duncan's multiple comparisons test using the

SPSS 16.0. Spearman's correlation coefficients of gut microbiota, DEGs, and metabolites were calculated with R software (v3.2.1; Mathsoft Company) and visualized with the Complex Heatmap package in R software (v3.2.1). All data are expressed as means $\pm$ SD.

3. Results

3.1. Identification of SISH Composition

SISH was obtained from sacha inchi oil press-cake upon trypsin treatment, and the molecular weight distribution and amino acid composition of SISH were determined (Table 1). The results indicated that most of the peptides in SISH were oligopeptides with a molecular weight of less than 1000 Da (90.70%). In addition, SISH contained 17 kinds of amino acids in total, mainly including glutamic acid (16.7%), aspartic acid (12.6%), and arginine (10.8%). Among them, the essential amino acid content was in the range of 30%, and the hydrophobic amino acid content was 24.9%. De novo sequencing was performed on the collected high-resolution mass spectrometry dataset, and 24 polypeptides were identified based on scores, peak areas, fragment ion error distribution, and other factors (Table 2). MVVKK, KVVL, RLLVWELER, WLPDVK, and TVLLPR were identified as the most abundant peptides in SISH.

Table 1. The molecular weight distribution and amino acid composition of SISH.

		SISH
Molecular weight distribution (%)		
	<1 kDa	90.7
	1~2 kDa	6.0
	2~5 kDa	1.9
	>5 kDa	1.4
Amino acid composition (%)		
	Glutamate	16.7
	Aspartic acid	12.6
	Arginine	10.8
	Glycine	6.6
	Serine	5.9
	Lysine	4.9
	Valine	5.8
	Threonine	5.0
	Proline	4.1
	Leucine	7.1
	Alanine	3.9
	Tyrosine	4.4
	Isoleucine	4.7
	Phenylalanine	2.2
	Histidine	2.2
	Cysteine	2.4
	Methionine	1.0
	Essential amino acid content	30.7
	Hydrophobic amino acid	24.9

3.2. SISH Attenuated Hyperuricemia

Established SD rat models were used to evaluate the anti-hyperuricemia activity of SISH, and elevated serum UA level was monitored as an indicator, which is a common clinical observation in patients with hyperuricemia. After the initial 10 days of high purine diet feeding, the levels of serum UA were above 110 μmoL/L in the MG, SISH, and DG groups, a significant increase ($p < 0.05$) as compared with the CD group, demonstrating the success in the establishment of the hyperuricemic rat model (Supplementary Figure S1). After four weeks of indicated SISH treatments, there were no significant differences in body weight among the four test groups ($p > 0.05$) (Figure 1a). However, the serum UA of hyperuricemic rats after SISH treatment significantly decreased by 38.56% compared to the

MG group (Figure 1b). As reported, XOD is highly expressed in the liver, and an increase in XOD activity would enhance UA production [32]. Therefore, we next examined the effect of SISH treatment on the XOD activity in hyperuricemic rats. As compared with the CD group, XOD activity in serum and hepatic was significantly increased in the MG group ($p < 0.001$), and its activity was reversed to normal levels upon SISH supplementation ($p < 0.01$) (Figure 1c,d). These results indicated that SISH could effectively rescue the induced hyperuricemia by inhibiting XOD activity.

Table 2. The peptides sequence identified in SISH by HPLC/MS-MS analysis.

Peptide Sequence	Mass-to-Charge Ratio	Retention Time	Relative Intensity (%)	Peptide Score
TGGWSPLK	423.2326	19.9	1.19	98
WKPW	308.6679	27.18	1.55	98
FLTMEPR	447.2341	19.97	5.29	97
VVLDVK	672.4325	16.53	4.1	97
KVVL	458.3361	21.18	7.86	97
MVVKK	302.6903	21.15	8.66	97
LTGLNKL	379.7451	20.72	3.76	97
RLLVWELER	607.3524	18.13	8.13	97
KLSLEWWLK	601.8453	17.98	2.9	96
FVKLL	310.2144	25.61	5.04	96
LGDLGTKL	408.7478	22.05	1.99	96
LTGLDKL	380.2367	22.14	3.34	96
LFAEMDK	427.2123	18.29	4.69	96
EADGTLR	381.1941	14.02	1.11	96
VVLFK	303.2065	19.5	2.1	95
T(+42.01)LLNPR	378.2258	17.03	1.04	95
AYLTGLK	383.2322	18.84	1.01	95
WLPDVK	379.2173	21.21	7.68	95
VLWLPR	392.2499	30.21	1.25	95
RWQVWEDR	587.7963	21.99	2.107	95
TVLLPR	349.7339	18.2	8.63	95
LVRFPK	380.2433	20.96	1.6	95
TLLFGDK	397.2278	22.5	1.56	95
WSELVK	381.2155	18.68	1.66	95

3.3. SISH Attenuated Renal Damage in Hyperuricemic Rats

Hyperuricemia would increase the risk of renal injury, and thus we needed to examine the effects of SISH on the rat kidney. As shown, SISH supplement indeed profoundly reduced the kidney index as compared with the MG group (Figure 2a). The BUN and Cr, as the indexes of kidney functions, were also evaluated in hyperuricemic rats (Figure 2b,c). The levels of BUN and Cr were substantially increased in the MG group, indicating that the renal functions of hyperuricemia rats were impaired. However, both the BUN and Cr levels were markedly decreased by 12.12% and 13.63% upon SISH treatment ($p < 0.05$). These results showed that SISH could ameliorate renal damage in hyperuricemic rats. Histologically, the sections of kidneys from the CD group showed normal architecture with intact glomerulus and renal tubules. However, in contrast to the normal group, glomerular atrophy, renal tubular dilation, and inflammatory cell infiltration were present in the MG group, suggesting that hyperuricemic rats had obvious renal damage (Figure 2d). Nevertheless, renal lesions were alleviated upon SISH treatment, which was manifested by a moderate decrease in glomerular volume, mild tubules dilation, and mild inflammatory cell infiltration in the interstitium. Simultaneously, injury scores in the SISH group were also significantly lower than that in the MG group (Figure 2e), and the glomerular counts were similarly reversed (Figure 2f). These results showed that SISH treatment effectively relieved the renal damage in hyperuricemic rats.

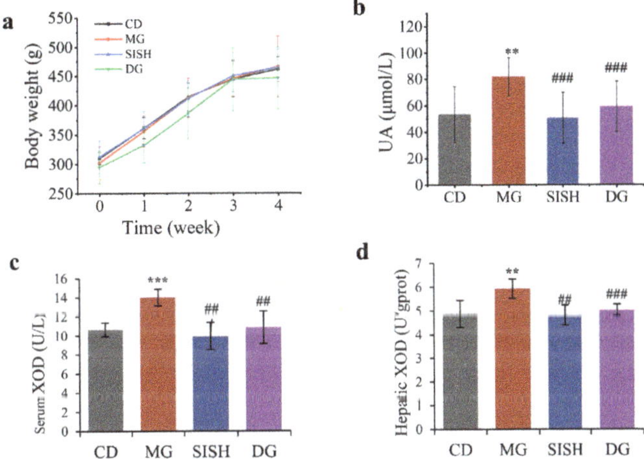

Figure 1. Evaluation of SISH anti-hyperuricemic activities in hyperuricemic rats. (**a**) Body weight. (**b**) Serum uric acid (UA) levels. (**c**) Serum xanthine oxidase (XOD) activity. (**d**) Hepatic XOD activity. All data are expressed as means $\pm$ SD. ** $p < 0.01$, and *** $p < 0.001$ vs. the CD group; ## $p < 0.01$, and ### $p < 0.001$ vs. the MG group.

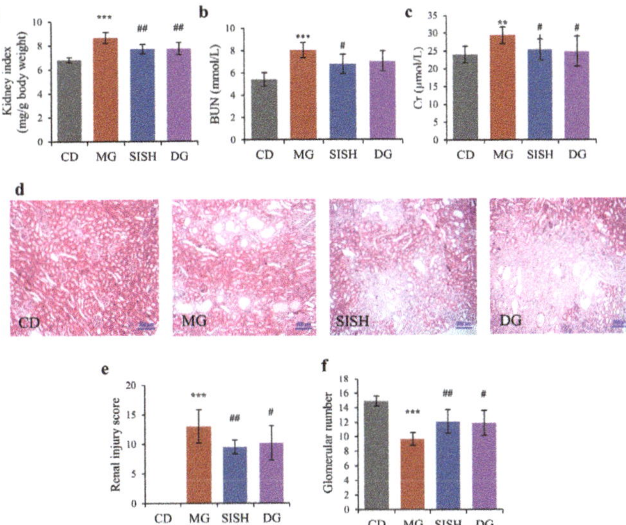

Figure 2. The effects of SISH on renal damage in hyperuricemic rats. (**a**) The kidney indexes. (**b**) Blood urea nitrogen (BUN) levels. (**c**) Serum creatinine (Cr) levels. (**d**) Photomicrographs of representative sections of kidneys. (**e**) Renal injury score. (**f**) Glomerular number in each group. All data are expressed as means $\pm$ SD. ** $p < 0.01$, and *** $p < 0.001$ vs. the CD group; # $p < 0.05$, and ## $p < 0.01$ vs. the MG group.

3.4. SISH Reshaped the Gut Microbiota Community in Hyperuricemic Rats

The gut microbiota alteration was analyzed by 16S rRNA gene sequencing. A total of 2,021,638 clean reads were generated based on sequencing the V4 region of 16S rRNA genes, and each colonic sample produced an average of 84,234 ± 4205 clean reads. The rarefaction curves were close to the saturation plateau (Figure 3a), indicating that the sequencing depth covered rare new phylotypes and represented most of the accessible microbiota. The Shannon index that indicates the diversity of microbial communities [33] was significantly decreased in the MG group as compared to that of the CD group ($p < 0.001$). Upon SISH and allopurinol treatments, the Shannon index was significantly increased in the SISH and DG groups as compared with the MG group (Figure 3b). Therefore, SISH clearly increased the diversity of gut microbiota in hyperuricemic rats. Meanwhile, PCoA analysis also showed that the community composition of the MG group was separated from the other three groups (Figure 3c). Anosim and MRPP algorithms further confirmed significant differences ($p < 0.05$) between the MG groups versus the CD, SISH, and DG groups (Figure 3d).

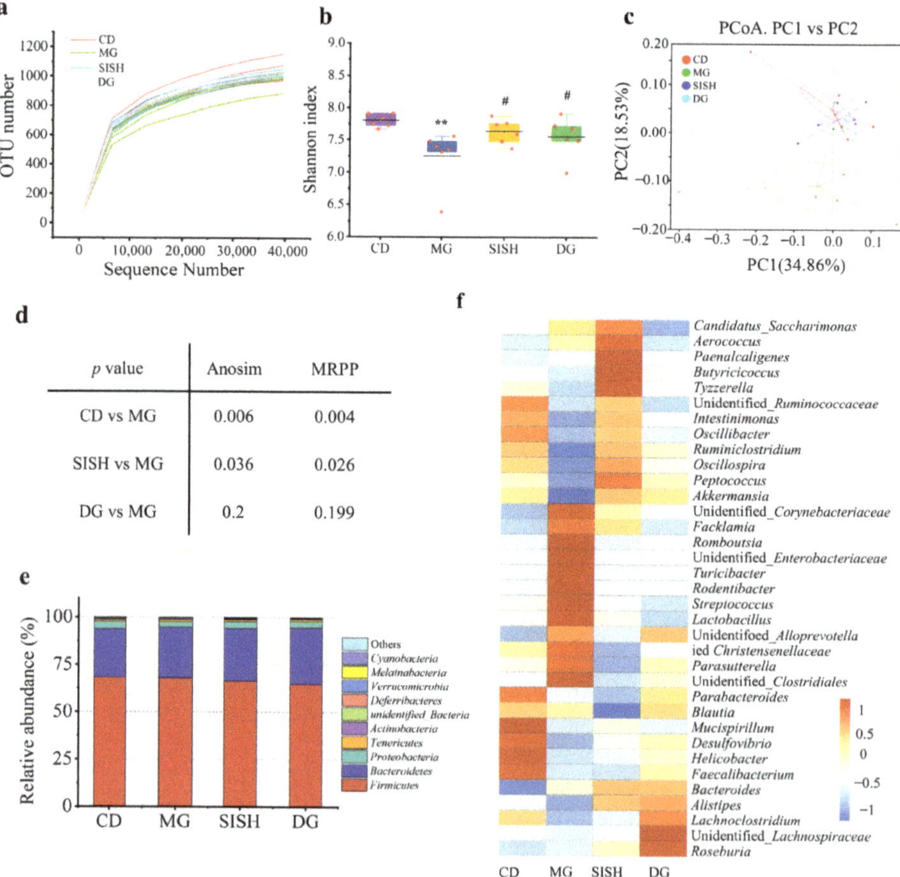

Figure 3. The effects of SISH on the diversity and integrity of the gut microbiota in hyperuricemic rats. (**a**) Rarefaction curves. (**b**) Shannon index. (**c**) Principal coordinate analysis (PCoA). (**d**) Anosim and multi-response permutation procedure (MRPP) analyses. (**e**) Relative abundance at phylum levels. (**f**) The heatmap analysis at genus levels. All data are expressed as means ± SD. ** $p < 0.01$ vs. the CD group; # $p < 0.05$ vs. the MG group.

Taxonomic profiling was also used to define SISH effects on the gut microbiota composition in hyperuricemic rats (Figure 3e). As shown, *Firmicutes*, *Bacteroidetes*, *Proteobacteria*, and *Tenericutes* were the dominant phylum in all test groups (Figure 3e). There was no significant difference among groups at the phylum levels. Nevertheless, heatmap analysis showed a dramatic alteration of gut microbiota distribution at the genus level (Figure 3f). Compared with the CD group, *Romboutsia* (1.64% vs. 6.71%) and unidentified_*Clostridiales* (0.42% vs. 0.80%) were markedly increased, while unidentified_*Ruminococcaceae* (9.63% vs. 6.16%), *Ruminiclostridium* (1.54% vs. 0.66%), *Oscillibacter* (1.92% vs. 0.88%), and *Intestinimonas* (1.01% vs. 0.52%) were significantly decreased in the MG group. Of importance, SISH intervention substantially rescued the relative abundance of those intestinal florae in hyperuricemic rats (Figure 3f, Supplementary Figure S2). Specifically, SISH enhanced the relative abundance of probiotics such as *Akkermansia* and *Alistipes* and reduced the *Streptococcus* (XOD-producing bacteria) population [10]. Taken together, these results evidenced that SISH could modulate the gut microbial organization and enhance the microbial diversity in hyperuricemic rats.

3.5. SISH Intervention Altered Serum Metabolome in Hyperuricemic Rats

The effects of SISH on serum metabolome in hyperuricemia rats were next determined by the UHPLC-MS/MS analyses. PCA score plotting showed a clear separation trend between the CD and the MG groups, with similar separation trends being observed between the SISH and the MG groups and between the DG and the MG groups (Figure 4a), respectively. In comparison with that of the MG group, a total of 58, 56, and 70 metabolites were significantly changed in the CD, the SISH, and the DG groups, respectively (Figure 4b,c). OPLS-DA analysis further showed that there was a distinct classification in the metabolite profiles between the MG and the SISH groups (Figure 4d). In addition, cluster analyses revealed a distinct separation of serum metabolites between the MG and the SISH groups (Figure 4e). We also found that UA levels were significantly increased, while 2′-deoxycytidine levels were decreased in the MG group. Other metabolites related to amino acid and energy metabolism, such as L-threonine, DL-carnitine, and tigliylcarnitine, were also substantially reduced in the MG group, indicating that hyperuricemia is extensively linked to amino acid and energy metabolic dysfunction [34,35]. In any case, upon SISH treatment, levels of the above five metabolites were all pronouncedly restored. In comparison, allopurinol treatment rescued the levels of UA, L-threonine, and tigliylcarnitine only (Figure 4f, Supplementary Table S1).

We further performed Spearman correlation analysis to visualize an association between the gut microbiota and metabolome (Figure 4g). As shown, UA was positively ($p < 0.05$) correlated with the gut genus of *Turicibacter* while negatively ($p < 0.05$) correlated with the gut genus of *Ruminiclostridium* and *Intestinimonas*. 2′-deoxycytidine was positively ($p < 0.05$) correlated with *Ruminiclostridium* while negatively ($p < 0.05$) correlated with *Romboutsia*, unidentified_*Clostridiales*, *Rodentibacter*, and *Faecalibacterium*. Metabolite of L-threonine was positively ($p < 0.05$) correlated with the gut genu of *Ruminiclostridium*, but negatively ($p < 0.05$) correlated with the gut genus of *Romboutsia*, unidentified_*Clostridiales*, *Rodentibacter*, and *Faecalibacterium*. Overall, these results delineated the impacts of SISH on the serum metabolome in hyperuricemic rats.

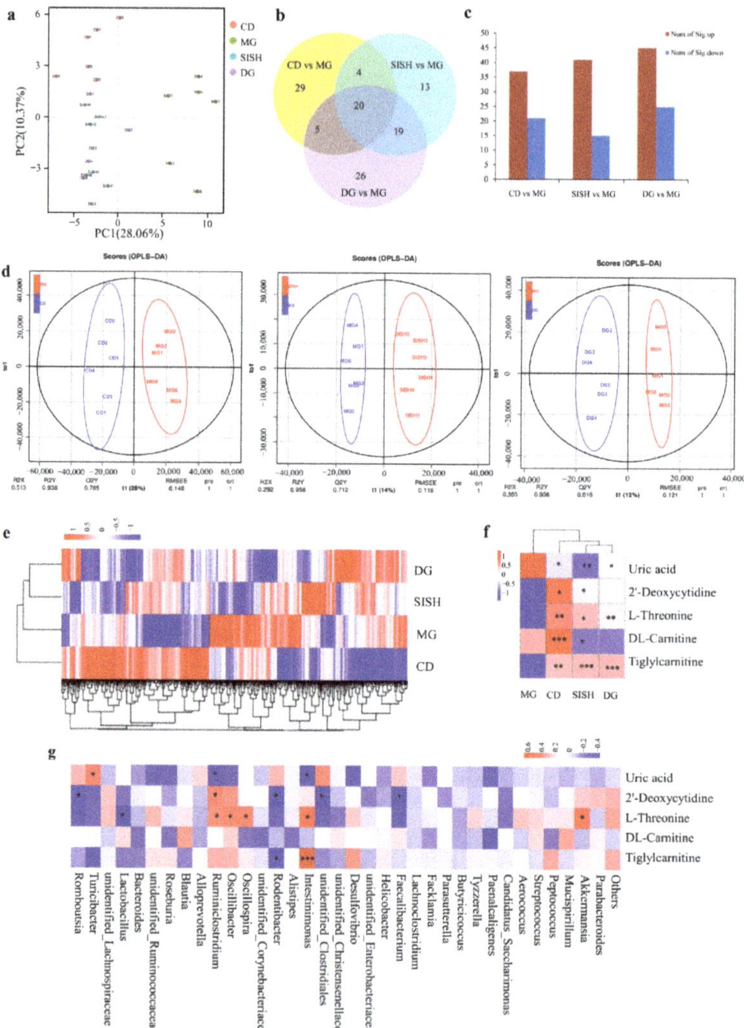

Figure 4. The impacts of SISH on the serum metabolome of hyperuricemic rats. (**a**) Principal component analysis (PCA) score plot. (**b**) Venn diagram. (**c**) The numbers of differential metabolites. (**d**) The Orthogonal partial least-squares discriminant analysis (OPLS-DA) scores plot. (**e**) A heatmap of differential metabolites potentially associated with hyperuricemia. (**f**) A heatmap of differential metabolites potentially associated with hyperuricemia disease for each treatment. (**g**) Spearman correlation analysis between the metabolites and the gut microbiota. Note: * $p < 0.05$, ** $p < 0.01$, and *** $p < 0.01$.

3.6. Identification of the Potential SISH Target Genes and Signaling Pathways Involved in Hyperuricemic Rats

In order to further define the molecular determinants underlying SISH beneficial effects in hyperuricemic rats, we then explored the renal transcriptome. As compared with the CD group, 2584 genes were differentially expressed in the MG group. Furthermore, a total of 1394 genes were differentially expressed between the SISH and the MG groups

(Figure 5a). Hierarchical clustering and PCA analysis revealed that DEGs in the SISH group were obviously distinct from those in the MG group (Figure 5b,c).

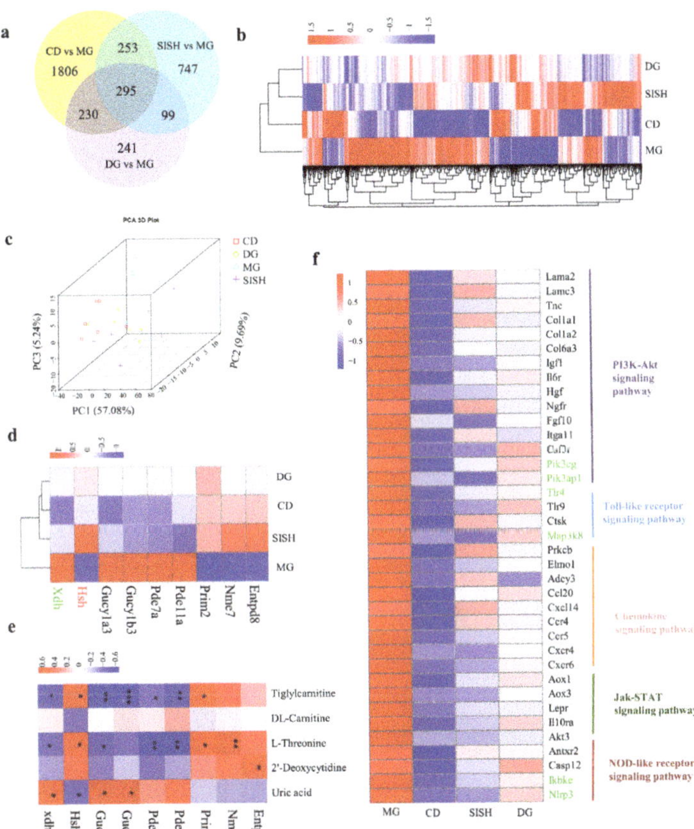

Figure 5. The Effects of SISH on the renal transcriptome in hyperuricemic rats. (**a**) Venn diagram. (**b**) Heatmap for hierarchical cluster analysis of differential expression genes (DEGs) (fold change of ≥ 1.5 or ≤ -1.5, $p < 0.05$). (**c**) The principal component analysis (PCA) score plot. (**d**) Heatmap for genes involved in purine metabolism. (**e**) Spearman correlation analysis between the metabolites and genes related to purine metabolism. Differences were assessed by ANOVA. (**f**) Heatmap for genes involved in inflammatory responses. Note: * $p < 0.05$, ** $p < 0.01$, and *** $p < 0.01$.

Next, we evaluated the expression of several crucial genes related to purine metabolism (Figure 5d, Supplementary Table S2). Compared with the CD group, Xdh, Gucy1α3, Gucy1β3, Pde7a, and Pde11a transcripts were markedly upregulated in the MG group, while Prim2, Nme7, and Entpd8 were substantially downregulated. Importantly, compared with the MG group, the expression levels of Xdh, Gucy1α3, Gucy1β3, Pde7a, and Pde11a were all downregulated, while Hsh was upregulated upon SISH treatment. Consistently, based on the Spearman correlation analysis, we found that the metabolite of UA was positively ($p < 0.05$) correlated with Xdh, Gucy1α3, and Gucy1β3, but negatively ($p < 0.05$) correlated with Hsh (Figure 5e). These results demonstrated that SISH could reduce UA levels by regulating genes functioning in purine metabolism.

Excessive levels of serum UA are known to increase the risk of kidney inflammation and damage [36]. In accordance, the transcriptomic analysis identified enrichment in a group of DEGs relevant to inflammation (Figure 5f). Compared with the MG group, the SISH group substantially reduced the expression of genes involved in the PI3K-Akt signaling pathway, Jak-STAT signaling pathway, Toll-like receptor signaling pathway, and NOD-like receptor signaling pathway. Among them, Tlr4, Map3k8, Ikbke, Nlrp3, Pik3cg, Pik3ap1, and Akt3 genes were significantly downregulated in the SISH group as compared with the MG group. According to GO analysis (Supplementary Figures S3 and S4), the significantly upregulated DEGs in the SISH group (as compared to the MG group) were functionally enriched in the transport and metabolic-related processes. In comparison, most of the significantly downregulated genes were associated with inflammatory responses, such as chemotaxis, ERK1 and ERK2 cascade, regulation of ERK1 and ERK2 cascade, and positive regulation of MAPK cascade. Together, these findings specified these inflammation-related genes that could mediate the beneficial effects of SISH on hyperuricemia.

4. Discussion

With the increasing prevalence of hyperuricemia worldwide, it is imperative to seek more effective alternatives for hyperuricemia intervention with fewer side effects. Plant protein hydrolysates are among the best options in terms of scientific, economic, social, environmental, and industrial values. In this study, we evidenced that sacha inchi oil press-cake is a good source of bioactive protein hydrolysates that have favorable anti-hyperuricemia effects. Instead of discarding the sacha inchi oil press-cake as industrial by-products and environmental biohazards, we could obtain SISH from it by hydrolyzing proteins. A series of tests in biochemical, biological, and physiological settings helped us to define the composition and anti-hyperuricemic effects of SISH, potentializing its use and application in various food and bio-industrial domains.

Hyperuricemia is characterized by high levels of UA in serum [37], and previous studies have shown that some small molecular weight peptides (<1 kDa) could lower the serum UA level [18,38,39]. In this study, SISH was confirmed to contain a large number of small peptides (up to 90.7%, <1 kDa) and shown to reduce serum UA levels in hyperuricemic rats, effects that were validated in serum metabolome analysis, as well. Purine metabolism disorders could also lead to an increase in serum UA, in which the XOD plays a central role by catalyzing the oxidation and conversion of hypoxanthine to UA [18]. Consistently, our study found that SISH inhibited the XOD activity in serum and liver in hyperuricemic rats, providing a mechanistic explanation. It should be noted that Trp residue in peptides, as identified in the SISH oligopeptides, acts as a critical factor for achieving XOD inhibitory activity [37]. Moreover, a mixture of peptides in the SISH would confer combinatory effects on UA metabolism, likely due to synergistic interactions among these peptides [40]. In line with these findings, renal transcriptome analysis indicated that SISH could downregulate genes relevant to UA production (such as Xdh, Gucy1α3, Gucy1β3, Pde7a, and Pde11a) [41] and UA catabolism (like Hsd) that promotes the degradation of UA into allantoin [42]. Collectively, we propose that the anti-hyperuricemic activities of SISH are derived from its inhibitory activity on XOD and its regulation of genes engaged in UA metabolism.

The imbalance between UA production and excretion leads to hyperuricemia, and approximately 2/3 of UA is excreted through the renal route [9]; however, a high level of UA in serum could cause renal injury [43]. Therefore, normal renal function is an essential factor in ensuring the balance of UA production and excretion. Serum BUN and Cr are important indicators of renal functions, and consistently, in our study, both levels of BUN and Cr were enhanced in hyperuricemic rats. On the contrary, SISH treatment significantly reversed the increases in BUN and Cr, in agreement with the findings by Shan et al. [44]. In addition, high UA in serums could induce renal inflammation [36], and histopathological examination did show inflammatory cell infiltration in the interstitium of the kidney in hyperuricemic rats, while SISH again alleviated the inflammation. Mechanistically, renal transcriptome analysis in hyperuricemic rats further demonstrated that SISH downregulated genes relevant to inflammatory signaling, namely the PI3K-Akt signaling pathway, Toll-like receptor signaling pathway, chemokine signaling pathway, Jak-STAT signaling pathway, and NOD-like receptor signaling pathway. Of significance, SISH repressed the

Tlr4 gene that functions as a UA receptor [45]. Upon Tlr4 activation by UA, its downstream signal cascade (like the IKK/IκB/NF-κB pathway) will be unleashed, ultimately leading to the induction of pro-inflammatory cytokines/mediators and the onset of tissue damage [36,46]. Together, we determined by multiple channels that SISH markedly alleviates the kidney damage bound with hyperuricemia.

In addition to the predominant excretion of UA through the kidneys, the remaining about 1/3 of UA is excreted from the intestines [9]; importantly, UA excretion will be mainly converted to the intestine when kidneys become dysfunctional [43]. Gut microbiota, as an important component of the intestine, plays an active role in UA excretion [9,47]. In accordance, previous studies indicated that the anti-hyperuricemic effects of some oligopeptides were mediated by the gut microbiota, such as tuna meat oligopeptides [40], Val-Val-Tyr-Pro peptide [48], and *Apostichopus japonicus* oligopeptide [49]. As reported, the diversity of intestinal flora in the hyperuricemia group is generally lower than that in the normal group [50]. Indeed, the alpha diversity analysis in the current study revealed that the Shannon diversity index was significantly decreased in the hyperuricemic rats, whereas it was substantially enhanced by SISH supplementation. Short-chain fatty acids (SCFAs), especially butyric acid, can relieve hyperuricemia by providing energy to intestinal wall cells to promote UA excretion [9]. Compared with the hyperuricemic group, the SISH group had a significantly increased abundance of *Ruminococcaceae* (unidentified_*Ruminococcaceae*, *Oscillibacter*, *Ruminiclostridium*, *Intestinimonas*) which are considered butyrate-producing prebiotics [51]. In addition, SISH also increased the abundance of probiotics (*Akkermansia*, *Alistipes*) and reduced the population of the XOD-producing bacteria (*Streptococcus*) [10]. *Lactobacillus* is known to sequentially degrade UA to urea by synthesizing uricase, allantoinase, and allantoicase, thereby improving UA excretion [9]. Unexpectedly, we found that the relative abundance of *Lactobacillus* was higher in the MG group as compared with the CD group, and SISH decreased *Lactobacillus* in hyperuricemic rats, which is in contradiction to some reports [9,50]. We speculated that the increase in *Lactobacillus* in the MG group could reflect the body's intrinsic compensatory system as a part of feedback at certain time points to consume excess UA in hyperuricemic rats [43]. Overall, these results clearly indicate that SISH supplementation could reverse gut microbiota dysbiosis in hyperuricemic rats.

By integrating the results of 16S rRNA gene sequencing, serum metabolome, and renal transcriptome analyses, we also found that UA level was negatively correlated with the *Ruminococcaceae* (unidentified_*Ruminococcaceae*, *Oscillibacter*, *Ruminiclostridium*, *Intestinimonas*) abundance and *Hsh* gene expression level, and an increase in those bacteria and the *Hsh* gene can enhance UA excretion [51]. In addition, UA levels positively correlated with *Streptococcus* abundance and *Xdh* gene expression, and increases in those bacteria and the *Xdh* gene would enhance UA production [10,41]. Consistently, SISH boosted the *Ruminococcaceae* abundance and Hsh gene expression and reduced the *Streptococcus* abundance and *Xdh* gene expression, consequentially leading to a decline in serum UA and relief of hyperuricemia. In summary, SISH attenuated hyperuricemia by ameliorating renal damage and maintaining intestinal flora integrity. Our findings provide new insights on the utilization of sacha inchi oil press-cake and the potential of derived SISH as a hyperuricemia therapeutic regime.

5. Conclusions

In conclusion, SISH could reduce the serum UA level and improve hyperuricemia. The underlying mechanisms could be related to the inhibition of the XOD activity, regulation of gut microbiota structure, and attenuation of the renal injury in SD rats. However, additional studies are required to investigate the deeper mechanisms underlying the anti-hyperuricemic activity of SISH; for instance, fecal microbiota transplantation could be used to further explore whether alterations in the gut microbiota are a key factor in the anti-hyperuricemic activity of SISH. Overall, these results suggest that the protein in sacha inchi oil pressed-cake is a valuable source of hydrolysates, and SISH could be used as a potential functional component for preventing hyperuricemia.

Supplementary Materials: The following supporting information can be downloaded at: https://www.mdpi.com/article/10.3390/foods11162534/s1, Table S1, Identification of the potential biomarkers in SISH treated hyperuricemic rats; Table S2, Identification of candidate functional genes in SISH treated hyperuricemic rats; Figure S1, The serum UA levels in rats upon adenine and potassium oxonate treatments for 10 days; Figure S2, Comparison of indicated specific taxa between the four test groups; Figure S3, GO enrichment analyses between the CD and the MG groups; Figure S4, GO enrichment analyses between the SISH and the MG groups.

Author Contributions: Conceptualization, K.W., S.W. and P.L.; data curation, P.L.; formal analysis, J.L. and S.L.; investigation, N.X. and X.C.; methodology, K.W., S.W. and J.W.; project administration, P.L. and B.D.; resources, J.W.; software, K.W., N.X. and Y.X.; writing—original draft, K.W.; writing—review and editing, S.W. and P.L.; supervision: B.D.; funding acquisition, B.D. All authors have read and agreed to the published version of the manuscript.

Funding: This research was funded by the China Agriculture Research System of MOF and MARA (CARS-21), the Natural Science Foundation of Guangdong Province (2020A1515011268), the Key-Area Research and Development Program of Guangdong Province (2020B020226008) and Du Bing Expert Workstation Project of Yunnan Province (202005AF150071).

Institutional Review Board Statement: The animal study protocol was approved by the Ethics Committee of Guangdong Medical Laboratory Animal Center (protocol code: 44007200058370; date of approval: 5 April 2019).

Informed Consent Statement: Not applicable.

Data Availability Statement: The data presented in this study are available on request from the corresponding author.

Acknowledgments: We would like to thank Novogene (Beijing, China) for sharing the Illumina platform. We also thank Pure U-Multitude Biological Resources Co., Ltd. (Puer, China) for the supply of the defatted and squeezed sacha inchi seed flour.

Conflicts of Interest: The authors declare no conflict of interest.

References

1. Guo, Z.; Zhang, J.C.; Wang, Z.L.; Ang, K.Y.; Huang, S.; Hou, Q.C.; Su, X.Q.; Qiao, J.M.; Zheng, Y.; Wang, L.F.; et al. Intestinal Microbiota Distinguish Gout Patients from Healthy Humans. *Sci. Rep.* **2016**, *6*, 20602. [CrossRef]
2. Zhang, R.; Zhan, S.Y.; Li, S.Y.; Zhu, Z.Z.; He, J.R.; Lorenzo, J.M.; Barba, F.J. Anti-hyperuricemic and nephroprotective effects of extracts from Chaenomeles sinensis (Thouin) Koehne in hyperuricemic mice. *Food Funct.* **2018**, *9*, 5778–5790. [CrossRef] [PubMed]
3. Li, X.Q.; Gao, X.X.; Zhang, H.; Liu, Y.Y.; Sarker, M.M.R.; Wu, Y.J.; Chen, X.H.; Zhao, C. The anti-hyperuricemic effects of green alga Enteromorpha prolifera polysaccharide via regulation of the uric acid transporters in vivo. *Food Chem. Toxicol.* **2021**, *158*, 112630. [CrossRef] [PubMed]
4. Mehmood, A.; Zhao, L.; Ishaq, M.; Xin, W.; Zhao, L.; Wang, C.T.; Hossen, I.; Zhang, H.M.; Lian, Y.H.; Xu, M.L. Anti-hyperuricemic potential of stevia (Stevia rebaudiana Bertoni) residue extract in hyperuricemic mice. *Food Funct.* **2020**, *11*, 6387–6406. [CrossRef] [PubMed]
5. Abeles, A.M. Hyperuricemia, gout, and cardiovascular disease: An update. *Curr. Rheumatol. Rep.* **2015**, *17*, 13. [CrossRef]
6. Ma, Z.M.; Wang, Y.F.; Xu, C.N.; Ai, F.L.; Huang, L.; Wang, J.P.; Peng, J.; Zhou, Y.M.; Yin, M.H.; Zhang, S.; et al. Obesity-Related Genetic Variants and Hyperuricemia Risk in Chinese Men. *Front. Endocrinol.* **2019**, *10*, 230. [CrossRef]
7. Grassi, D.; Desideri, G.; Di Giacomantonio, A.V.; Di Giosia, P.; Ferri, C. Hyperuricemia and cardiovascular risk. *High Blood Press. Cardiovasc. Prev.* **2014**, *21*, 235–242. [CrossRef]
8. Su, H.Y.; Yang, C.; Liang, D.; Liu, H.F. Research Advances in the Mechanisms of Hyperuricemia-Induced Renal Injury. *BioMed Res. Int.* **2020**, *2020*, 5817348. [CrossRef]
9. Wang, J.; Chen, Y.; Zhong, H.; Chen, F.; Regenstein, J.; Hu, X.S.; Cai, L.Y.; Feng, F.Q. The gut microbiota as a target to control hyperuricemia pathogenesis: Potential mechanisms and therapeutic strategies. *Crit. Rev. Food Sci. Nutr.* **2021**, *62*, 3979–3989. [CrossRef]
10. Gao, Y.; Sun, J.; Zhang, Y.; Shao, T.J.; Li, H.C.; Wang, M.J.; Zhang, L.; Bian, H.; Wen, C.P.; Xie, Z.J.; et al. Effect of a Traditional Chinese Medicine Formula (CoTOL) on Serum Uric Acid and Intestinal Flora in Obese Hyperuricemic Mice Inoculated with Intestinal Bacteria. *Evid. Based Complementary Altern. Med.* **2020**, *2020*, 8831937. [CrossRef]

11. Dalbeth, N.; Merriman, T.R.; Stamp, L.K. Gout. *Lancet* **2016**, *388*, 2039–2052. [CrossRef]
12. Kohagura, K.; Tana, T.; Higa, A.; Yamazato, M.; Ishida, A.; Nagahama, K.; Sakima, A.; Iseki, K.; Ohya, Y. Effects of xanthine oxidase inhibitors on renal function and blood pressure in hypertensive patients with hyperuricemia. *Hypertens. Res.* **2016**, *39*, 593–597. [CrossRef] [PubMed]
13. Bailén, R.; González Senac, N.M.; López, M.M.; Luisa Llena, M.; Migoya, M.; Teresa Rodriguez, M.; de Miguel, E.; Torres, R.J.; Puig, J.G. Efficacy and Safety of a Urate Lowering Regimen in Primary Gout. *Nucleos. Nucleot. Nucl.* **2014**, *33*, 174–180. [CrossRef]
14. Abd El-Maksoud, A.A.; Korany, R.M.S.; Abd El-Ghany, I.H.; El-Beltagi, S.H.; de Gouveia, G.M.A.F. Dietary solutions to dyslipidemia: Milk protein–polysaccharide conjugates as liver biochemical enhancers. *J. Food Biochem.* **2020**, *44*, e13142. [CrossRef]
15. Qi, X.F.; Chen, H.R.; Guan, K.F.; Wang, R.C.; Ma, Y. Anti-hyperuricemic and nephroprotective effects of whey protein hydrolysate in potassium oxonate induced hyperuricemic rats. *J. Sci. Food Agric.* **2021**, *101*, 4916–4924. [CrossRef]
16. Li, Y.J.; Kang, X.Y.; Li, Q.Y.; Shi, C.C.; Lian, Y.Y.; Yuan, E.D.; Zhou, M.; Ren, J.Y. Anti-hyperuricemic peptides derived from bonito hydrolysates based on in vivo hyperuricemic model and in vitro xanthine oxidase inhibitory activity. *Peptides* **2018**, *107*, 45–53. [CrossRef] [PubMed]
17. Wu, Y.Q.; He, H.; Hou, T. Purification, identification, and computational analysis of xanthine oxidase inhibitory peptides from kidney bean. *J. Food Sci.* **2021**, *86*, 1081–1088. [CrossRef]
18. He, W.W.; Su, G.W.; Sun-Waterhouse, D.; Waterhouse, G.I.N.; Zhao, M.M.; Liu, Y. In vivo anti-hyperuricemic and xanthine oxidase inhibitory properties of tuna protein hydrolysates and its isolated fractions. *Food Chem.* **2019**, *272*, 453–461. [CrossRef]
19. Li, Q.; Kang, X.; Shi, C.; Li, Y.; Majumder, K.; Ning, Z.; Ren, J. Moderation of hyperuricemia in rats via consuming walnut protein hydrolysate diet and identification of new antihyperuricemic peptides. *Food Funct.* **2018**, *9*, 107. [CrossRef]
20. Chirinos, R.; Zuloeta, G.; Pedreschi, R.; Mignolet, E.; Larondelle, Y.; Campos, D. Sacha inchi (*Plukenetia volubilis*): A seed source of polyunsaturated fatty acids, tocopherols, phytosterols, phenolic compounds and antioxidant capacity. *Food Chem.* **2013**, *141*, 1732–1739. [CrossRef]
21. Torres Sánchez, E.G.; Hernández-Ledesma, B.; Gutiérrez, L.F. Sacha Inchi Oil Press-cake: Physicochemical Characteristics, Food-related Applications and Biological Activity. *Food Rev. Int.* **2021**, *5*, 1–12. [CrossRef]
22. Chirinos, R.; Aquino, M.; Pedreschi, R.; Campos, D. Optimized Methodology for Alkaline and Enzyme-Assisted Extraction of Protein from Sacha Inchi (*Plukenetia volubilis*) Kernel Cake. *J. Food Process. Eng.* **2017**, *40*, e12412. [CrossRef]
23. Lai, X.; Pan, S.; Zhang, W.; Sun, L.; Li, Q.; Chen, R.; Sun, S. Properties of ACE inhibitory peptide prepared from protein in green tea residue and evaluation of its anti-hypertensive activity. *Process. Biochem.* **2022**, *92*, 277–287. [CrossRef]
24. Chirinos, R.; Pedreschi, R.; Campos, D. Enzyme-assisted hydrolysates from sacha inchi (*Plukenetia volubilis*) protein with in vitro antioxidant and antihypertensive properties. *J. Food Process. Preserv.* **2020**, *44*, e14969. [CrossRef]
25. Rawdkuen, S.; Rodzi, N.; Pinijsuwan, S. Characterization of sacha inchi protein hydrolysates produced by crude papain and Calotropis proteases. *Lwt-Food Sci. Technol.* **2018**, *98*, 18–24. [CrossRef]
26. Perez-Galvez, R.; Morales-Medina, R.; Espejo-Carpio, F.; Guadix, A.; Guadix, E.M. Modelling of the production of ACE inhibitory hydrolysates of horse mackerel using proteases mixtures. *Food Funct.* **2016**, *7*, 3890–3901. [CrossRef]
27. Benjakul, S.; Oungbho, K.; Visessanguan, W.; Thiansilakul, Y.; Roytrakul, S. Characteristics of gelatin from the skins of bigeye snapper, Priacanthus tayenus and Priacanthus macracanthus. *Food Chem.* **2009**, *116*, 445–451. [CrossRef]
28. Li, G.S.; Zhan, J.Q.; Hu, L.P.; Yuan, C.H.; Takaki, K.; Ying, X.G.; Hu, Y.Q. Identification of a new antioxidant peptide from porcine plasma by in vitro digestion and its cytoprotective effect on H_2O_2 induced HepG2 model. *J. Funct. Foods* **2021**, *86*, 104679. [CrossRef]
29. Liao, W.W.; Chen, H.; Jin, W.G.; Yang, Z.N.; Cao, Y.; Miao, J.Y. Three Newly Isolated Calcium-Chelating Peptides from Tilapia Bone Collagen Hydrolysate Enhance Calcium Absorption Activity in Intestinal Caco-2 Cells. *J. Agric. Food Chem.* **2020**, *68*, 2091–2098. [CrossRef]
30. Morgan, X.C.; Tickle, T.L.; Sokol, H.; Gevers, D.; Devaney, K.L.; Ward, D.V.; Reyes, J.A.; Shah, S.A.; LeLeiko, N.; Snapper, S.B.; et al. Dysfunction of the intestinal microbiome in inflammatory bowel disease and treatment. *Genome Biol.* **2012**, *13*, R79. [CrossRef]
31. Kanehisa, M.; Goto, S.; Sato, Y.; Furumichi, M.; Tanabe, M. KEGG for integration and interpretation of large-scale molecular data sets. *Nucleic Acids Res.* **2012**, *40*, 109–114. [CrossRef] [PubMed]
32. Chen, G.; Tan, M.L.; Li, K.K.; Leung, P.C.; Ko, C.H. Green tea polyphenols decreases uric acid level through xanthine oxidase and renal urate transporters in hyperuricemic mice. *J. Ethnopharmacol.* **2015**, *175*, 14–20. [CrossRef] [PubMed]
33. Tian, B.M.; Zhao, J.H.; Zhang, M.; Chen, Z.F.; Ma, Q.Y.; Liu, H.C.; Nie, C.X.; Zhang, Z.Q.; An, W.; Li, J.X. Lycium ruthenicum Anthocyanins Attenuate High-Fat Diet-Induced Colonic Barrier Dysfunction and Inflammation in Mice by Modulating the Gut Microbiota. *Mol. Nutr. Food Res.* **2021**, *65*, 2000745. [CrossRef]
34. Wang, Y.N.; Zhao, M.; Xin, Y.; Liu, J.J.; Wang, M.; Zhao, C.J. ^{1}H NMR and MS based metabolomics study of the therapeutic effect of Cortex Fraxini on hyperuricemic rats. *J. Ethnopharmacol.* **2016**, *185*, 272–281. [CrossRef]
35. Wang, C.; Pan, Y.; Zhang, Q.Y.; Wang, F.M.; Kong, L.D. Quercetin and allopurinol ameliorate kidney injury in STZ-treated rats with regulation of renal NLRP3 inflammasome activation and lipid accumulation. *PLoS ONE* **2012**, *7*, e38285. [CrossRef] [PubMed]
36. Zhou, Y.; Fang, L.; Jiang, L.; Wen, P.; Cao, H.D.; He, W.C.; Dai, C.S.; Yang, J.W. Uric acid induces renal inflammation via activating tubular NF-kappaB signaling pathway. *PLoS ONE* **2012**, *7*, e39738. [CrossRef]

37. Li, Q.Y.; Shi, C.C.; Wang, M.; Zhou, M.; Liang, M.; Zhang, T.; Yuan, E.D.; Wang, Z.; Yao, M.J.; Ren, J.Y. Tryptophan residue enhances in vitro walnut protein-derived peptides exerting xanthine oxidase inhibition and antioxidant activities. *J. Funct. Foods* **2019**, *53*, 276–285. [CrossRef]
38. Liu, N.X.; Wang, Y.; Zeng, L.; Yin, S.G.; Hu, Y.; Li, S.S.; Fu, Y.; Zhang, X.P.; Xie, C.; Shu, L.J.; et al. RDP3, A Novel Antigout Peptide Derived from Water Extract of Rice. *J. Agric. Food Chem.* **2020**, *68*, 7143–7151. [CrossRef]
39. Wei, L.Y.; Ji, H.W.; Song, W.K.; Peng, S.; Zhan, S.H.; Qu, Y.S.; Chen, M.; Zhang, D.; Liu, S.C. Hypouricemic, hepatoprotective and nephroprotective roles of oligopeptides derived from Auxis thazard protein in hyperuricemic mice. *Food Funct.* **2021**, *12*, 11838–11848. [CrossRef]
40. Han, J.J.; Wang, X.F.; Tang, S.S.; Lu, C.Y.; Wan, H.T.; Zhou, J.; Li, Y.; Ming, T.H.; Wang, Z.J.; Su, X.R. Protective effects of tuna meat oligopeptides (TMOP) supplementation on hyperuricemia and associated renal inflammation mediated by gut microbiota. *FASEB J.* **2020**, *34*, 5061–5076. [CrossRef]
41. Wang, C.H.; Zhang, C.; Xing, X.H. Xanthine dehydrogenase: An old enzyme with new knowledge and prospects. *Bioengineered* **2016**, *7*, 395–405. [CrossRef] [PubMed]
42. French, J.B.; Ealick, S.E. Structural and kinetic insights into the mechanism of 5-hydroxyisourate hydrolase from Klebsiella pneumoniae. *Acta. Crystallogr. D* **2011**, *67*, 671–677. [CrossRef] [PubMed]
43. Pan, J.; Shi, M.; Guo, F.; Ma, L.; Fu, P. Pharmacologic inhibiting STAT3 delays the progression of kidney fibrosis in hyperuricemia-induced chronic kidney disease. *Life Sci.* **2021**, *285*, 119946. [CrossRef] [PubMed]
44. Shan, B.; Chen, T.; Huang, B.X.; Liu, Y.; Chen, J. Untargeted metabolomics reveal the therapeutic effects of Ermiao wan categorized formulas on rats with hyperuricemia. *J. Ethnopharmacol.* **2021**, *281*, 114545. [CrossRef]
45. Ru, L.B.; Scott, P.; Sydlaske, A.; Rose, D.M.; Terkeltaub, R. Innate immunity conferred by Toll-like receptors 2 and 4 and myeloid differentiation factor 88 expression is pivotal to monosodium urate monohydrate crystal-induced inflammation. *Arthritis Rheumatol.* **2005**, *52*, 2936–2946. [CrossRef]
46. Bi, F.F.; Chen, F.; Li, Y.N.; Wei, A.; Cao, W.S. Klotho preservation by Rhein promotes toll-like receptor 4 proteolysis and attenuates lipopolysaccharide-induced acute kidney injury. *J. Mol. Med.* **2018**, *96*, 915–927. [CrossRef]
47. Mehmood, A.; Zhao, L.; Wang, C.T.; Hossen, I.; Raka, R.N.; Zhang, H.M. Stevia residue extract increases intestinal uric acid excretion via interactions with intestinal urate transporters in hyperuricemic mice. *Food Funct.* **2019**, *10*, 7900–7912. [CrossRef]
48. Xie, X.S.; Zhang, L.; Yuan, S.; Li, H.L.; Zheng, C.J.; Xie, S.S.; Sun, Y.B.; Zhang, C.H.; Wang, R.K.; Jin, Y. Val-Val-Tyr-Pro protects against non-alcoholic steatohepatitis in mice by modulating the gut microbiota and gut-liver axis activation. *J. Cell. Mol. Med.* **2021**, *25*, 1439–1455. [CrossRef]
49. Lu, C.Y.; Tang, S.S.; Han, J.J.; Fan, S.Q.; Huang, Y.M.; Zhang, Z.; Zhou, J.; Ming, T.H.; Li, Y.; Su, X.R. Apostichopus japonicus Oligopeptide Induced Heterogeneity in the Gastrointestinal Tract Microbiota and Alleviated Hyperuricemia in a Microbiota-Dependent Manner. *Mol. Nutr. Food Res.* **2021**, *65*, 2100147. [CrossRef]
50. Yu, Y.R.; Liu, Q.P.; Li, H.C.; Wen, C.P.; He, Z.X. Alterations of the Gut Microbiome Associated With the Treatment of Hyperuricaemia in Male Rats. *Front. Microbiol.* **2018**, *9*, 2233. [CrossRef]
51. Solano-Aguilar, G.I.; Jang, S.; Lakshman, S.; Gupta, R.; Beshah, E.; Sikaroodi, M.; Vinyard, B.; Molokin, A.; Gillevet, P.M.; Urban, J.F. The Effect of Dietary Mushroom Agaricus bisporus on Intestinal Microbiota Composition and Host Immunological Function. *Nutrients* **2018**, *10*, 1721. [CrossRef] [PubMed]

Article

Crosslinking Mechanism on a Novel *Bacillus cereus* Transglutaminase-Mediated Conjugation of Food Proteins

Hongbin Wang [1,†], Yuanfu Zhang [1,†], Zhaoting Yuan [1], Xiaotong Zou [1], Yuan Ji [1], Jiayi Hou [1], Jinfang Zhang [2,*], Fuping Lu [1] and Yihan Liu [1,*]

1   Key Laboratory of Industrial Fermentation Microbiology, Ministry of Education, The College of Biotechnology, Tianjin University of Science and Technology, Tianjin 300457, China
2   Tianjin Key Laboratory of Industrial Microbiology, Tianjin 300457, China
*   Correspondence: jfzhang1992@sohu.com (J.Z.); lyh@tust.edu.cn (Y.L.); Tel.: +86-022-6060-1958 (J.Z.); +86-022-6060-2949 (Y.L.)
†   These authors contributed equally to this work.

Abstract: Until now, *Streptoverticillium mobaraense* transglutaminase (TG) is the only commercialized TG, but limited information is known about its selection tendency on crosslinking sites at the protein level, restricting its application in the food industry. Here, four recombinant *Bacillus* TGs were stable in a broad range of pH (5.0–9.0) and temperatures (<50 °C), exhibiting their maximum activity at 50–60 °C and pH 6.0–7.0. Among them, TG of *B. cereus* (BCETG) demonstrated the maximal specific activity of 177 U/mg. A structural analysis indicated that the Ala147-Ala156 region in the substrate tunnel of BCETG played a vital role in catalytic activity. Furthermore, bovine serum albumin, as well as nearly all protein ingredients in soy protein isolate and whey protein, could be cross-linked by BCETG, and the internal crosslinking paths of three protein substrates were elucidated. This study demonstrated *Bacillus* TGs are a candidate for protein crosslinking and provided their crosslinking mechanism at the protein level for applications in food processing.

Keywords: transglutaminase; enzymatic characterization; structural analysis; crosslinking properties; protein substrate crosslinking mechanism

Citation: Wang, H.; Zhang, Y.; Yuan, Z.; Zou, X.; Ji, Y.; Hou, J.; Zhang, J.; Lu, F.; Liu, Y. Crosslinking Mechanism on a Novel *Bacillus cereus* Transglutaminase-Mediated Conjugation of Food Proteins. *Foods* **2022**, *11*, 3722. https://doi.org/10.3390/foods11223722

Academic Editor: Cristina Martínez-Villaluenga

Received: 23 September 2022
Accepted: 14 November 2022
Published: 19 November 2022

Publisher's Note: MDPI stays neutral with regard to jurisdictional claims in published maps and institutional affiliations.

1. Introduction

Proteins perform a vital role in food quality as a critical fraction of food products. Nowadays, protein modification technologies attract more and more attention due to their importance in satisfying the consumers' various food requirements [1]. Compared to chemical modification, protein modification using enzymatic techniques demonstrates lots of advantages, such as a low frequency of side reactions, high reaction specificity, and the lack of a need for chemical solvents (environmental friendliness) or high-pressure and high-temperature conditions [2]. In this context, protein crosslinking, which is described as "the process of joining protein molecules by inter or intra-molecular covalent bonds", plays an important role in determining the functional characteristics of foods [3].

Transglutaminase (EC 2.3.2.13, TG) leads to inter- and intra-molecular crosslinking through catalyzing the generation of ε-(g-glutamyl) lysine "isopeptide" covalent bonds in proteins. Currently, TGs have been discovered and characterized in prokaryotes and eukaryotes. Comparing with the eukaryotic counterparts, bacterial TGs demonstrate lots of obvious advantages, such as cofactor independency, small size, high stability and improved performance [4], and they have been chiefly discovered in *Bacillus* and *Streptoverticillium* strains. TG derived from *Streptoverticillium mobaraense* (MTG) is the sole transglutaminase commercialized for food processing [5]. It is mainly used for changing proteins' elasticity and solubility, producing textured products, improving protein encapsulation capacity, emulsifying properties, and water-holding capacity, improving the nutritious value of food products by the incorporation of essential amino acids, forming heat and water-resistant films, avoiding gelation by thermal treatment, and, recently, to reduce allergenicity [6]. MTG demonstrates its stability below 40 °C from pH 5.0 to 8.0 and its maximum activity at 50 °C between pH 6.0 and 7.0 [7]. However, TG from *B. subtilis* (BSUTG) exhibits its maximum activity at pH 8.0 and 60 °C, which are more stable under wide ranges of pH (pH 5.0–9.0) and temperature (30–60 °C). Meanwhile, BSUTG and MTG have only ~10% homology in amino acid sequences, and they also show lots of differing preference in substrates such as soy protein isolate (SPI) and whey protein (WP). Obviously, it would be helpful to meet the large number of needs of various food substrates and processing due to the diversity of TGs in nature. Hence, it would offer a novel challenge for utilizing TGs in the rising food field by exploring novel TGs. In addition, our previous study investigated BSUTG's crosslinking characteristics on peptide levels [8]. Nevertheless, selecting food protein crosslinking sites might be affected by both the neighboring amino acids and the protein spatial structure of BSUTG. So far, limited information on the favorite amino acid crosslinking sites is known on the food protein level.

In this study, we characterized TGs derived from *B. amyloliquefaciens*, *B. cereus*, *B. safensis*, and *B. aryabhattai*. Then, molecular dynamics simulation was performed to analyze the *B. cereus* TG's tertiary structure to explore the molecular mechanism of its catalytic activity under different temperatures. Finally, we estimated the abilities of *B. cereus* TG towards various substrates and the preferred crosslinked amino acid sites. This study facilitates the applications of *Bacillus* TGs in the food industry and the depth our understanding of them.

2. Materials and Methods

2.1. Strains, Plasmids, Materials, and Growth Media

B. amyloliquefaciens TCCC 111018, *B. cereus* TCCC 111006, *B. safensis* TCCC 111022, and *B. aryabhattai* TCCC 11368 were kept in our laboratory, and they were incubated at 37 °C in liquid LB medium. *Escherichia coli* BL21 (DE3) was utilized for expressing the recombinant enzyme, and kanamycin was added when necessary. The pET-28a(+) was employed as an expression vector.

Reagents were provided by the following sources: *N,N*-dimethylcasein and dansyl cadaverine [N-(5-aminopentyl)-5-dimethylamino-1-naphthalenesulfonamide, MDC] were ordered from Sigma-Aldrich Co. (St. Louis, MO, USA) and Aladdin Chemical Co., Ltd. (Shanghai, China), respectively. Bovine serum albumin (BSA, ≥98%), soy protein isolate (SPI, ≥88%), and whey protein (WP, ≥80%) were provided by Solarbio Science & Technology Co., Ltd. (Beijing, China), Macklin Biochemical Co., Ltd. (Shanghai, China), and Sigma-Aldrich (St Louis, MO, USA), respectively.

2.2. Strains Construction

The TG genes encoding *bamtg*, *bcetg*, *bsatg*, and *bartg* were obtained using the forward and reverse primers presented in Table S1 and genomic DNA from *B. amyloliquefaciens* TCCC 111018, *B. cereus* TCCC 111006, *B. safensis* TCCC 111022, and *B. aryabhattai* TCCC 11368 as templates. After digestion with *Nco*I and *Hind*III, the PCR products *bamtg*, *bsatg*, and *bartg* were inserted into the *Nco*I and *Hind*III-linearized pET-28a (+). However, the PCR product *bcetg* was ligated to the *Nco*I–*Not*I-linearized pET28a(+) due to the *Hind*III restriction site in the *bcetg* gene. The sequences of *bamtg*, *bcetg*, *bsatg*, and *bartg* were deposited in GenBank, respectively (Accession No: MN537144, MW281568, MW292483, and MW916358). *E. coli* BL21 (DE3) cells harboring the recombinant vectors pET-*bamtg*, pET-*bcetg*, pET-*bsatg*, and pET-*bartg* were utilized for protein expression. Their protein sequences were analyzed using DNAMAN for multiple protein sequence alignment, and the signal peptides were predicted by the Signal P 3.0 server (http://www.cbs.dtu.dk/services/SignalP-3.0 (accessed on 8 June 2019)).

2.3. Production and Purification of TGs

Each single colony of BL21/pET-*bamtg*, BL21/pET-*bcetg*, BL21/pET-*bsatg*, and BL21/pET-*bartg* was grown at 37 °C for 12 h in 5mL of LB medium containing 50 µg/mL kanamycin with shaking at 200 rpm. Subsequently, the seed culture (1 mL) was inoculated into 50 mL of fresh LB medium with kanamycin (50 µg/mL). When OD600 of the cells reached 0.6–0.8, 0.5 mM isopropyl-β-D-1-thiogalactopyranoside (IPTG) was utilized to induce the production of recombinant TGs (rTGs) at 16 °C for 20 h.

Cells were centrifugally collected at $8000 \times g$ and 4 °C for 15 m, followed by resuspension in the lysis buffer (20 mmol/L CAPS, 0.5 mol/L NaCl, 10% glycerol, 10 mmol/L imidazole, pH 8.0). Then, the cells were sonicated at 260 W with 2 s strokes and 4 s intervals. After centrifugation at $12,000 \times g$ for 30 min, the resulting supernatant was loaded onto a nickel–nitrilotriacetic acid (Ni-NTA) agarose column (Qiagen, Germany) which was preequilibrated with lysis buffer. After removing non-target proteins with washing buffer (20 mmol/L CAPS, 0.5 mol/L NaCl, 10% glycerol, 300 mmol/L imidazole, pH 8.0), the purified rTGs were collected after eluting with elution buffer (20 mmol/L CAPS, 0.5 mol/L NaCl, 10% glycerol, 60 mmol/L imidazole, pH 8.0). Sodium dodecyl sulfate–polyacrylamide gel electrophoresis (SDS-PAGE) was performed to analyze the purified protein, and the Bradford method was utilized toevaluate the protein concentration, with bovine serum albumin as the standard.

2.4. TGs Activity Assays

TG activities were measured based on the protocol described by Liu et al. [9] with some modifications. After mixing 200 µL of phosphate buffer (50 mM, pH 6.0 or 7.0, optimal pH for various TGs) and 100 µL of enzyme-incorporating MDC (12.5 µM) and *N,N*-dimethylcasein (0.2%), the enzymatic reaction was performed for 20 min at 50 °C or 60 °C (the optimum temperature for different TGs). The fluorescent intensity was recorded by the Fluorescence Spectrophotometer-Infinite 200PRO (Tecan, Austria) based on a previous study [8].

2.5. Characterization of TGs

The effect of pH on the activities of TGs was determined over the pH ranging from 5.0 to 9.0 under their optimum temperatures. The following buffer systems were applied in this work: citrate-phosphate buffer (50 mM, pH 5.0), phosphate buffer (50 mM, pH 6.0–8.0), and Tris-HCl buffer (50 mM, pH 9.0). After incubating the enzyme in the above-mentioned buffers at 4 °C for different times, the influence of pH on the stability of TGs was tested at pH = 5.0–9.0 by calculating the retained enzyme activity under standard conditions.

The optimum temperatures of TGs were ascertained by measuring the activities in the buffer of their optimal pH at differing temperatures of 30–80 °C. Thermostability was detected by keeping TGs in phosphate buffer (50 mM, pH 7.0) for various periods at 30–80 °C, and then calculating the residual activity.

2.6. Molecular Dynamics Simulations

The docking of TG (BCETG) from *B. cereus* TCCC 111006 with the substrate was conducted using AutoDock4.2.6 software [10]. After inputting the protein sequence of BCETG, the searching results were obtained on SWISS-MODEL (http://www.swissmodel.expasy.org (accessed on 7 January 2020)). We selected the 3D structure of TG from *B. subtilis* (BSUTG, PDB accession No. 4P8I) as the template for building the BCETG structure. In addition, we obtained the 3D structure of the MDC substrate from the PubChem Web site (http://pubchem.ncbi.nlm.nih. gov (accessed on 7 January 2020)). By observing the location of the BCETG active site (Cys117, Glu116, and Glu189), the volume of 60 Å × 60 Å × 60 Å was chosen for the center of the grid box to fit the ligand easily and cover the whole pocket of the active center. Then, MD simulation was carried out using the best conformation exhibiting the minimum binding energy for protein–substrate docking.

MD simulation was conducted with GROMACS 5.1.4 software in combination with the GROMOS96 54a7 force field parameters [11], using almost the same simulation parameters as those reported in detail [12,13]. The minimum distance of BCETG and the cube box edge was 15 Å under the periodic boundary conditions, and then water molecules filled in the box. The SPC/E model was utilized to describe water. The system was charged zero to replace the equivalent number of ions by using random water molecules. The simulation systems were optimized using the steepest decent method of 50,000 steps. Both short-range van der Waals (vdW) and electrostatic interaction (ELE) were easily truncated at 10 Å. Different temperature gradients (323 K and 353 K) were repeated in the isochoricisothermal (NVT) ensemble three times. The size of the mesh was 0.16 Å with the intercept at 12 Å. The external temperature and pressure baths were to couple MDC, protein and water molecules. In the end, each simulation was carried out for 100 ns while recording the coordinates every 2 ps.

The trajectory of the MD simulation was evaluated with the supplementary program in the GROMACS 5.1.4 software package (SciLifeLab, Stockholm, Sweden). The typical snapshots of the BCETG tertiary structures were obtained using the Visual Molecular Dynamics (VMD) 1.9.4 software [14].

2.7. Protein Substrates Crosslinked by BCETG

The BSA (1 mg/mL), SPI (8 mg/mL), and WP (1 mg/mL) solutions were made in sodium phosphate buffer (50 mM, pH 6.0). Then, the crosslinking was performed at 50 °C by mixing BCETG with protein in the concentrations of substrate (0.1, 0.03, 0.1 mg/mg) containing DTT (2 mM) for various periods (1–6 h). Finally, SDS-PAGE was utilized to characterize BCETG that catalyzed the crosslinking conditions of BSA, SPI, and WP for differing periods, as noted by Liu et al. [5,7,15].

2.8. Analysis of Protein Substrate's Crosslinked Amino Acid Sites by BCETG

The crosslinked sites in the crosslinked protein samples by BCETG were identified by liquid chromatography linked to the tandem mass spectrometry (LC-MS/MS) through analyzing the crosslinked peptides. After being boiled for 10 min to stop crosslinking, the protein samples that were crosslinked for 1 h were then denatured, reduced and digested with trypsin based on the previous method [8]. LC-MS/MS analysis was performed by a 1260 series HPLC coupled to an ESI-Q/TOF mass spectrometry (Agilent Technologies, Palo Alto, CA, USA) using the previously reported conditions [8]. The crosslinked peptides were identified and quantified using the software PLink 2 (The Institute of Computing Technology of the Chinese Academy of Sciences, Beijing, China) and the Masshunter B.08.00 of Agilent (Palo Alto, CA, USA), respectively, by processing the mgf data. The conditions used for the identification by the software PLink were as the following. Protein database: BSA for BSA protein, soybean proteins for SPI sample, and whey proteins for WP sample; protease: trypsin; variable modifications: methionine oxidation; fixed modification: cysteine carbamidomethylation; missed cleavages: no more than two; fragment ion tolerance: 50 ppm; precursor mass tolerance: 50 ppm.

3. Results and Discussion

3.1. Molecular Cloning and Genes Sequence of bamtg, bcetg, bsatg, and bartg Genes

The *bamtg*, *bcetg*, *bsatg*, and *bartg* genes showed an open reading frame of 735, 828, 738, and 813 bp, coding for a full-length protein of 244, 275, 245, and 270 amino acids with a theoretical molecular mass of 28.3, 31.4, 28.4, and 31.3 kDa (BAMTG, BCETG, BSATG, and BARTG), respectively. No signal peptide was found to locate at the N-terminus of the BAMTG, BCETG, BSATG, and BARTG after analysis using the Signal P program. The gene sequences were stored in the NCBI database, and the accession numbers of the *bamtg*, *bcetg*, *bsatg*, and *bartg* genes were MN537144, MW281568, MW292483, and MW916358, respectively. BAMTG, BCETG, BSATG, and BARTG showed 71.4%, 34.7%, 51.8%, and 38.3% of protein sequence identities with the reported BSUTG, respectively, after multiple amino acid sequence alignments (Figure S1). Meanwhile, they only shared ~10% identity with MTG. It suggested that they were probably the novel members of TG. As the first TG isolated from *Bacillus* species, BSUTG is produced to crosslink the highly resilient spore surface during sporulation [16]. In addition, it is the smallest TG produced as an active single-domain protein. Furthermore, it functions via a peculiar partly redundant catalytic dyad containing Glu115 and Cys116 or Glu187. Comparative analysis of the primary sequences indicated that BAMTG, BCETG, BSATG, and BARTG contained the conserved amino acids correspondent to Glu115, Cys116, and Glu187 of BSUTG, which were considered to play a catalytic role. However, their amino acid sequences showed a significant homology with BSUTG, and especially MTG, which could cause differing characteristics and substrate preference from MTG and BSUTG.

3.2. Heterologous Production and Purification of TGs

The genes of *bamtg*, *bcetg*, *bsatg*, and *bartg* from *B. amyloliquefaciens* TCCC 111018, *B. cereus* TCCC 111006, *B. safensis* TCCC 111022, and *B. aryabhattai* TCCC 11368 in *E. coli* with a C-terminal 6 × His tag. The clear bands corresponding to the purified BAMTG, BCETG, BSATG, and BARTG were observed on the SDS-PAGE gel (Figure S2)

3.3. Biochemical Characterization of BAMTG, BCETG, BSATG, and BARTG

The profile of BARTG activity as a function of temperature variation indicated its optimum condition around 60 °C (Table 1), and BARTG displayed > 40% of its highest activity between 40 °C and 70 °C (Figure 1a). MTG only showed about 60% of its maximal activity at 60 °C and showed no activity at 70 °C, though it was most active at 55 °C [9]. Additionally, BAMTG, BCETG, and BSATG showed their maximal activity at 50 °C (Table 1, Figure 1a). MTG retained merely ~20% of its initial activity after incubating at 50 °C for 100 min [9], but BAMTG, BCETG, BSATG, and BARTG retained about 100%, 100%, 60%, and 50% of their original activities after 120 min incubation at 50 °C, respectively (Figure 2a–d). These results verify that four TGs had better thermostabilities than MTG.

Table 1. The summarized biochemical characterization of recombinant TGs.

Enzymes	Optimum Temperature	Optimum pH	Thermostability Range	pH Stability Range
BAMTG	50 °C	7.0	≤50 °C	5.0–9.0
BCETG	50 °C	6.0	≤50 °C	5.0–9.0
BSATG	50 °C	6.0	≤50 °C	5.0–9.0
BARTG	60 °C	6.0	≤50 °C	5.0–9.0

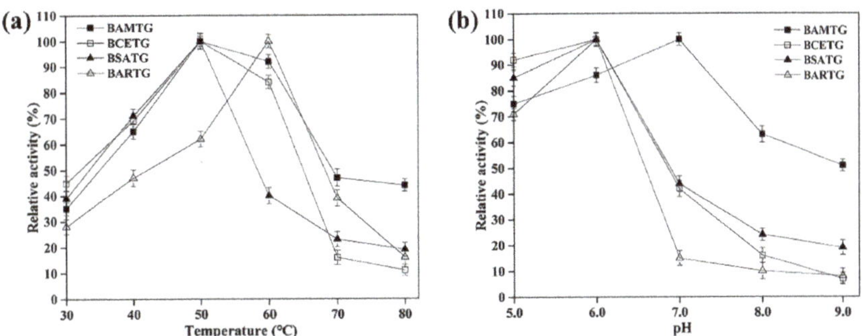

Figure 1. The optimum temperature and pH of recombinant TGs. (**a**) The influence of temperature on the BAMTG, BCETG, BSATG, and BARTG activities; (**b**) The influence of pH (5.0–9.0) on the activities of BAMTG, BCETG, BSATG, and BARTG. Three independent experiments were performed to calculate the values described by means ± SD.

BAMTG showed over 50% activity from pH 5.0 to 9.0, and its optimal pH was 7.0, but the optimal pH of BARTG, BCETG, and BSATG was 6.0 (Table 1, Figure 1b). By contrast, MTG showed no activity at pH 9.0 [9]. Moreover, BAMTG, BARTG, BCETG, and BSATG kept stable under an extensive pH range of 5.0 to 9.0, retaining approximately

100%, > 90%, > 80%, > 70% of the initial activity after incubation for 5 d between pH 5.0 and 9.0, respectively (Figure 3a–d), which was superior to MTG with a pH stability range of 5.0–8.0 [9]. These data indicate that these TGs could be a potential candidate for various biotechnological processes under acidic–neutral–alkaline conditions.

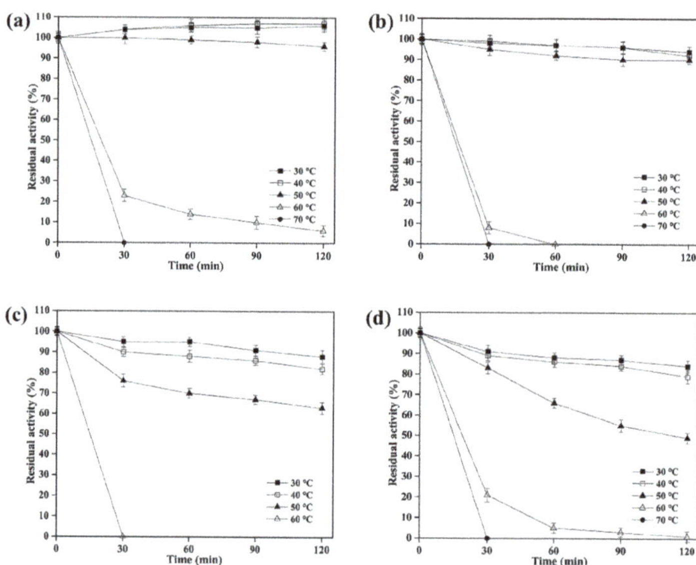

Figure 2. The thermostability of recombinant TGs. The effect of temperature on the stability of (**a**) BAMTG, (**b**) BCETG, (**c**) BSATG, and (**d**) BARTG. Three independent experiments were performed to calculate the values described by means ± SD.

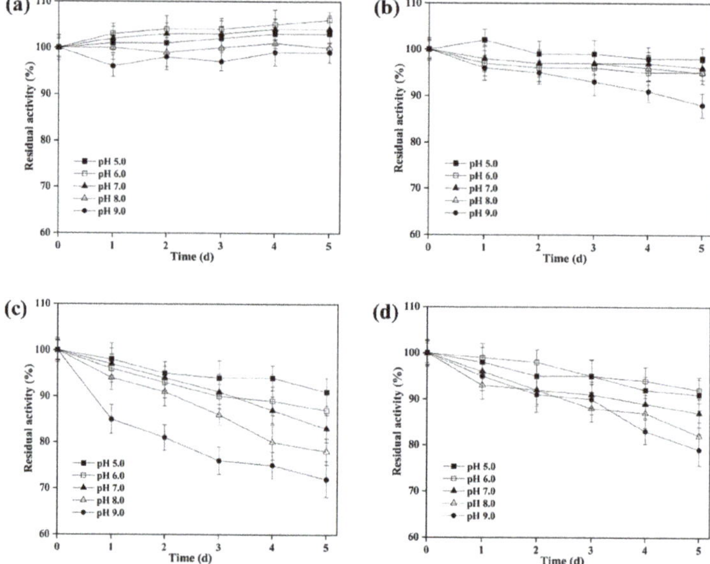

Figure 3. The pH stability of recombinant TGs. The effect of pH on the stability of (**a**) BAMTG, (**b**) BCETG, (**c**) BSATG, and (**d**) BARTG. Three independent experiments were performed to calculate the values described by means ± SD.

As the most common enzyme applied for protein crosslinking, MTG is the only commercially available enzyme utilized in the food industry. However, the biological approach might be limited by the enzymatic properties. Hence, it could be essential to provide an extensive range of available TGs for crosslinking to the food processors. These four TGs demonstrated various temperature and pH ranges for their activities and stabilities, and these properties were clearly unlike that of MTG. Thus, they were beneficial for covering the disadvantages of MTG's properties in different types of food processing.

3.4. Molecular Dynamics Simulation

According to the optimum pH and temperature, the specific activities of BAMTG, BCETG, BSATG, and BARTG were determined to be 4.3, 176.8, 2.5, and 9.1 U/mg, respectively. BCETG exhibited the highest specific activity; however, its optimum temperature (50 °C) was unremarkable. Therefore, in order to investigate the reason for the low catalytic activity of BCETG at high temperatures, MD simulation was used to analyze the structural changes of BCETG at 323 K (50 °C) and 353 K (80 °C). Figure 4a shows the change in BCETG's root-mean-square deviation (RMSD) values at 50 °C and 80 °C as the function of simulation time (100 ns). It demonstrates that BCETG at 50 °C and 80 °C showed stabilized RMSD values after 80 ns, and BCETG showed a ~0.05 Å lower RMSD value at 50 °C than that at 80 °C, finally. This indicates that the structure of BCETG at 50 °C was more stable than that at 80 °C.

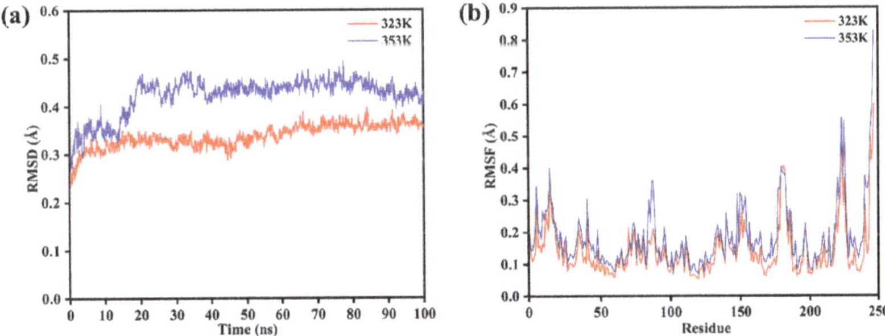

Figure 4. The structural changes of BCETG at 323 K and 353 K. (**a**) The root-mean-square deviation (RMSD) of BCETG at 323 K and 353 K. (**b**) The root-mean-square fluctuation (RMSF) value for each residue in BCETG at 323 K and 353 K.

During MD simulation, the root-mean-square fluctuation (RMSF) values of every amino acid residue at 50 °C and 80 °C were analyzed and represented to determine the sensitive part of BCETG with the increasing temperature (Figure 4b). Comparing to BCETG at 50 °C, the two regions of Trp83-Phe91 and Ala147-Ala156 exhibited an observable increase in RMSF values at 80 °C, suggesting that the two regions demonstrated enhanced flexibility from 50 °C to 80 °C. The Ala147-Ala156 region, which was adjacent to the Trp83-Phe91 region, was situated at one side of the substrate channeling tunnel of BCETG (Figure 5a,b). Leu88 and Trp150, which were both in the loop region, were the two typical residues demonstrating the largest enhancement of RMSF values in each region. As exhibited in Figure 5a,b, the two regions of Trp83-Phe91 and Ala147-Ala156 were closer to each other at 80 °C than 50 °C. However, the two regions were far from another region, Val177-Trp186 (the other side of the tunnel), resulting in the expansion of the tunnel entrance of the substrate MDC and the change in interaction between the active sites of BCETG and substrate MDC (Figure 5a,b). As suggested in Figure 5c, a short distance was discovered between the active center of BCTG and substrate MDC at the beginning of the MD simulation. Substrate MDC showed similar positions at 50 °C (Figure 5d) in the end. By contrast, substrate MDC was far from the active center of BCETG at 80 °C and moved into the inside of BCETG in the end (Figure 5e). In addition, it was clear that the distance between MDC and BCETG's active center was enhanced with the temperature raising from 50 °C to 80 °C during 15–100 ns of MD simulation (Figure S3). At 80 °C, the distance was quickly enhanced to approximately 0.29 Å from an initial 0.25 Å, and subsequently fluctuated close to 0.36 Å, suggesting the removal of the substrate from the active center. Hence, the WT activity might be remarkably destroyed at 80 °C. Nevertheless, the distance kept stable in the process of the whole simulation at 50 °C, demonstrating the stable manifold structure of the BCETG and MDC. Therefore, we speculated that the conformation change of critical regions Trp83-Phe91 and Ala147-Ala156 at the high temperature indeed enlarged the substrate tunnel entrance, leading to the substrate MDC being 'beyond control'. This indicated that substrate MDC not only was far away from the active center of BCETG, but also entered into the BCETG's structure interior, leading to the decrease in catalytic activity of BCETG and the impediment of substrate MDC release and enzyme regeneration. Thus, the two Trp83-Phe91 and Ala147-Ala156 regions of BCETG could be modified to improve its high-temperature activity in future research.

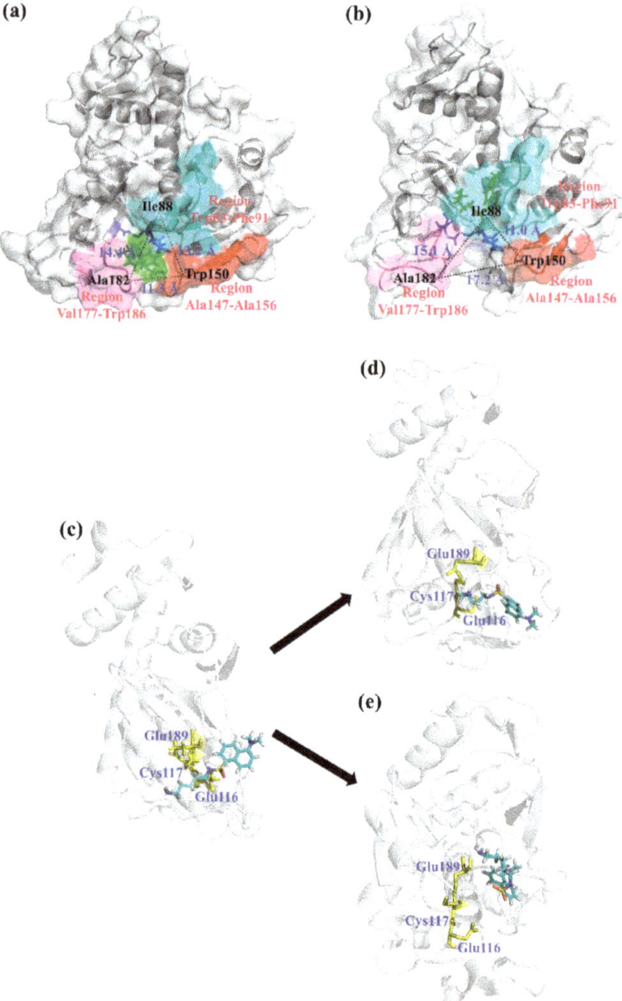

Figure 5. The structural illustration of BCETG-MDC at (**a**) 323 K and (**b**) 353 K. (**a**) The front view for the conformation of BCETG-MDC at 323 K at the end of molecular dynamics simulation. (**b**) The

front view for the conformation of BCETG-MDC at 353 K at the end of molecular dynamics simulation. The Trp83-Phe91, Ala147-Ala156, and Val177-Trp186 regions are labeled in cyan, red, and magenta, respectively. The structure of BCETG is exhibited in gray. (**c**) The side view for the conformation of BCETG-MDC at the beginning of molecular dynamics simulation. (**d**) The side view for the conformation of BCETG-MDC at 323 K at the end of molecular dynamics simulation. (**e**) The side view for the conformation of BCETG-MDC at 353 K at the end of molecular dynamics simulation. The active center is colored in yellow. The structure of BCETG and MDC are shown in gray and cyan, respectively.

3.5. Electrophoresis Analysis of Protein Substrates Crosslinked by BCETG

SDS-PAGE was utilized to monitor the changes that occurred in the course of BSA, SPI, and WP treated with BCETG (Figure 6) for different times (1–6 h). A remarkable difference was noticed among the samples treated with various times. It was observed that a gradual loss of monomers simultaneously occurred with intensifying higher molecular-weight bands. However, some crosslinked proteins still stayed in the stacking gel and could not get into the resolving gel (Figure 6). As demonstrated in Figure 6a, the SDS-PAGE profiles of the BCETG-treated BSA samples were similar to that of MTG-treated samples [17–19]. The primary components of SPI (Figure 6b, lane 2) are 7S globulin, which is a trimer containing three subunits such as α, α′, and β subunits (45 kDa to 116 kDa), and 11S globulin chiefly composed of basic subunits and acidic subunits (18.4–45 kDa). The results demonstrate that BCETG could crosslink almost all components in SPI (Figure 6b, lane 4–9). However, no BCETG preference was observed in the crosslinking reaction for the protein ingredients of SPI. However, the BS in 11S globulin was hardly crosslinked by MTG [15]. The untreated WP showed three main bands (~14 kDa, ~18 kDa, and ~67 kDa) on the SDS-PAGE gel, matching α-la, β-lg, and BSA monomers, respectively (Figure 6c, lane 2). WPs crosslinked by BCETG resulted in the slow reduction in α-la, β-lg, and BSA amounts with the increasing incubation time, suggesting that BCETG could effectively catalyze the three components in WP (Figure 6c, lane 4–9). Nevertheless, α-la was the major substrate in WPI for MTG, which showed a weak capability of crosslinking β-lg and BSA [17,18,20]. These results indicate BCETG showed differing substrate preferences from MTG, which could be owing to the various structures of proteins. Hence, it suggested that BCETG and other *Bacillus* TGs would be potential candidates for supplementing MTG functions in the food-processing industry.

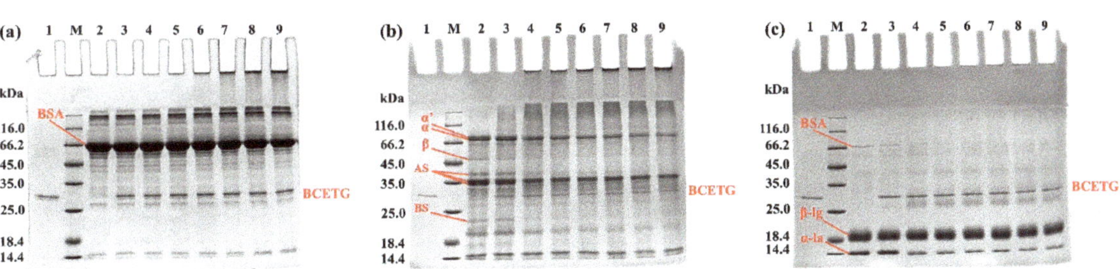

Figure 6. The electrophoretic arrangements of the native and BCETG-treated (**a**) BSA, (**b**) SPI, and (**c**) WP. Lane M, protein standard ladder; Lane 1, the purified BCETG; Lane 2, the native BSA, SPI, or WP; Lane 3–9, BCETG-treated BSA, SPI, or WP for 0–6 h.

3.6. Crosslinked Amino Acid Preference in Protein Substrates by BCETG Treatment

So far, the crosslinking mechanism of TGs at the protein level has not been investigated; thus, it is difficult to obtain the real abilities of TGs to promote the crosslinking of proteins with different characteristics, as well as rationally modulate and control the functional properties of protein matrices. Hence, the application of TG in food processing could be benefited by revealing the crosslinked sites in proteins. LC-MS/MS was utilized to determine the crosslinked sites in BSA, SPI, and WP by BCETG through identifying the crosslinked peptides. BCETG-treated BSA was taken as an example, and the characterization results of ADEKKFWGK and QNCDQFEK peptides are demonstrated in Figure S4. Figure S4a shows the extraction ion chromatogram of the crosslinked peptide ADEKKFWGK-QNCDQFEK and the uncrosslinked QNCDQFEK. However, uncrosslinked ADEKKFWGK was not observed. The mass spectra of the ADEKKFWGK-QNCDQFEK and QNCDQFEK peptides are presented in Figure S4b,c, respectively. The mass obtained from the experiment was consistent with the calculated theoretical monoisotopic mass.

Only a few Q and K sites were crosslinked in BSA, SPI, and WP. Figure 7 exhibits the investigation of crosslinked sites of BCETG-treated BSA, SPI, and WP. Figure 7a–c demonstrates the five most plentiful crosslinked peptides from BCETG-treated proteins. It was obvious that only a smaller number of them showed higher abundance than others, indicating that BCETG had remarkable site preference on crosslinking these proteins. Figure 7d–f presents the most plentiful crosslinked sites in the protein materials. As suggested in Figure 7d, the favorite cross-linking sites of BCETG were Q (118), K (156), Q (413), K (235) and K (245) on BSA. As exhibited in Figure 7e, the five favorite crosslinking sites of BCETG on soybean proteins were Seed linoleate 13S-lipoxygenase-1's K (250), Seed linoleate 13S-lipoxygenase-1's Q (322), P24 oleosin isoform B's Q (180), Glycinin G4's K (535), and P34 probable thiol protease's K (119), which were from differing SPI protein ingredients. As demonstrated in Figure 7f, the four favorite crosslinking sites of BCETG on WP were found in BSA, such as K (155), K (244), Q (603), and K (160), and some were distinct from those from BSA. The possible reason should be that BSA could be crosslinked with other protein constituents in WP. Additionally, one preferred crosslinked site of BCETG on WP were K (133), which was from α-lactalbumin. Figure 7g–i suggested that these sites were mostly on the surface of BSA, SPI, and WP. Based on the above results, the internal crosslinking mode of BCETG at the protein level could help to estimate its crosslinking ability and regulate its applied manner.

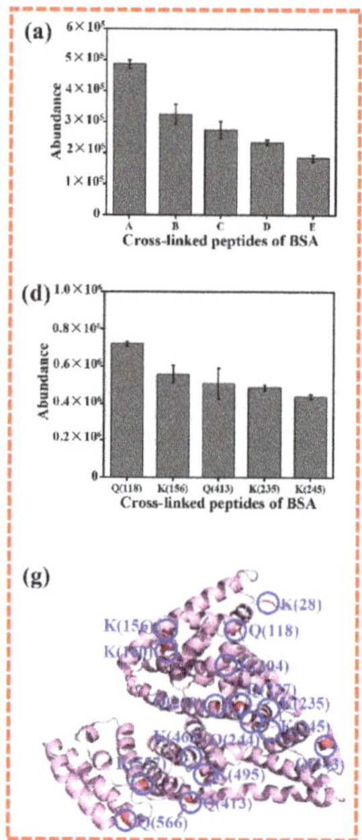
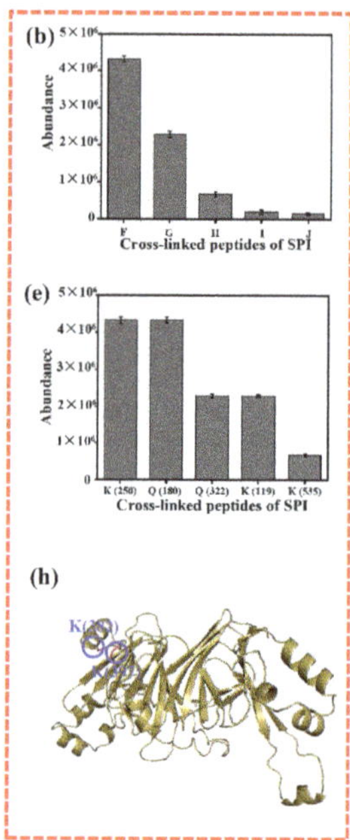
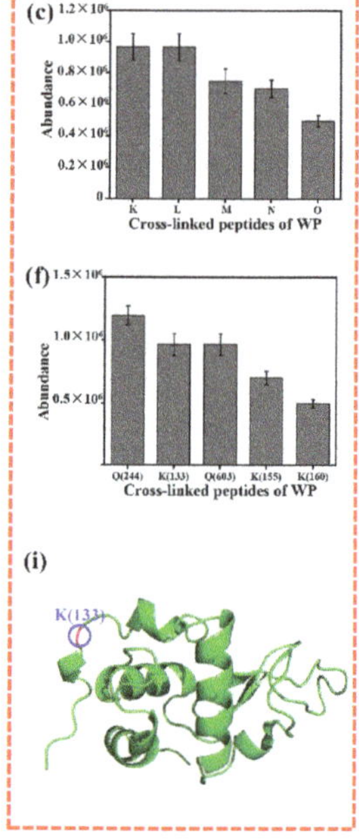

Figure 7. The crosslinking sites in the crosslinked BSA, SPI, and WP were analyzed by BCTG. (**a**) Five of the most plentiful peptides crosslinked in BSA were A (QEPER-FGERALKAWSVAR),

B (ADEKKFWGK-QNCDQFEK), C (LSQKFPK-ATEEQLK), D (QEPER-ADEKKFWGK), and E (QNCDQFEK-TPVSEKVTK). (**b**) The five most plentiful peptides crosslinked in SPI were F (Glycinin G4/QQLQDSHQKIR-Seed linoleate 9S-lipoxygenase-2/KDQNSEK), G (Seed linoleate 13S-lipoxygenase-1/TDGQHILK-P34 probable thiol protease/DVSQQIKMANKK), H (Glycinin G4/AKSSSR-β-conglycinin alpha subunit 1/QSQVSELK), I (Glycinin G2/GRSQRPQDRHQK-Seed linoleate 9S-lipoxygenase/PLANGKGKVGK), and J (Glycinin G5/SQQQLQDSHQK-β-conglycinin alpha subunit 1/SSSRKTISSEDKPFNLR). (**c**) The five most plentiful peptides crosslinked in WP were: K (BSA/QEPER-ADEKKFWGK), L (α-lactalbumin/ALCSEKLDQWLCEK-BSA/LVVSTQTALA), M (β-lactoglobulin/TPEVDDEALEKFDK-BSA/NYQEAK), N (BSA/ADEKK-ALKAWSVARLSQK), and O (BSA/FWGKYLYEIARR-AWSVARLSQK). (**d**) The five most plentiful sites crossllinked in BSA: Q (118), K (156), Q (413), K (235), and K (245). (**e**) The five most plentiful sites crosslinked in SPI: Seed linoleate 13S-lipoxygenase-1/K (250), P24 oleosin isoform B/Q(180), Seed linoleate 13S-lipoxygenase-1/Q (322), P34 probable thiol protease/K (119), Glycinin G4/K (535). (**f**) The five most plentiful sites crosslinked in WP sample: BSA/K (244), α-lactalbumin/K (133), BSA/Q (603), BSA/K (155), and BSA/K (160). (**g**) The sites crosslinked in the tertiary structure of BSA containing K (28), Q (118), K (156), K (160), K (204), Q (219), Q (227), K (235), Q (244), K (245), K (303), Q (343), Q (413), K (463), K (495), K (557), and Q (566). (**h**) The sites crosslinked in the SPI-β-conglycinin alpha subunit 1 containing K (382) and K (397) (**i**) The sites crosslinked in the tertiary structure of WP-α-lactalbumin containing K (133).

4. Conclusions

In brief, four novel TGs demonstrating better thermostability and pH stability than MTG were exploited from *Bacillus* strains, which could offer different options for various food-processing processes. A structural analysis indicated that the Ala147-Ala156 region in the substrate tunnel of BCETG was related to its high temperature adaptability, which could be modified in further research. Furthermore, BCETG exhibited better preference for the protein constituents in SPI and WP than MTG, and the crosslinking mode of BCETG towards BSA, SPI, and WP displayed the intramolecular and intermolecular crosslinked path of glutamine and lysine in food protein substrates. Hence, novel protein crosslinking enzymes were provided in this study, which enhanced the understanding of the structure–activity relationship in new TGs associated with a high temperatures and further promoted its applications in food industry.

Supplementary Materials: The following supporting information can be downloaded at https://www.mdpi.com/article/10.3390/foods11223722/s1. Table S1. Primers used in this study. Figure S1. Multiple alignments analysis of protein sequences for BAMTG, BCETG, BSATG, and BARTG with software DNAMAN. The similar amino acids were labeled in grey and the identical amino acids were highlighted in solid black. The active sites were presented in yellow color. Figure S2. The purified rTGs analyzed using SDS-PAGE. (a) Purified BAMTG. (b) Purified BCETG. (c) Purified BSATG. (d) Purified BARTG. Lane M: Standard protein ladder; Lane 1: purified BAMTG, BCETG, BSATG, and BARTG. Figure S3. Distance between the active center of BCETG and MDC as a function of time (100 ns) at 323 K and 353 K. Figure S4. LC-MS evaluation of the cross-linked peptides of ADEKKFWGK and QNCDQFEK from cross-linked BSA. (a) EIC of the cross-linked peptides ADEKKFWGK-QNCDQFEK and QNCDQFEK. (b) Mass spectrum of cross-linked peptide ADEKKFWGK-QNCDQFEK. (c) Mass spectrum of peptide QNCDQFEK.

Author Contributions: H.W.: Conceptualization, Validation, Formal analysis, Writing—original draft. Y.Z.: Formal analysis, Investigation, Writing—original draft. Z.Y.: Investigation. X.Z.: Investigation. Y.J.: Investigation. J.H.: Investigation. J.Z.: Conceptualization, Investigation. F.L.: Conceptualization, Formal analysis. Y.L.: Conceptualization, Formal analysis, Resources, Writing—review and editing, Supervision, Project administration, Funding acquisition. All authors have read and agreed to the published version of the manuscript.

Funding: This work was supported by the National Key Research and Development Program of China (2021YFC2100300) and the Tianjin Key-Training Program of "Project and Team" of China (XC202032).

Data Availability Statement: The data used to support the findings of this study can be made available by the corresponding author upon request.

Conflicts of Interest: There are no conflicts to declare.

References

1. Miwa, N. Innovation in the food industry using microbial transglutaminase: Keys to success and future prospects. *Anal. Biochem.* **2020**, *597*, 113638. [CrossRef] [PubMed]
2. Feeney, R.E.; Yamasaki, R.B.; Geoghegan, K.F. Chemical Modification of Proteins: An Overview. *Modif. Proteins* **1982**, *98*, 3–55. [CrossRef]
3. Heck, T.; Faccio, G.; Richter, M.; Thöny-Meyer, L. Enzyme-catalyzed protein crosslinking. *Appl. Microbiol. Biotechnol.* **2013**, *97*, 461–475. [CrossRef] [PubMed]
4. Deweid, L.; Avrutina, O.; Kolmar, H. Microbial transglutaminase for biotechnological and biomedical engineering. *Biol. Chem.* **2019**, *400*, 257–274. [CrossRef] [PubMed]
5. Liu, Y.; Liu, Y.; Xu, Z.; Shan, M.; Ge, X.; Zhang, Y.; Shao, S.; Huang, L.; Wang, W.; Lu, F. Effects of Bacillus subtilis transglutaminase treatment on the functional properties of whey protein. *LWT* **2019**, *116*, 108559. [CrossRef]
6. Mostafa, H.S. Microbial transglutaminase: An overview of recent applications in food and packaging. *Biocatal. Biotransform.* **2020**, *38*, 161–177. [CrossRef]
7. Liu, Y.; Huang, L.; Zheng, D.; Fu, Y.; Shan, M.; Li, Y.; Xu, Z.; Jia, L.; Wang, W.; Lu, F. Characterization of transglutaminase from *Bacillus subtilis* and its cross-linking function with a bovine serum albumin model. *Food Funct.* **2018**, *9*, 5560–5568. [CrossRef] [PubMed]
8. Wang, H.; Wang, Y.; Yuan, Z.; Wang, Y.; Li, X.; Song, P.; Lu, F.; Liu, Y. Insight into the cross-linking preferences and characteristics of the transglutaminase from Bacillus subtilis by in vitro RNA display. *LWT* **2021**, *151*, 112152. [CrossRef]
9. Liu, Y.; Lin, S.; Zhang, X.; Liu, X.; Wang, J.; Lu, F. A novel approach for improving the yield of *Bacillus subtilis* transglutaminase in heterologous strains. *J. Ind. Microbiol. Biotechnol.* **2014**, *41*, 1227–1235. [CrossRef]
10. Morris, G.M.; Huey, R.; Lindstrom, W.; Sanner, M.F.; Belew, R.K.; Goodsell, D.S.; Olson, A.J. AutoDock4 and AutoDockTools4: Automated docking with selective receptor flexibility. *J. Comput. Chem.* **2009**, *30*, 2785–2791. [CrossRef] [PubMed]
11. Abraham, M.J.; Murtola, T.; Schulz, R.; Páll, S.; Smith, J.C.; Hess, B.; Lindahl, E. GROMACS: High performance molecular simulations through multi-level parallelism from laptops to supercomputers. *SoftwareX* **2015**, *1–2*, 19–25. [CrossRef]
12. Li, Y.; Hu, X.; Sang, J.; Zhang, Y.; Zhang, H.; Lu, F.; Liu, F. An acid-stable β-glucosidase from Aspergillus aculeatus: Gene expression, biochemical characterization and molecular dynamics simulation. *Int. J. Biol. Macromol.* **2018**, *119*, 462–469. [CrossRef] [PubMed]
13. Wang, J.; Zhang, Y.; Wang, X.; Shang, J.; Li, Y.; Zhang, H.; Lu, F.; Liu, F. Biochemical characterization and molecular mechanism of acid denaturation of a novel α-amylase from *Aspergillus niger*. *Biochem. Eng. J.* **2018**, *137*, 222–231. [CrossRef]
14. Humphrey, W.; Dalke, A.; Schulten, K. VMD: Visual molecular dynamics. *J. Mol. Graph.* **1996**, *14*, 33–38. [CrossRef]
15. Liu, Y.; Zhang, Y.; Guo, Z.; Wang, C.; Kang, H.; Li, J.; Wang, W.; Li, Y.; Lu, F.; Liu, Y. Enhancing the functional characteristics of soy protein isolate via cross-linking catalyzed by *Bacillus subtilis* transglutaminase. *J. Sci. Food Agric.* **2021**, *101*, 4154–4160. [CrossRef] [PubMed]
16. Fernandes, C.G.; Plácido, D.; Lousa, D.; Brito, J.A.; Isidro, A.; Soares, C.M.; Pohl, J.; Carrondo, M.A.; Archer, M.; Henriques, A.O. Structural and Functional Characterization of an Ancient Bacterial Transglutaminase Sheds Light on the Minimal Requirements for Protein Cross-Linking. *Biochemistry* **2015**, *54*, 5723–5734. [CrossRef] [PubMed]
17. Agyare, K.K.; Damodaran, S. pH-Stability and Thermal Properties of Microbial Transglutaminase-Treated Whey Protein Isolate. *J. Agric. Food Chem.* **2010**, *58*, 1946–1953. [CrossRef] [PubMed]
18. Damodaran, S.; Agyare, K.K. Effect of microbial transglutaminase treatment on thermal stability and pH-solubility of heat-shocked whey protein isolate. *Food Hydrocoll.* **2013**, *30*, 12–18. [CrossRef]
19. Sun, X.D.; Arntfield, S.D. Gelation properties of salt-extracted pea protein isolate catalyzed by microbial transglutaminase cross-linking. *Food Hydrocoll.* **2011**, *25*, 25–31. [CrossRef]
20. Truong, V.-D.; Clare, D.A.; Catignani, G.L.; Swaisgood, H.E. Cross-Linking and Rheological Changes of Whey Proteins Treated with Microbial Transglutaminase. *J. Agric. Food Chem.* **2004**, *52*, 1170–1176. [CrossRef] [PubMed]

Article

The Loading of Epigallocatechin Gallate on Bovine Serum Albumin and Pullulan-Based Nanoparticles as Effective Antioxidant

Zikun Li [1], Xiaohan Wang [1], Man Zhang [1], Hongjun He [1], Bin Liang [2,*], Chanchan Sun [1,*], Xiulian Li [3] and Changjian Ji [4]

[1] College of Life Sciences, Yantai University, Yantai 264005, China
[2] College of Food Engineering, Ludong University, Yantai 264025, China
[3] School of Pharmacy, Binzhou Medical University, Yantai 264003, China
[4] Department of Physics and Electronic Engineering, Qilu Normal University, Jinan 250200, China
* Correspondence: liangbin1989311@163.com (B.L.); sunchan88@126.com (C.S.)

Abstract: Due to its poor stability and rapid metabolism, the biological activity and absorption of epigallocatechin gallate (EGCG) is limited. In this work, EGCG-loaded bovine serum albumin (BSA)/pullulan (PUL) nanoparticles (BPENs) were successfully fabricated via self-assembly. This assembly was driven by hydrogen bonding, which provided the desired EGCG loading efficiency, high stability, and a strong antioxidant capacity. The encapsulation efficiency of the BPENs was above 99.0%. BPENs have high antioxidant activity in vitro, and, in this study, their antioxidant capacity increased with an increase in the EGCG concentration. The in vitro release assays showed that the BPENs were released continuously over 6 h. The Fourier transform infrared spectra (FTIR) analysis indicated the presence of hydrogen bonding, hydrophobic interactions, and electrostatic interactions, which were the driving forces for the formation of the EGCG carrier nanoparticles. Furthermore, the transmission electron microscope (TEM) images demonstrated that the BSA/PUL-based nanoparticles (BPNs) and BPENs both exhibited regular spherical particles. In conclusion, BPENs are good delivery carriers for enhancing the stability and antioxidant activity of EGCG.

Keywords: bovine serum albumin; pullulan; epigallocatechin gallate; nanoparticles; antioxidant activity

Citation: Li, Z.; Wang, X.; Zhang, M.; He, H.; Liang, B.; Sun, C.; Li, X.; Ji, C. The Loading of Epigallocatechin Gallate on Bovine Serum Albumin and Pullulan-Based Nanoparticles as Effective Antioxidant. *Foods* **2022**, *11*, 4074. https://doi.org/10.3390/foods11244074

Academic Editors: Fuping Lu and Wenjie Sui

Received: 12 October 2022
Accepted: 8 December 2022
Published: 16 December 2022

Publisher's Note: MDPI stays neutral with regard to jurisdictional claims in published maps and institutional affiliations.

1. Introduction

Previous studies have shown that catechins have antioxidant effects [1] and that they are highly effective natural antioxidants with low toxicity [2]. Furthermore, they also have anticancer [3], anti-obesity [4], and anti-inflammatory properties [5]. Catechins contain three basic ring nuclei—A, B and C—that form a D ring after esterification. According to the different groups connected to the B and C rings, common catechins can be divided into four forms, which can be divided as follows: epicatechin (EC), epigallocatechin (EGC), epicatechin gallate (ECG), and epigallocatechin gallate (EGCG) [6]. Among them, EGCG has the highest content, accounting for approximately 70% of the total catechins. The molecular structure of EGCG contains eight monophenolic hydroxyl groups and it is the most active catechin.

Although EGCG has various beneficial health effects, its application in pharmaceuticals and nutritional supplements is very limited, mainly due to its low bioavailability [7]. In addition, reactions such as oxidation and polymerization can occur easily under conditions such as high temperatures and high humidity, and it can be easily decomposed in neutral and alkaline physiological environments [8]. The hydrophobic benzene ring and more than five hydrophilic hydroxyl groups in the molecule make EGCG both hydrophilic and lipophilic. However, the solubility of EGCG is poor. Its low solubility and poor stability lead to its low bioavailability [9]; however, these problems can be effectively solved by nano-delivery systems.

Nanoparticles are fine microparticles with more than one dimension, and they are between 1–100 nm in size. Furthermore, they are generally composed of natural macromolecular substances or synthetic macromolecular substances. They can be used as carriers for conducting or delivering active substances and drugs. Embedding small-molecule active substances or drugs in nanoparticles can alter their release rate, increase the permeability of their biofilms, and improve their bioavailability [10].

The use of bovine serum albumin (BSA) to prepare nanoparticles as carriers has been gradually gaining attention. Due to the special heart-shaped structure of BSA, a large number of hydrophobic amino acids are concentrated inside it, which form a hydrophobic inner core and a hydrophilic outer shell. This structure allows BSA to interact with both hydrophilic and hydrophobic small molecules [11]. BSA is widely used as a drug carrier due to its non-antigenicity, its degradability, and its nontoxicity [12], in spite of its allergenic disadvantage [13].

Pullulan (PUL) is a natural, degradable macromolecular polymer that is nonhazardous and has been widely used in many fields [14]. PUL contains three hydroxyl groups in each glucose unit and has good solubility in aqueous solutions. Its aqueous solution is stable and less dense than other water-soluble polysaccharides and is not affected by temperature, pH, or most metal ions. Thus, PUL is often used as a food additive [15]. It is widely used in food processing because of its high non-immunogenicity [16], non-carcinogenicity [17], and biocompatibility [18].

In this context, an attempt was made to encapsulate EGCG into a BSA-PUL nanoparticulate (BPNs) system to retain or enhance its antioxidant potential. The objectives of this study were as follows: (1) to investigate the particle size, the polydispersity index (PDI), zeta potential, and thermal stability of BSA or BSA-PUL complexes prepared at different ratios of BSA and PUL or with different pH solutions; (2) to study the encapsulation efficiency and loading rate of EGCG-loaded BPNs (BPENs); (3) to determine the effect of EGCG concentrations on the antioxidant activity and to release the capacity of BPENs in vitro; and (4) to characterize the structures of BPNs and BPENs using Fourier transform infrared spectra (FTIR) and a transmission electron microscope (TEM).

2. Materials and Methods

2.1. Materials

Bovine serum albumin (BSA), epigallocatechin gallate (EGCG), anhydrous ferric chloride, potassium ferricyanide, ferrous ammonium sulfate hexahydrate, salicylic acid, trichloroacetic acid, 1,1-diphenyl-2-Picrylhydrazine (DPPH), and 2,2′-Diaza-bis-3-ethylbenzothiazoline-6-sulfonic acid (ABTS) were obtained from Shanghai Macklin Biochemical Co., Ltd. (Shanghai, China). Pullulan (PUL) was purchased from Shanghai Yien Chemical Technology Co., Ltd. (Shanghai, China). Hydrogen peroxide (purity $\geq$ 30%), potassium dihydrogen phosphate, dipotassium hydrogen phosphate, absolute ethanol, hydrochloric acid, and sodium hydroxide were purchased from China National Medicines Corporation Ltd. (Shanghai, China).

2.2. Preparation and Characterization of BPNs

2.2.1. Preparation of BPNs

First, 0.25 g of BSA and 0.25 g of PUL were dissolved in 50 mL of deionized water. After complete dissolution, the solution was placed in a refrigerator at 4 °C overnight to fully hydrate.

The 5 mg/mL PUL stock solution was diluted to the following three concentrations: 2.5 mg/mL, 1 mg/mL, and 0.5 mg/mL. Equal volumes of 5 mg/mL BSA solution and four concentrations of PUL (5 mg/mL, 2.5 mg/mL, 1 mg/mL, and 0.5 mg/mL) were mixed. Finally, the ratios of BSA/PUL were 1:1, 2:1, 5:1, and 10:1 (all of the BSA concentrations were 2.5 mg/mL). The mixtures of each ratio were adjusted to pH 4.5, 5.0, 5.5, 6.0, 6.5, and 7.0, with 0.01 M hydrochloric acid and 0.01 M sodium hydroxide solution. Next, the solution was incubated overnight to enable it to become completely mixed and homogeneous. The preparation of BPNs is shown in Figure 1(1).

Samples with different BSA-PUL ratios were called BPNs (1:1), BPNs (2:1), BPNs (5:1), and BPNs (10:1).

2.2.2. Determination of Particle Size, PDI, and Zeta-Potential of BPNs

The particle size and distribution of the samples were determined with a particle size analyzer (90Plus Zeta, BIC, NYC, USA). All determinations were performed using three batches of the samples, and the results are expressed as mean $\pm$ standard error.

2.2.3. Thermal Stability

The BSA and BPNs solutions with different ratios and pH levels were heated at 80 °C for 30 min and then cooled to room temperature in ice water [19]. The turbidity (OD_{600}) of the BSA and BPNs solutions with and without a heating treatment was measured using a UV spectrophotometer (UV-2500, Shimadzu, Shanghai, China). The turbidity curves were plotted with the change in turbidity (OD_{600}) as the dependent variable and the change in pH as the independent variable to characterize the thermal stability of the BSA and BPNs.

2.3. Preparation and Characterization of EGCG-Loaded BPNs (BPENs)

2.3.1. Preparation of BPENs

Different concentrations (0.5 mg/mL, 1.0 mg/mL, 1.5 mg/mL, 2.0 mg/mL, and 2.5 mg/mL) of EGCG solutions were added to the BPNs solution, which was prepared under optimum conditions. The EGCG solution was mixed thoroughly with the protein solution, and then it was mixed with the polysaccharide solution. After adjusting the dispersion to the desired pH, using a 0.1 M hydrochloric acid solution or sodium hydroxide solution, the samples were left overnight and lyophilized. The preparation of BPENs is shown in Figure 1(2).

In this paper, the samples loaded with different concentrations of EGCG are called BPENs (0.5), BPENs (1.0), BPENs (1.5), BPENs (2.0), and BPENs (2.5).

2.3.2. Encapsulation Efficiency and Loading Rate

The BPEN solutions were centrifuged at $3500\times g$ for 30 min. The free EGCG in the clear supernatant was measured according to the vanillin–hydrochloric acid colorimetric method. The measurement of EGCG was determined by the absorbance method at 285 nm, and the standard curve is presented in Figure S1. All analyses were performed using three batches of the samples. The results were calculated according to Equations (1) and (2) [20]:

$$\text{Encapsulation efficiency (\%)} = \left(1 - \frac{\text{The amount of free EGCG}}{\text{The amount of total EGCG}}\right) \times 100\% \qquad (1)$$

$$\text{Loading rate (mg EGCG/mg BSA)} = \frac{\text{Encapsulated EGCG amount}}{\text{Total BSA amount}} \qquad (2)$$

2.3.3. DPPH Scavenging Assay

We modified this assay slightly according to the method used by Tang et al. [21]. We added 2 mL of the sample solutions to the 2 mL of 0.2 mM DPPH% ethanol solution. The reaction mixture was incubated for 30 min on a shaker at room temperature in a dark place. The absorbance of the reaction solution was recorded at 517 nm, using a UV spectrophotometer (UV-2500, Shimadzu, Shanghai, China). The DPPH scavenging activity (%) was calculated using Equation (3), as follows:

$$\text{DPPH scavenging activity (\%)} = \frac{(A_0 - A_i)}{A_0} \times 100\% \qquad (3)$$

where A_0 is the absorbance of the DPPH solution after reacting with deionized water, and A_i is the absorbance of the DPPH solution after reacting with the sample.

2.3.4. ABTS$^+$ Scavenging Assay

The method was slightly changed with reference to Wang et al. [22]. ABTS powder was dissolved in deionized water to prepare a 7 mM ABTS solution. Potassium persulfate was dissolved in a 75 mM phosphate buffer solution (PBS) at pH 7.4 to prepare a 4.9 mM potassium persulfate solution. Next, ABTS$^+$ was generated by the equal volume oxidation of the ABTS solution and the aforementioned potassium persulfate solution. Next, the ABTS$^+$ solution was diluted with PBS to the absorbance value of 0.70 ± 0.02 ($\lambda = 734$ nm). Finally, 0.2 mL of the sample solutions were added to 4 mL of the ABTS$^+$ solution and mixed thoroughly. Immediately after incubation in the dark for 15 min, the absorbance of the reaction mixture was recorded at 734 nm. The ABTS$^+$ scavenging activity was calculated using Equation (4), as follows:

$$\text{ABTS}^+ \text{ scavenging assay } (\%) = \frac{A_0 - A_i}{A_0} \times 100\% \tag{4}$$

where A_0 represents the absorbance value of the ABTS$^+$ solution after reacting with deionized water, and A_i represents the absorbance of the ABTS$^+$ solution after reacting with the samples.

2.3.5. Ferric-Reducing Antioxidant Power (FRAP) Assay

This experiment was carried out with reference to the method of Liu et al. [23]. We added 30 μL of the BPENs solution and 2.5 mL of 10 g/L potassium ferricyanide to 2.5 mL of PBS (0.2 mol/L, pH 6.6). The mixed samples were put into thermostatic water for 15 min at 37 °C. After adding 2.5 mL of 100 g/L trichloroacetic acid, 2.5 mL of the supernatant was taken after standing. Next, 2.5 mL of deionized water and 0.5 mL of 1 g/L $FeCl_3$ were added to the supernatant, and, subsequently, the absorbance was measured at 700 nm.

2.3.6. Hydroxyl Radicals (OH) Scavenging Assay

The hydroxyl radical (OH) was generated by a Fenton reaction model system, and the scavenging activity of the nanoparticles was determined following the procedures of Yang et al. [24]. We mixed 1.0 mL of 9 mmol/L ammonium ferrous sulfate and 1.0 mL of a 9 mmol/L salicylic acid–ethanol solution and left the solution for 10 min. We added 1.0 mL of an 8.8 mmol/L H_2O_2 solution and 1.0 mL of the BPENs solutions to the above solution. After mixing evenly, the mixture was placed at 37 °C to react for 1 h. Subsequently, the absorbance was measured at 510 nm. The ·OH scavenging activity was calculated using Equation (5), as follows:

$$\text{Hydroxyl radical scavenging rate } (\%) = A_0 - \frac{A_i - A_{i0}}{A_0} \times 100\% \tag{5}$$

where A_0 is the absorbance of the ·OH solution after reacting with deionized water, A_i is the absorbance of the ·OH solution after reacting with the BPENs solutions, and A_{i0} represents the absorbance of the ·OH solution after reacting with the BPENs solutions, without H_2O_2.

2.3.7. In Vitro Release Behavior

The in vitro release behaviors of BPENs with different EGCG contents were studied using the dynamic dialysis method in deionized water [25]. Briefly, the dispersion of BPENs in distilled water was placed into dialysis tubing (molecular weight cut-off 8–14 kDa, Solarbio Science and Technology Co., Ltd., Beijing, China) and dialyzed against deionized water at 37 ± 0.2 °C in an air-bath shaker at 120 rpm. At predefined time intervals, 4 mL of the release media were collected, and the fresh release media were added. The cumulative percentage of EGCG released was determined at 285 nm and calculated using Equation (6), as follows:

$$\text{The cumulative percentage of EGCG released} = \frac{\sum_{i=1}^{13} C_i [V_2 - (i-1)V_1]}{W} \qquad (6)$$

where C_i is the concentration of EGCG in the released medium obtained in time i, mg/mL; V_1 is the volume of the release media, 4 mL; and V_2 is the volume of dialysis deionized water, 200 mL.

2.3.8. Fourier Transform Infrared Spectra (FTIR) Analysis

All samples (BPNs, BPENs, BSA, and PUL) were freeze-dried and fully mixed with 200 mg of KBr and pelletized [26]. Spectrum scanning was taken in the wavelength range of 4000–400 cm^{-1}, at a resolution of 4 cm^{-1}, and with a scan speed of 2 mm/s, using a Fourier transform infrared spectrometer (FTIR, VERTEX70, Bruker, Germany).

2.3.9. Morphological Observation of BPNs and BPENs

The freshly prepared samples were dropped onto the copper mesh, and the morphology of the BPNs and BPENs were observed using a transmission electron microscope (TEM, TECNAI F20, FEI Company, HIO, OSU, Columbus, OH, USA) after drying.

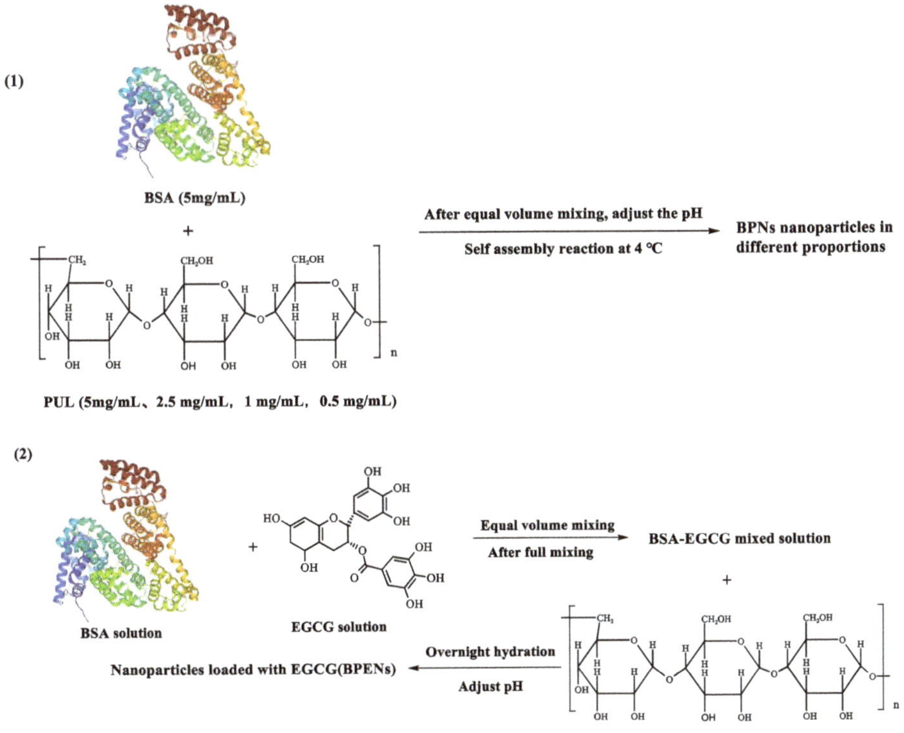

Figure 1. The preparation process of BPNs (**1**) and BPENs (**2**).

2.3.10. Statistical Analysis

All the tests were repeated in triplicate, and the data are expressed as the means $\pm$ SDs (standard deviations). A one-way analysis of variance (ANOVA) and significant difference tests were performed using SPSS 27.0 software. Significant differences were determined with an independent sample *t*-test or Duncan's multiple-range tests, at $p = 0.05$. Mean differences were considered significant when $p < 0.05$.

3. Results and Discussion

3.1. Particle Size and Zeta-Potential of BPNs

The results of the particle size of the nanoparticles are summarized in Table 1. It can be seen from Table 1 that the pH level had a significant effect on the particle size of the BPNs. As the pH deviated from the isoelectric point, the particle size first decreased and then increased, probably because they were in a neutral environment [27], which led to some particle aggregation. In addition, as shown in Table 1, the different BSA/PUL ratios had a significant effect on the particle size of the BPNs. The particle size of the BPNs at each pH was the smallest when the BSA/PUL ratio was 1:1. As the BSA/PUL ratio increased, the particle size of the BPNs also increased. This can be mainly attributed to the decrease in the concentration of PUL, which led to the rapid nucleation of the BSA at different rates, which, in turn, resulted in the formation of larger and more uneven particles [27]. Therefore, the optimal ratio of BSA/PUL was 1:1.

The electrostatic repulsion between nanoparticles is an important factor in determining the stability of the particles. Zeta potential is an important indicator in characterizing the stability of a colloidal dispersion [28]. It represents the coefficient of change in the surface charge and its absolute value represents the stability. It can be seen from Figure 2 that the potential of a solution was negative when the pH was 4.5–7.0, and that the potential gradually decreased with the increase in the pH. With the increase in the PUL concentration, the potential was at its maximum at a 1:1 ratio of PUL/BSA. This was because the surface of the BSA was negatively charged, whereas the surface of the PUL was not charged [29]. The decrease in the zeta potential of all the BPNs was obvious as the pH level increased from 4.5 to 5.5. This was because the charge of the PUL and BSA was neutralized; therefore, the electrostatic repulsion decreased. The aggregation of the BSA and PUL was also proved by the trend of the particle size increasing (Table 1).

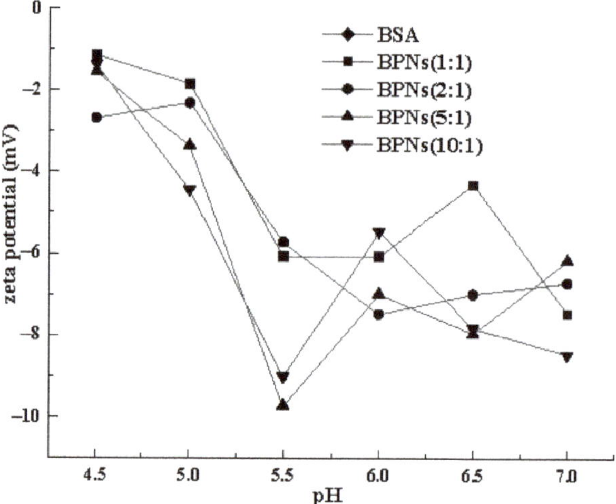

Figure 2. Zeta potential of BPNs as a function of pH.

Table 1. Particle size and polymer dispersity index (PDI) of BPNs.

pH	1:1		2:1		5:1		10:1	
	Particle Size (nm)	PDI	Particle Size (nm)	PDI	Particle Size (nm)	PDI	Particle Size (nm)	PDI
4.5	341 ± 5 [a]	0.369	382 ± 5 [a]	0.516	558 ± 8 [a]	0.521	462 ± 45 [a]	0.375
5.0	219 ± 6 [b]	0.392	335 ± 26 [b]	0.372	325 ± 9 [b]	0.350	644 ± 48 [b]	0.330
5.5	180 ± 8 [c]	0.369	226 ± 1 [c]	0.218	251 ± 21 [c]	0.122	620 ± 57 [bc]	0.312
6.0	139 ± 13 [d]	0.355	447 ± 32 [d]	0.341	606 ± 56 [ad]	0.300	224 ± 21 [d]	0.321
6.5	251 ± 11 [e]	0.265	488 ± 37 [de]	0.190	670 ± 3 [e]	0.402	257 ± 13 [de]	0.434
7.0	347 ± 28 [af]	0.449	362 ± 18 [abf]	0.546	2011 ± 62 [f]	0.503	749 ± 50 [f]	0.303

Values that do not bear the same letter in the same column were significantly different ($p < 0.05$).

3.2. Turbidity Determination and Thermal Stability

Nanoparticles may be affected by various pH environments during the manufacturing process, in storage, and during their passage through the human digestive tract. Therefore, it is necessary to assess the stability of a drug delivery system under different pH conditions. In this mixed solution system, the OD_{600} (Figure 3A) of the unheated BPNs sample over the entire pH range was close to 0, and the naked-eye view of the solution was transparent and translucent. This means that BSA and PUL form soluble complexes at different ratios and under different pH conditions. In general, the BPNs were stable over a wide pH range, which indicates a broad application prospect.

Thermal treatment—an important step in food preservation and production—can influence the structural and functional properties of proteins, resulting in the denaturation of proteins by disrupting the forces that stabilize their natural conformation [30]. The single BSA solution and the BPN samples with different pH levels were heated, and the results are shown in Figure 3B. We can see that when the pH was 4.5 in the four ratios, the sample solution had slightly flocculent precipitates. When the pH was 5.0, the sample solution had obviously larger flocs. When the pH was 5.5, the sample solution presented a more obvious milky-white colloidal solution state, and no precipitated flocs were formed. This was attributed to the thermal denaturation of the protein [31]. However, at pH 6.0, 6.5, and 7.0, this behavior was not observed. It is possible that the nanoparticle structure formed by the protein and polysaccharide was more stable at these pH conditions, demonstrating that PUL could improve the anti-agglomeration ability of nanoparticles and ensure that they are uniformly dispersed [32]. The denaturation temperature of BSA was reported to be between 62–65 °C [33], at which point it had a weak electrostatic repulsion and its interior hydrophobic groups were exposed [32] which, in turn, led to the formation of BSA aggregates.

In Figure 3B, the sample solution with a pH of 6.0 presented a translucent milky-white colloidal solution state. The sample solutions with a pH of 6.5 and 7.0 were clear and transparent. As the pH gradually increased, the solution state gradually became clear. The reason for this was that the isoelectric point of BSA is approximately 4.7, and the solution was easy to coagulate and flocculate near the isoelectric point. The more it deviated from the isoelectric point, the more stable the solution was [34].

As shown in Figure 3, it was observed that no turbidity or aggregates were formed at the three pH values of 6.0, 6.5, and 7.0, and this was true in both the heated group and the unheated group. However, the solution formed a state of suspension without precipitation at a pH of 6.0, which was more stable than the complex system prepared at pH values of 4.5–5.5. Therefore, it was indicated that BSA and PUL basically completed the self-assembly and presented a uniform and stable colloidal solution system under the pH 6.5–7.0 condition. Based on their smaller particle size and higher thermal stability, the BPNs (1:1) prepared at pH 6.5 were selected for the encapsulation of EGCG, and a series of characterization analyses were carried out.

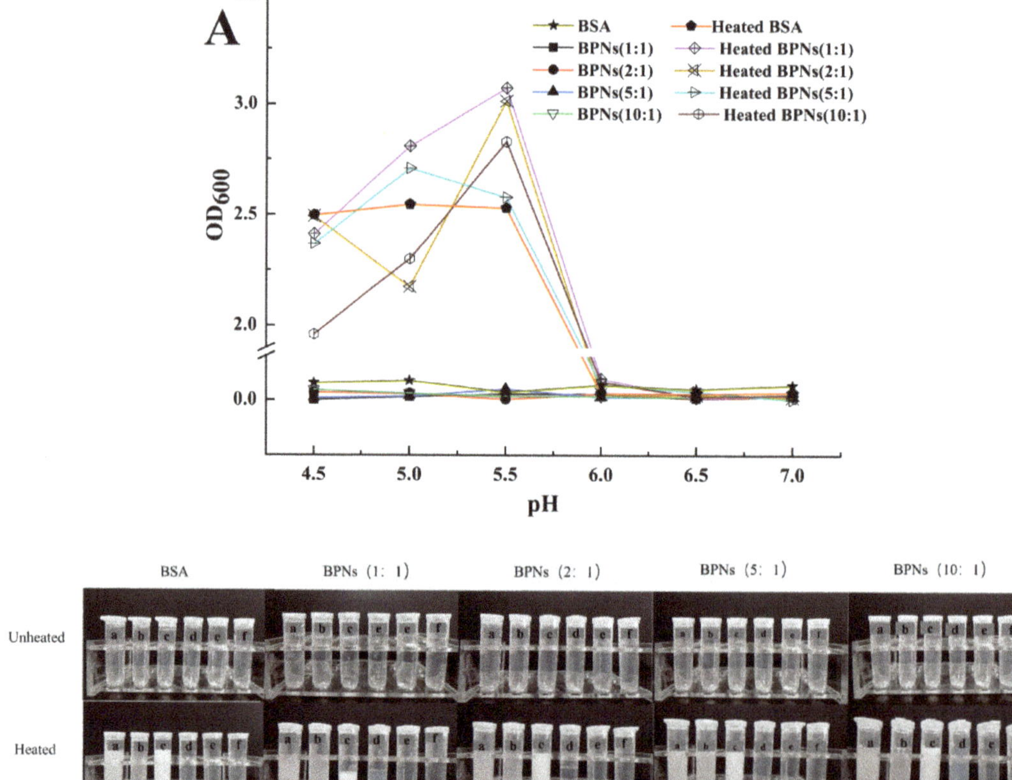

Figure 3. Turbidity curves (**A**) and visual appearance (**B**) of BSA and BPNs as a function of pH, without being heated and heated at 80 °C for 30 min. a–f represents pH 4.5, 5.0, 5.5, 6.0, 6.5, and 7.0, respectively.

3.3. BPENs' Encapsulation Efficiency and Loading Rate

The encapsulation efficiency and loading rate of the BPENs loaded with EGCG are shown in Figure 4. As can be seen from Figure 4A, the encapsulation efficiency of BPENs (0.5) was 99.67%. With the increase in the EGCG loading concentration, the encapsulation efficiency of the BPENs did not change significantly.

In Figure 4B, we can see that the loading rate of BPENs (0.5) was 17.4%, and the loading rate also changed significantly with the increase in the concentration of EGCG. The loading of BPENs (2.5) reached 53.4%, which was 36.0% higher than that of BPENs (0.5).

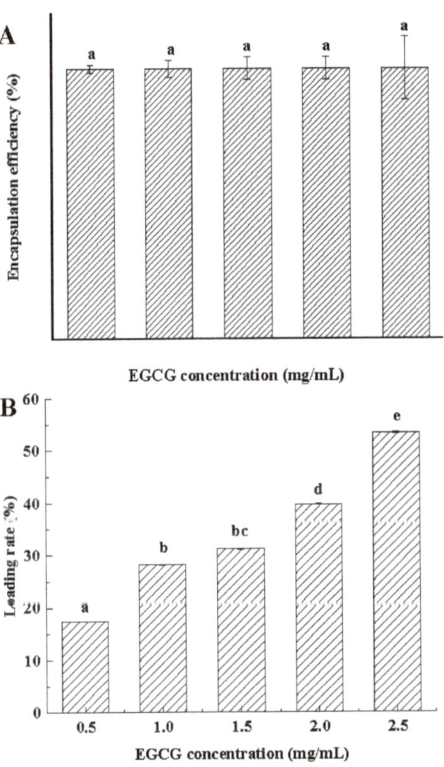

Figure 4. Encapsulation efficiency (**A**) and loading rate (**B**) of BPENs. Histograms followed by different letters were significantly different at $p < 0.05$.

3.4. In Vitro Antioxidant Activities

DPPH, DPPH, ABTS, FRAP, and hydroxyl radical scavenging capacity assays were carried out to evaluate the in vitro antioxidant capacity of the BPENs that were loaded with various concentrations of EGCG, and the results are shown in Figure 5.

DPPH is a very stable and classic nitrogen-containing free radical. EGCG can react with a DPPH free radical to form a phenoxy free radical. This reaction is stable, and it is not easy to cause new reactions. An analysis of the ability to eliminate DPPH· can determine EGCG antioxidant activity [35]. As shown in Figure 5A, the DPPH radicals scavenging activity was, in general, positively correlated with the sample concentration. The radical scavenging activities of the BPENs and EGCG (free) were significantly higher than those of the BPNs at all concentrations, suggesting that EGCG is a much more effective antioxidant than proteins and polysaccharides. The DPPH scavenging activity increased significantly with the increase in the antioxidant concentration. The spherical structure of the BPENs enables them to have a high surface area, which increases the interaction between the DPPH radicals and EGCG, thereby enhancing the antioxidant activity [36]. In addition, improving the water solubility of EGCG by nanoencapsulation can significantly enhance the scavenging ability of DPPH radicals. However, the DPPH scavenging ability of EGCG was significantly higher than that of the BPENs. This may be accounted for by the loss of exposed hydroxyl groups, since hydroxyl groups make a significant contribution to superior antioxidant activity [37]. However, when the EGCG was added to the protein, a much slower decrease in its antioxidant activity was observed. This was attributed to the protective effect provided by the protein, which was in line with previous studies [38].

Figure 5B shows that the ABTS$^+$ free radical scavenging activity of the free EGCG was not very different from the activities of the free EGCG and BPENs that were loaded with different concentrations of EGCG. This shows that the nanoparticles can effectively improve the stability of EGCG and maintain its antioxidant activity. In line with the ideas of Laura et al. [39], we concluded that hydroxyl radicals are the most active oxygen-containing radicals, so the ability of BPENs to scavenge ·OH is an important indicator of their antioxidant activity [40]. FRAP is measured by monitoring the change in Fe^{3+} to Fe^{2+}, which results from the electrons in the reducing substances [41]. As shown in Figure 5C, D, the FRAP scavenging activity and the hydroxyl radicals scavenging activity were also, in general, positively correlated with the sample concentration. Moreover, the free radical scavenging activity of free EGCG was not significantly different from that of the BPENs. When the concentration was increased, the free radical scavenging activity was significantly improved. This was mainly attributed to the higher surface area of the BPENs and the water solubility of EGCG after nanoencapsulation. The experimental results of the DPPH radicals scavenging activity and the ABTS$^+$ scavenging activity were consistent, and it was shown that nanoparticles can enhance or retain the antioxidant activity of catechins [42].

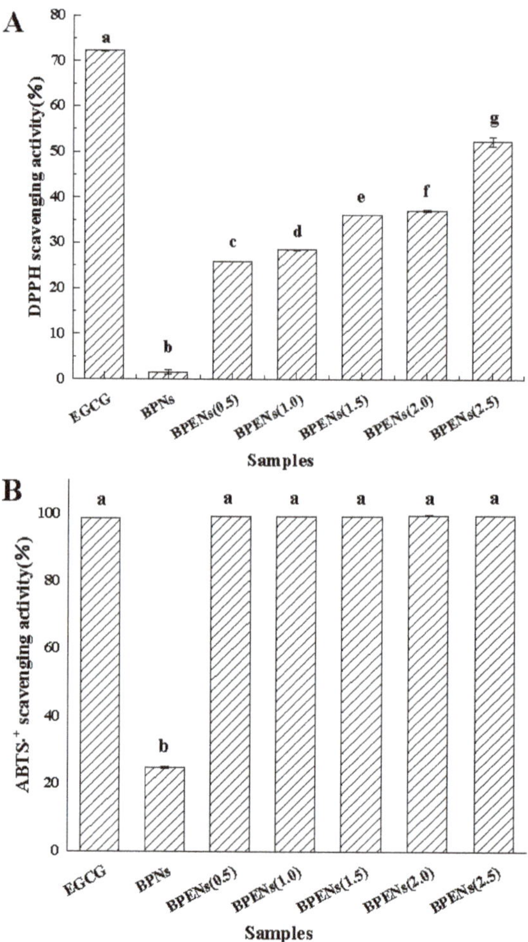

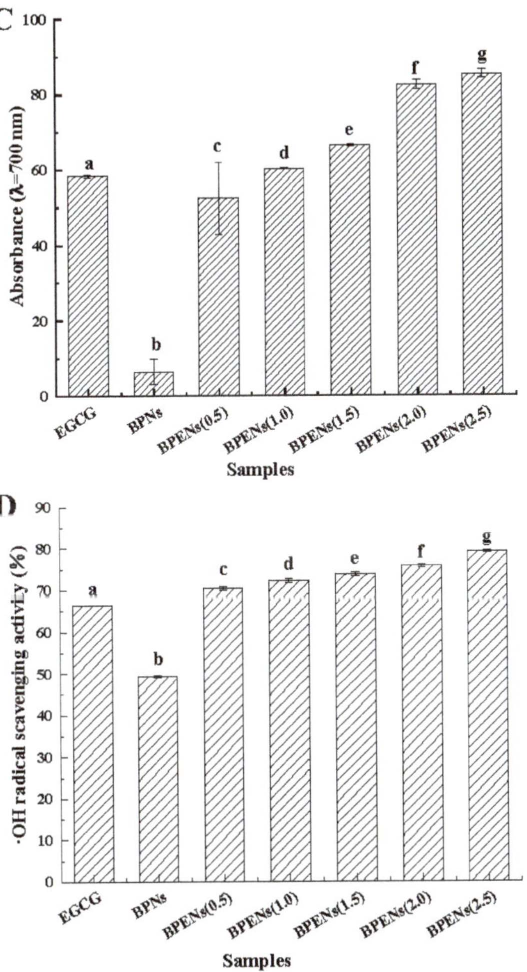

Figure 5. The in vitro antioxidant capacity of BPENs, as determined by the DPPH (**A**), ABTS (**B**), FRAP (**C**), and hydroxyl radical scavenging capacity (**D**), respectively. Histograms followed by different letters were significantly different at $p < 0.05$.

3.5. In Vitro Release Test

The in vitro release behavior of the BPENs loaded with different concentrations of EGCG was investigated at 37 °C, and the results are shown in Figure 6A. The results of the in vitro release properties of EGCG showed that the BPENs loaded with different concentrations of EGCG showed a sustainable release ability. Similar release behavior characteristics were also observed in the carriers of soybean protein/EGCG nanoparticles [43]. In BPENs (0.5), EGCG was slowly released during the first hour. It is most likely that the EGCG release corresponded to the EGCG being closer to the nanoparticle surface. After 1 h, a sudden release of EGCG was observed, which lasted until after the second hour. The slow and sustained release of EGCG may be due to the diffusion of the EGCG embedded within the cores of the BPENs.

In BPENs (1.0), BPENs (1.5), BPENs (2.0), and BPENs (2.5), the release rate and the total release amount of EGCG increased with the increase in the initial concentration of EGCG. During the whole process, the BPENs exhibited a rapid but limited release of EGCG. Within the first hour, EGCG was rapidly released from the nanoparticles. The amount of EGCG released by the BPENs increased significantly with the increase in the EGCG-loaded concentration. After 1 h, the release of EGCG had significantly slowed and reduced. This release of EGCG was thought to be due to the partial adsorption of EGCG near the surface of the BPENs. The entire release process lasted for more than 5 h. The cumulative release rate of BPENs (0.5) at 6 h reached 37.19%, while the cumulative release rate of BPENs (2.5) reached 89.04%. It indicated that the cumulative release rate of the BPENs was related to the EGCG-loaded concentration, and the higher the concentration, the better the release effect was.

3.6. FTIR Analysis

The molecular characterization of the BPNs and BPENs was carried out using FTIR spectra. This technique has been used to evaluate the chemical and conformational changes that occur when nanoparticles are formed or when they interact with other compounds through the slight shift in the characteristic bands of the spectral regions of amide I and amide II. Figure 6B shows the FTIR spectra of BSA, PUL, BPNs, and BPENs. In Figure 6B, the FTIR spectra of PUL had a broad absorption band, at 3316 cm^{-1}, which was caused by the -OH stretching vibration. Furthermore, an absorption peak appeared near 2925 cm^{-1} and 1640 cm^{-1}, which was caused by the -CH$_2$ stretching vibration and the O-C-O bond stretching vibration connecting the glucose units [44]. In the FTIR spectra of BSA, 3285 cm^{-1} was related to amide A, which was related to the N-H stretching; 2957 cm^{-1} was related to amide B, which was associated with the N-H stretching of the NH^{3+} free ion; 1643 cm^{-1} was the peak of amide I and the C=O stretching vibration; the absorption peak at 1530 cm^{-1} was related to amide II, which was related to the C-N stretching and the N-H bending vibration; and 1392 cm^{-1} was related to amide III, which was associated with the C-N stretching and the N-H bending [45]. In the case of the BPNs, the presence of the protein enhanced the intensity of the 1643 and 1530 cm^{-1} bands, due to superposition with the amide I and amide II bands, respectively [46]. Furthermore, small shifts in the observation bands were observed, compared to BSA and the BPNs [45]. The spectral bands of the BPNs showed the characteristic peaks of these proteins and polysaccharides, which indicates that BSA and PUL produced BPNs through noncovalent binding, which is the self-assembly effect. A comparison of the FTIR spectrum of the BPENs and BPNs showed that both have similar profiles, suggesting that the structure of the nanoparticles does not change significantly with the addition of EGCG.

In the Fourier transform infrared spectra of the BPENs, 1284 cm^{-1} was related to the stretching vibration of the C-O bond of EGCG. Therefore, the BPENs had one more absorption peak as a result of the stretching vibration of the C-O bond of the catechins, compared with the BPNs. This indicates that the nanoparticles successfully embedded in the EGCG. Meanwhile, the presence of the aromatic ring quadrant, the -OH deformation of the aromatic alcohols, and the C-O stretching of the aromatic alcohols and aliphatic diols were shown at 1078 cm^{-1} [47]. This indicates that no bonds were formed between the BPNs and EGCG.

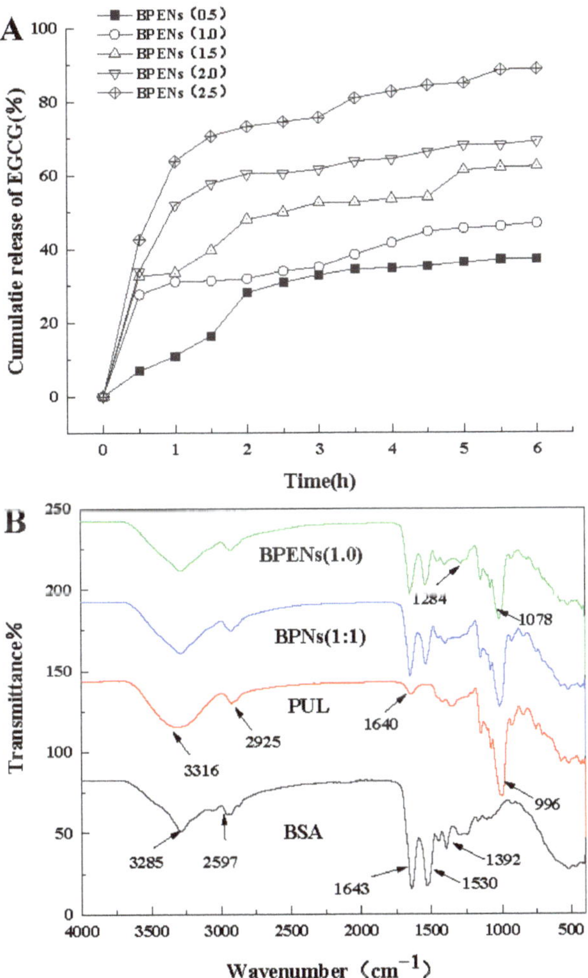

Figure 6. In vitro release curves (**A**) of EGCG and FTIR spectrum (**B**) of BSA, PUL, BPNs, BPENs.

3.7. TEM Observation of Microscopic Morphology

TEM images can be used to directly observe the morphology and structure of samples. As shown in Figure 7A, it can be seen that the BPNs (BSA: PUL = 1:1 and pH = 6.5) are spherical and that the particle size is approximately 350 nm, which is roughly the same as the results measured by the particle size analyzer. The particle morphology was similar to the shape of woolen balls. The inner and outer contrasts were different, indicating that PUL and BSA self-assembled to form spherical nanoparticles. The images demonstrate a spherical morphology for the nanoparticles, which corroborates the literature that has been published [48].

According to Figure 7B, it can be seen that the shape of the EGCG-encapsulated composite particles changed, indicating that EGCG might be successfully encapsulated inside the BPNs.

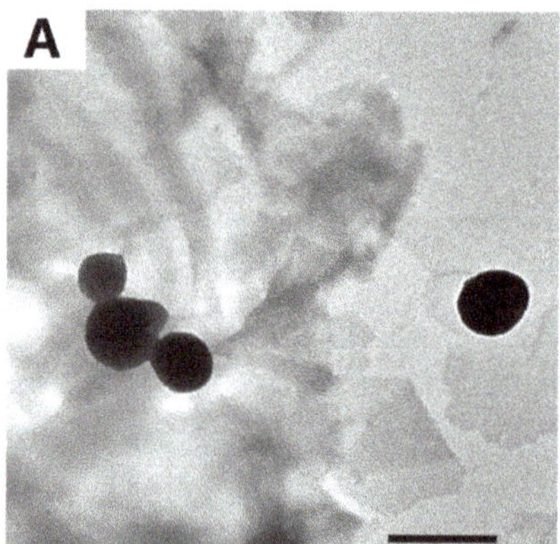

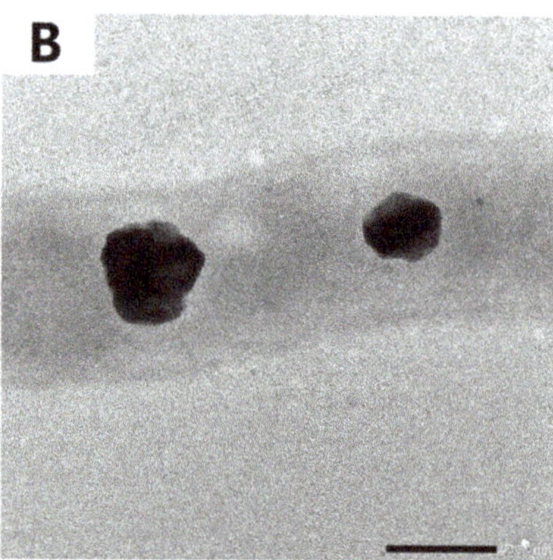

Figure 7. TEM images of BPNs (1:1) (**A**) and BPENs (1.0) (**B**).

4. Conclusions

In conclusion, EGCG-loaded BSA/PUL nanoparticles were successfully fabricated via self-assembly. This was driven by hydrogen bonding, which provided the desirable EGCG-loading efficiency, high stability, and a strong antioxidant capacity. The encapsulation efficiency of the BPENs was above 99.0%. The BPENs had high antioxidant activity in vitro, and their antioxidant capacity increased with the increase in the EGCG concentration. The in vitro release assays showed that the BPENs were released continuously over 6 h. An FTIR analysis indicated the presence of hydrogen bonding, hydrophobic interactions, and electrostatic interactions, which were the driving force for the formation of the EGCG carrier nanoparticles. The TEM images demonstrated that the BPNs and BPENs both exhibited regular spherical particles. In conclusion, BPENs are good delivery carriers for enhancing the stability and the antioxidant activity of EGCG, which is of great importance in the development of functional foods.

Supplementary Materials: The following supporting information can be downloaded at: https://www.mdpi.com/article/10.3390/foods11244074/s1, Figure S1: Standard curve of EGCG.

Author Contributions: Data curation, X.W.; investigation, B.L.; methodology, X.L.; project administration, C.S.; resources, H.H. and C.J.; writing—original draft, Z.L.; writing—review and editing, Z.L. and M.Z. All authors have read and agreed to the published version of the manuscript.

Funding: The authors are grateful for the support of the Natural Science Foundation of Shandong Province (ZR2020QC219, ZR2019BCE036) and the National Natural Science Foundation of China (32101904).

Institutional Review Board Statement: Not applicable.

Informed Consent Statement: Not applicable.

Data Availability Statement: The data are contained within the article.

Acknowledgments: We gratefully acknowledge the four anonymous reviewers who provided valuable assistance for our analysis, feedback for our results, and constructive comments which greatly helped improve this manuscript. The authors thank the reviewers and editors for their careful review of this manuscript.

Conflicts of Interest: The authors declare that they have no known competing financial interests or personal relationships that could have appeared to influence the work reported in this paper.

References

1. Ebrahimian, J.; Khayatkashani, M.; Soltani, N.; Yousif, Q.A.; Salavati-Niasari, M. Catechin mediated green synthesis of Au nanoparticles: Experimental and theoretical approaches to the determination HOMO-LUMO energy gap and reactivity indexes for the (+)-epicatechin (2S, 3S). *Arab. J. Chem.* **2022**, *15*, 103758. [CrossRef]
2. Song, H.; Wang, Q.; He, A.; Li, S.; Guan, X.; Hu, Y.; Feng, S. Antioxidant activity, storage stability and in vitro release of ep-igallocatechin-3-gallate (EGCG) encapsulated in hordein nanoparticles. *Food Chem.* **2022**, *388*, 132903. [CrossRef] [PubMed]
3. Ye, Q.; Li, T.; Li, J.; Liu, L.; Dou, X.; Zhang, X. Development and evaluation of tea polyphenols loaded water in oil emulsion with zein as stabilizer. *J. Drug Deliv. Sci. Technol.* **2020**, *56*, 101528. [CrossRef]
4. Li, J.; Wang, Y.; Suh, J.H. Multi-omics approach in tea polyphenol research regarding tea plant growth, development and tea processing: Current technologies and perspectives. *Food Sci. Hum. Wellness* **2022**, *11*, 524–536. [CrossRef]
5. Han, Y.; Jia, F.; Bai, S.; Xiao, Y.; Meng, X.; Jiang, L. Effect of operating conditions on size of catechin/β-cyclodextrin nanopar-ticles prepared by nanoprecipitation and characterization of their physicochemical properties. *LWT* **2022**, *153*, 112447. [CrossRef]
6. Dube, A.; Ng, K.; Nicolazzo, J.A.; Larson, I. Effective use of reducing agents and nanoparticle encapsulation in stabilizing catechins in alkaline solution. *Food Chem.* **2010**, *122*, 662–667. [CrossRef]
7. Ahmad, M.; Mudgil, P.; Gani, A.; Hamed, F.; Masoodi, F.A.; Maqsood, S. Nano-encapsulation of catechin in starch nanopar-ticles: Characterization, release behavior and bioactivity retention during simulated in-vitro digestion. *Food Chem.* **2019**, *270*, 95–104. [CrossRef]
8. Wang, R.; Zhou, W.; Jiang, X. Mathematical modeling of the stability of green tea catechin epigallocatechin gallate (EGCG) during bread baking. *J. Food Eng.* **2008**, *87*, 505–513. [CrossRef]
9. Chanphai, P.; Tajmir-Riahi, H.A. Conjugation of tea catechins with chitosan nanoparticles. *Food Hydrocoll.* **2018**, *84*, 561–570. [CrossRef]
10. Laha, B.; Maiti, S. Design of Core-Shell Stearyl Pullulan Nanostructures for Drug Delivery. *Mater. Today Proc.* **2019**, *11*, 620–627. [CrossRef]
11. Zhou, R.; Dong, X.; Song, L.; Jing, H. Interaction mode and nanoparticle formation of bovine serum albumin and anthocyanin in three buffer solutions. *J. Lumin.* **2014**, *155*, 244–250. [CrossRef]
12. Zhou, D.; Xu, Z.; Li, Y.; Chen, L.; Liu, Y.; Xu, Y.; Meng, K.; Cheng, L.; Sun, J. Preparation and characterization of thermosen-sitive hydrogel system for dual sustained-release of chlorhexidine and bovine serum albumin. *Mater. Lett.* **2021**, *300*, 130121. [CrossRef]
13. Rostamnezhad, F.; Hossein Fatemi, M. Comprehensive investigation of binding of some polycyclic aromatic hydrocarbons with bovine serum albumin: Spectroscopic and molecular docking studies. *Bioorg. Chem.* **2022**, *120*, 105656. [CrossRef]
14. Roy, S.; Rhim, J.-W. Effect of chitosan modified halloysite on the physical and functional properties of pullulan/chitosan biofilm integrated with rutin. *Appl. Clay Sci.* **2021**, *211*, 106205. [CrossRef]
15. Yuzbasioglu, D.; Mamur, S.; Avuloglu-Yilmaz, E.; Erikel, E.; Celebi-Keskin, A.; Unal, F. Evaluation of the genotoxic and anti-genotoxic effects of exopolysaccharide pullulan in human lymphocytes in vitro. *Mutat. Res./Genet. Toxicol. Environ. Mutagen.* **2021**, *870–871*, 503391. [CrossRef]
16. Tiwari, S.; Patil, R.; Dubey, S.K.; Bahadur, P. Derivatization approaches and applications of pullulan. *Adv. Colloid Interface Sci.* **2019**, *269*, 296–308. [CrossRef]
17. Silva, N.H.; Vilela, C.; Almeida, A.; Marrucho, I.M.; Freire, C.S.R. Pullulan-based nanocomposite films for functional food packaging: Exploiting lysozyme nanofibers as antibacterial and antioxidant reinforcing additives. *Food Hydrocoll.* **2018**, *77*, 921–930. [CrossRef]
18. Zhang, Z.; Qiu, C.; Li, X.; McClements, D.J.; Jiao, A.; Wang, J.; Jin, Z. Advances in research on interactions between polyphenols and biology-based nano-delivery systems and their applications in improving the bioavailability of polyphenols. *Trends Food Sci. Technol.* **2021**, *116*, 492–500. [CrossRef]
19. Zhu, L.; Yang, F.; Li, D.; Wu, G.; Zhang, H. Preparation, structure and stability of protein-pterostilbene nanocomplexes coated by soybean polysaccharide and maltodextrin. *Food Biosci.* **2022**, *49*, 101899. [CrossRef]

20. Li, W.; Li, W.; Wan, Y.; Wang, L.; Zhou, T. Preparation, characterization and releasing property of antibacterial nano-capsules composed of ε-PL-EGCG and sodium alginate-chitosan. *Int. J. Biol. Macromol.* **2022**, *204*, 652–660. [CrossRef] [PubMed]
21. Tang, D.-W.; Yu, S.-H.; Ho, Y.-C.; Huang, B.-Q.; Tsai, G.-J.; Hsieh, H.-Y.; Sung, H.-W.; Mi, F.-L. Characterization of tea cat-echins-loaded nanoparticles prepared from chitosan and an edible polypeptide. *Food Hydrocoll.* **2013**, *30*, 33–41. [CrossRef]
22. Wang, Z.; Huang, J.; Yun, D.; Yong, H.; Liu, J. Antioxidant packaging films developed based on chitosan grafted with different catechins: Characterization and application in retarding corn oil oxidation. *Food Hydrocoll.* **2022**, *133*, 107970. [CrossRef]
23. Liu, Q.; Sun, Y.; Cui, Q.; Cheng, J.; Killpartrik, A.; Kemp, A.H.; Guo, M. Characterization, antioxidant capacity, and bioac-cessibility of Coenzyme Q10 loaded whey protein nanoparticles. *LWT* **2022**, *160*, 113258. [CrossRef]
24. Yang, J.; Wang, Y.; Yin, R.; Pang, J.; Cong, Y.; Yang, S. Water molecule attachment mode on the dried polysaccharide influences its free radical scavenging ability. *Process Biochem.* **2020**, *91*, 15–22. [CrossRef]
25. Jan, N.; Madni, A.; Rahim, M.A.; Khan, N.U.; Jamshaid, T.; Khan, A.; Jabar, A.; Khan, S.; Shah, H. In vitro anti-leukemic assessment and sustained release behaviour of cytarabine loaded biodegradable polymer based nanoparticles. *Life Sci.* **2021**, *267*, 118971. [CrossRef] [PubMed]
26. Tan, Y.; Li, S.; Liu, S.; Li, C. Modification of coconut residue fiber and its bile salt adsorption mechanism: Action mode of insoluble dietary fibers probed by microrheology. *Food Hydrocoll.* **2023**, *136*, 108221. [CrossRef]
27. Lu, J.; Xie, L.; Wu, A.; Wang, X.; Liang, Y.; Dai, X.; Cao, Y.; Li, X. Delivery of silybin using a zein-pullulan nanocomplex: Fabrication, characterization, in vitro release properties and antioxidant capacity. *Colloids Surf. B Biointerfaces* **2022**, *217*, 112682. [CrossRef]
28. Karami, K.; Jamshidian, N.; Hajiaghasi, A.; Amirghofran, Z. BSA nanoparticles as controlled release carriers for isophethalal-doxime palladacycle complex; synthesis, characterization, in vitro evaluation, cytotoxicity and release kinetics analysis. *New J. Chem.* **2020**, *44*, 4394–4405. [CrossRef]
29. Lin, P.; Zhang, W.; Chen, D.; Yang, Y.; Sun, T.; Chen, H.; Zhang, J. Electrospun nanofibers containing chitosan-stabilized bovine serum albumin nanoparticles for bone regeneration. *Colloids Surf. B: Biointerfaces* **2022**, *217*, 112680. [CrossRef]
30. Zhang, W.; Zhong, Q. Microemulsions as nanoreactors to produce whey protein nanoparticles with enhanced heat stability by thermal pretreatment. *Food Chem.* **2010**, *119*, 1318–1325. [CrossRef]
31. Seo, J.A.; Hedoux, A.; Guinet, Y.; Paccou, L.; Affouard, F.; Lerbret, A.; Descamps, M. Thermal Denaturation of Be-ta-Lactoglobulin and Stabilization Mechanism by Trehalose Analyzed from Raman Spectroscopy Investigations. *J. Phys. Chem. B* **2010**, *114*, 6675–6684. [CrossRef] [PubMed]
32. De Oliveira, R.C.; Benevides, C.A.; Rodrigues, G.C.P.; Tenório, R.P. Thermal Denaturation and γ-Irradiation effects on the Crack Patterns of Bovine Serum Albumin (BSA) Dry Droplets. *Colloid Interface Sci. Commun.* **2019**, *28*, 15–19. [CrossRef]
33. Fan, Y.T.; Yi, J.; Zhang, Y.Z.; Yokoyama, W. Fabrication of curcumin-loaded bovine serum albumin (BSA)-dextran nanopar-ticles and the cellular antioxidant activity. *Food Chem.* **2018**, *239*, 1210–1218. [CrossRef] [PubMed]
34. Zhang, X.; Wei, Z.; Wang, X.; Wang, Y.; Tang, Q.; Huang, Q.; Xue, C. Fabrication and characterization of core-shell glia-din/tremella polysaccharide nanoparticles for curcumin delivery: Encapsulation efficiency, physicochemical stability and bio-accessibility. *Curr. Res. Food Sci.* **2022**, *5*, 288–297. [CrossRef]
35. Hu, H.; Yong, H.; Yao, X.; Yun, D.; Huang, J.; Liu, J. Highly efficient synthesis and characterization of starch aldehyde-catechin conjugate with potent antioxidant activity. *Int. J. Biol. Macromol.* **2021**, *173*, 13–25. [CrossRef] [PubMed]
36. Fan, Y.; Liu, Y.; Gao, L.; Zhang, Y.; Yi, J. Improved chemical stability and cellular antioxidant activity of resveratrol in zein nanoparticle with bovine serum albumin-caffeic acid conjugate. *Food Chem.* **2018**, *261*, 283–291. [CrossRef] [PubMed]
37. Su, J.; Guo, Q.; Chen, Y.; Mao, L.; Gao, Y.; Yuan, F. Utilization of β-lactoglobulin- (−)-Epigallocatechin- 3-gallate(EGCG) composite colloidal nanoparticles as stabilizers for lutein pickering emulsion. *Food Hydrocoll.* **2020**, *98*, 105293. [CrossRef]
38. Chen, W.; Lv, R.; Muhammad, A.I.; Guo, M.; Ding, T.; Ye, X.; Liu, D. Fabrication of (−)-epigallocatechin-3-gallate carrier based on glycosylated whey protein isolate obtained by ultrasound Maillard reaction. *Ultrason. Sonochem.* **2019**, *58*, 104678. [CrossRef]
39. Gómez-Mascaraque, L.G.; Soler, C.; Lopez-Rubio, A. Stability and bioaccessibility of EGCG within edible micro-hydrogels. Chitosan vs. gelatin, a comparative study. *Food Hydrocoll.* **2016**, *61*, 128–138. [CrossRef]
40. Wang, Y.-Q.; Zhuang, G.; Li, S.-J. Multiple on-line screening and identification methods for hydroxyl radical scavengers in Yudanshen. *J. Pharm. Biomed. Anal.* **2018**, *156*, 278–283. [CrossRef]
41. Tan, C.; Xie, J.; Zhang, X.; Cai, J.; Xia, S. Polysaccharide-based nanoparticles by chitosan and gum arabic polyelectrolyte complexation as carriers for curcumin. *Food Hydrocoll.* **2016**, *57*, 236–245. [CrossRef]
42. Li, F.; Jin, H.; Xiao, J.; Yin, X.; Liu, X.; Li, D.; Huang, Q. The simultaneous loading of catechin and quercetin on chitosan-based nanoparticles as effective antioxidant and antibacterial agent. *Food Res. Int.* **2018**, *111*, 351–360. [CrossRef]
43. Li, J.; Chen, Z. Fabrication of heat-treated soybean protein isolate-EGCG complex nanoparticle as a functional carrier for curcumin. *LWT* **2022**, *159*, 113059. [CrossRef]
44. Kaith, B.S.; Sharma, K.; Kumar, V.; Kalia, S.; Swart, H.C. Fabrication and characterization of gum ghatti-polymethacrylic acid based electrically conductive hydrogels. *Synth. Met.* **2014**, *187*, 61–67. [CrossRef]
45. Solanki, R.; Patel, K.; Patel, S. Bovine Serum Albumin Nanoparticles for the Efficient Delivery of Berberine: Preparation, Characterization and In vitro biological studies. *Colloids Surf. A Physicochem. Eng. Asp.* **2021**, *608*, 125501. [CrossRef]

46. Dionísio, M.; Cordeiro, C.; Remuñán-López, C.; Seijo, B.; Rosa da Costa, A.M.; Grenha, A. Pullulan-based nanoparticles as carriers for transmucosal protein delivery. *Eur. J. Pharm. Sci.* **2013**, *50*, 102–113. [CrossRef]

47. Bera, H.; Abosheasha, M.A.; Ito, Y.; Ueda, M. Hypoxia-responsive pullulan-based nanoparticles as erlotinib carriers. *Int. J. Biol. Macromol.* **2021**, *191*, 764–774. [CrossRef]

48. Das, R.P.; Singh, B.G.; Kunwar, A.; Ramani, M.V.; Subbaraju, G.V.; Hassan, P.A.; Priyadarsini, K.I. Tuning the binding, release and cytotoxicity of hydrophobic drug by Bovine Serum Albumin nanoparticles: Influence of particle size. *Colloids Surf. B Biointerfaces* **2017**, *158*, 682–688. [CrossRef]

Article

Bacillus coagulans Alleviates Intestinal Damage Induced by TiO₂ Nanoparticles in Mice on a High-Fat Diet

Qingying Shi [1], Chen Yang [1], Bingjie Zhang [1], Dongxiao Chen [1], Fuping Lu [1,2] and Huabing Zhao [1,2,*]

[1]   College of Biotechnology, Tianjin University of Science and Technology, 9 TEDA 13th Street, Tianjin 300457, China
[2]   Key Laboratory of Industrial Fermentation Microbiology, Ministry of Education, Tianjin University of Science and Technology, Tianjin 300450, China
*   Correspondence: zhaohuabing@tust.edu.cn

Abstract: Titanium dioxide nanoparticles (TiO₂ NPs) are generally added in considerable amounts to food as a food additive. Oral exposure to TiO₂ NPs could induce intestinal damage, especially in obese individuals with a high-fat diet. The probiotic *Bacillus coagulans* (*B. coagulans*) exhibits good resistance in the gastrointestinal system and is beneficial to intestinal health. In this study, *B. coagulans* was used to treat intestinal damage caused by TiO₂ NPs in high-fat-diet mice via two intervention methods: administration of TiO₂ NPs and *B. coagulans* simultaneously and administration of TiO₂ NPs followed by that of *B. coagulans*. The intervention with *B. coagulans* was found to reduce the inflammatory response and oxidative stress. A 16S rDNA sequencing analysis revealed that *B. coagulans* had increased the diversity of gut microbiota and optimized the composition of gut microbiota. Fecal metabolomics analysis indicated that *B. coagulans* had restored the homeostasis of sphingolipids and amino acid metabolism. The intervention strategy of administering TiO₂ NPs followed by *B. coagulans* was found to be more effective. In conclusion, *B. coagulans* could alleviate intestinal damage induced by TiO₂ NPs in high-fat-diet mice TiO₂ *B. coagulans*. Our results suggest a new avenue for interventions against intestinal damage induced by TiO₂ NPs.

Keywords: TiO₂ NPs; *Bacillus coagulans*; intestinal microorganisms; metabolism; high-fat diet

Citation: Shi, Q.; Yang, C.; Zhang, B.; Chen, D.; Lu, F.; Zhao, H. *Bacillus coagulans* Alleviates Intestinal Damage Induced by TiO₂ Nanoparticles in Mice on a High-Fat Diet. *Foods* **2022**, *11*, 3368. https://doi.org/10.3390/foods11213368

Academic Editor: Evandro Leite de Souza

Received: 13 September 2022
Accepted: 24 October 2022
Published: 26 October 2022

Publisher's Note: MDPI stays neutral with regard to jurisdictional claims in published maps and institutional affiliations.

1. Introduction

Titanium dioxide nanoparticles (TiO₂ NPs) are promising nanomaterials that have been widely used in a variety of fields, such as the cosmetic industry, the field of medicine, and the food industry [1]. TiO₂ NPs are generally added in considerable amounts to food, as colorants to improve the sensory properties of food and as food preservatives due to their antibacterial properties [2]. For example, sweets and candies contain high levels of TiO₂ NPs, even as much as 2.5 mg of titanium per g of food [1,3]. A vital way for people to consume TiO₂ NPs is orally. It is estimated that the dietary consumption of TiO₂ NPs was about 2–3 mg/kg/day for children and 1 mg/kg/day for adults in the United Kingdom (UK) [4]. It is worth noting that the Food and Drug Administration (FDA) allows TiO₂ NPs to be used as a food additive in the United States, but its content cannot exceed 1% of the total quantity of the product [5]. In addition, in 2021, the European Food Safety Agency (EFSA) issued a safety assessment of TiO₂ NPs, indicating that TiO₂ NPs are no longer considered safe as a food additive [6].

Numerous studies have documented intestinal damage, as, after ingestion, TiO₂ NPs stay for extended periods in the intestinal tract and interact with intestinal microorganisms and intestinal epithelial cells [7–10]. Gut microbiota is an ecosystem with complex interactions, which plays a vital role in human health. The intestinal epithelial barrier can protect the body from commensal bacteria, pathogens, and foreign particles. It is believed that oxidative stress plays a vital role in TiO₂ NP toxicityTiO₂, and the increase in the oxidative stress level has been shown to lead to disturbed serum biochemical parameters and elevated inflammatory factor levels [11,12]. Ruiz et al. indicated that TiO_2 NPs aggravate colitis by increasing the production of reactive oxygen species (ROS) and inflammasomes [13]. Moreover, our previous study presented that exposure to TiO_2 NPs affects the intestinal microecology in mice: for example, gut microbiota diversity and composition, short-chain fatty acid (SCFA) production, inflammatory response levels, and gut-associated metabolism [10].

Some high-fat foodsTiO_2, such as chocolates, puffed foods, jams, and candies, contain large amounts of TiO_2 NPs. Obese people who consume large amounts of high-fat foods can be exposed to more TiO_2 NPs, which may cause more pronounced health effects. Moreover, obese populations have been shown to be more vulnerable to the potentially harmful effects of TiO_2 NPs. Cao et al. found that the effects of TiO_2 NPs, such as intestinal microbial disorders, decreased levels of SCFAs, abnormal changes in cytokines, and negative histopathology, were greater on obese mice on a high-fat diet compared to non-obese mice [14]. As per one study, about 1.9 billion adults worldwide are overweight, and more than 600 million people are obese [15]. Considering that obesity is still increasing globally and has become a major socio-economic burden [16], the impact of TiO_2 NPs on obese individuals deserves attention, and it is important to actively seek corresponding preventive or therapeutic intervention programs.

Bacillus coagulans (*B. coagulans*), with good heat resistance, strong resistance to acids, and a high survival rate after reaching the gastrointestinal system, has been reported as safe by the FDA and the EFSA and included in the Generally Recognized As Safe (GRAS) and Qualified Presumption of Safety (QPS) lists [17]. Microbial preparations made from *B. coagulans* have been widely used in food, medicine, animal husbandry, etc., enjoying broad application prospects. B.Coagulans MTCC 5856, a kind of *Bacillus coagulans*, has been found to exhibit excellent resistance to gastric acid and can withstand high temperatures of up to 90 °C [18]. Tanvi et al. [19] used *B. coagulans* MTCC 5856 to intervene in mice with inflammatory bowel disease (IBD) and found that *B. coagulans* MTCC 5856 has the ability to maintain intestinal epithelial integrity, promote SCFA production, and reduce colonic inflammation.

Here, taking into consideration its good properties, *Bacillus coagulans* was employed as a probiotic to treat intestinal damage caused by TiO_2 NPs. A mouse model of obesity induced by a high-fat diet was established in this study. These obese mice were fed with TiO_2 NPs daily for eight weeks. Two kinds of intervention strategies with *Bacillus coagulans* were conducted to explore both the regulatory effect of *Bacillus coagulans* and the influence of intervention methods on its effectiveness. This study will provide new insights into the prevention of the negative effects of TiO_2 NPs on people with chronic diseases.

2. Materials and Methods

2.1. Animals and Treatments

Six-week-old C57BL/6J mice (male, weight 20–24 g) were purchased from SPF (Beijing, China) Biotechnology Co., Ltd. All animal experiments were approved by the Animal Ethics Committee of JAK BIO company (JKX-2106-01). The animals were kept in a temperature-controlled room (25 °C, 45% humidity) exposed to a 12 h/12 h dark/light cycle and had free access to water and food. After a 1-week adaptation period, the mice were randomly divided into a normal-diet-fed group (5 kcal%) and a high-fat-diet-fed group (60 kcal%). The high-fat-diet mice were further divided into four subgroups ($n = 5$). One subgroup was given only a high-fat diet for eight weeks, one subgroup was subjected to a high-fat diet containing TiO_2 NPs for eight weeks, one subgroup was fed a high-fat diet containing TiO_2 NPs and *B. coagulans* MTCC 5856 simultaneously for eight weeks, and one subgroup was given a high-fat diet containing TiO_2 NPs for four weeks followed by *B. coagulans* MTCC 5856 for four weeks. Table 1 presents the specific groupings. TiO_2 NPs equivalent to 0.2% of the body weight of the mice were fed incorporated in a high-fat diet, a dose corresponding to one-fifth of the allowable maximum daily intake prescribed by the FDA and 50 times the estimated average daily human intake [14]. The *B. coagulans* MTCC 5856 was administered as a gavage daily with 10^9 CFUs suspended in 1 mL of the medium to the mice of the treatment groups, and an equal volume of normal saline was gavaged daily to the mice of the control group, and the HFD group. All the mice were weighed weekly, and the colorectal lengths of the mice were also measured. The mice were sacrificed by cervical dislocation after eight weeks, and the appropriate tissues were harvested.

Table 1. Animal groups and treatments.

Groups, $n = 10$	Treatments *	Duration
Control	Normal diet	8 weeks
HFD	High-fat diet	8 weeks
HFD + NPs	High-fat diet + TiO$_2$ NPs	8 weeks
HFD + NPs + BC	High-fat diet + TiO$_2$ NPs + *B. coagulans*	8 weeks
HFD + NPs&BC	High-fat diet + TiO$_2$ NPs, followed by a high-fat diet + *B. coagulans*	High-fat diet + TiO$_2$ NPs for 4 weeks, followed by a high-fat diet + *B. coagulans* MTCC 5856 for 4 weeks

* The daily dose was equivalent to 0.2% of the body weight of the mice for TiO$_2$ NPs and 10^9 CFUs for *B. coagulans* MTCC 5856.

The probiotic bacterium *B. coagulans* MTCC 5856 used in this study is a patented strain of Sabinsa Corporation/Sami Labs Limited and deposited to Microbial Type Culture Collection and Gene Bank (MTCC), Chandigarh, India. *B. coagulans* MTCC 5856 was manufactured by Sami Labs Limited as per a proprietary method in a good manufacturing practices (GMP) facility in Bangalore, India. *B. coagulans* MTCC 5856 spores were administered as a gavage daily with 10^9 CFU suspended in 1 mL of medium to the mice in this study, a dose corresponding to the safe and tolerable level for human and effective in diarrhea-predominant irritable bowel syndrome patients according to previous research [20–22]. A viable spore count of *B. coagulans* MTCC 5856 was determined as per the method described previously [23]. Briefly, 1.0 g of *B. coagulans* MTCC 5856 was mixed in sterile saline (0.9% NaCl, *w/v*) and then incubated in a water bath for 30 min at 75 °C, followed by immediate cooling to below 45 °C. The suspension was further serially diluted in sterile saline, and the viable count was enumerated by plating on glucose yeast extract agar by pour plate method. The plates were incubated at 37 °C for 48–72 h. Analysis was performed twice in triplicate. The average means of viable spore counts were expressed in cfu/g.

2.2. Measurement of the Blood Glucose

The blood glucose in the tail vein blood of the mice was measured by a blood glucose meter (Sinocare Inc., Changsha, China).

2.3. Analysis of the Lipid Profiles in Serum

The blood samples were collected by eyeball extraction and then transferred into 1.5 mL microcentrifuge tubes. After the blood samples were centrifuged at 800× *g* for 15 min, the levels of total cholesterol (TC) and triglyceride (TG) in the upper layer serum were assessed using the TC assay kit and the TG assay kit (Nanjing Jiancheng Bioengineering Institute, Nanjing, China) according to the manufacturer's instructions.

2.4. Assessment of Pro-Inflammatory Cytokines

To determine the inflammatory responses, the expression levels of interleukin 1β (IL-1β), interleukin 6 (IL-6), and tumor necrosis factor (TNF-α) in serums and colon tissues of the mice were measured by using enzyme-linked immunosorbent assay (ELISA) kits (Shanghai Enzyme-linked Biotechnology Co., Ltd., Shanghai, China).

2.5. Evaluation of Antioxidant Enzyme

The oxidative stress response was investigated. The total antioxidant capacity assay kit (T-Aoc), and the superoxide dismutase (SOD) assay kit (Nanjing Jiancheng Bioengineering Institute, Nanjing, China) were used to detect the levels of T-Aoc and SOD in the sera and colon tissues of the mice according to the manufacturer's instructions.

2.6. Analysis of Gut Microbiota

Before the mice were sacrificed, their individual fecal samples were collected and placed in 1.5 mL sterilized tubes. Then, all the samples were snap-frozen on dry ice and stored at $-80\,^{\circ}\text{C}$. The fecal genomic DNA was extracted using a fecal genomic DNA extraction kit (Tiangen Biotech Co., Ltd., Beijing, China). The Illumina HiSeq platform (Nuohe Zhiyuan Bio-Information Technology Co., Ltd., Tianjin, China) was used to analyze the 16S ribosomal RNA genes (16S rRNA) in the fecal samples. Reactions were conducted in triplicate, and the V3-V4 region of genomic DNA was amplified using the specific primers 341F (5′-ACTCCTACGGGAGGCAGCAG-3′) and 806R (5′-GGACTACHVGGGTWTCTAAT-3′). The obtained data were analyzed with the help of QIME (Version 1.8.0) and R (Version 4.0.5) software packages.

2.7. Quantification of Short-Chain Fatty Acids

In the cecal contents of mice, the SCFAs, including acetic acid (AA), propionic acid (PA), butyric acid (BA), valeric acid (VA), and isovaleric acid (IVA), were detected by gas chromatography-mass spectrometry (GC–MS, Agilent, Santa Clara, CA, USA). Briefly, cecal contents (0.1 g) were suspended in 0.1 mL of 20% phosphoric acid solution, homogenized adequately for 2 min by vortex, and centrifuged for 10 min at $14,500\times g$. After centrifugation, the supernatants were extracted with 500 µL of ethyl acetate and centrifuged for 10 min at $14,500\times g$. 4-Methylvaleric acid at a final concentration of 500 µM was added as the internal standard. The coefficient of determination for a standard curve, which included five concentrations, 0.05, 0.10, 0.15, 0.20, and 0.25 µL/mL, was greater than 0.99. The key parameters for GC–MS analysis are shown in Table S1 in Supplementary Information. The Agilent Mass Hunter software was used to process the data.

2.8. Metabolomic Analysis

Approximately 20 mg of each fecal sample was weighed and added to 400 µL of pre-cooled $\text{MeOH}/\text{H}_2\text{O}$ (8/2, *v/v*) buffer. The samples were subjected to ultrasonic for 10 min and allowed to stand for 30 min at $-20\,^{\circ}\text{C}$. Then, after centrifugation at $12,400\times g$ for 10 min at $4\,^{\circ}\text{C}$, 300 µL of the supernatants were transferred into new tubes. After further centrifugation at $12,400\times g$ for 3 min at $4\,^{\circ}\text{C}$, 200 µL of the supernatants were placed into liner pipes as the test solutions.

The specific detection information is presented in Tables S2 and S3 in Supplementary Information. The obtained raw data were converted into mzML format by ProteoWizard software; and the XCMS program was used for peak extraction, alignment, and retention time correction. The peak areas were corrected using the SVR method, and peaks with deletion rates > 50% were filtered from each group of samples. The metabolites were identified by matching the information with the metabolic database. Further statistical analysis was performed by MetaboAnalyst.

2.9. Statistical Analysis

Statistical analysis was performed using the SPSS statistical software (version 26.0). Multiple comparisons were made using a one-way analysis of variance (ANOVA). A nonparametric Kruskal–Wallis test was applied to analyze the statistical differences for the data that had failed the normality test. The web-based platform MetaboAnalyst 5.0 (https://www.metaboanalyst.ca/ (accessed on 21 March 2022)) was used to perform metabolic pathway analysis of the differentially expressed metabolites. All the samples were analyzed in the cation and negative ion mode. Normalized peak areas were used for quantification, and their values were log-transformed before statistical analyses. The data were pre-processed by Pareto scaling. PLS–DA, heatmap, and KEGG pathway analyses were conducted after normalizing the sample (sum normalization) and scaling the data (auto-scaling). Data were expressed as mean $\pm$ standard error of the mean (SEM). Differences were noted as significant at $p < 0.05$. The groups marked by different letters have significant differences, and the groups marked by the same letters have no significant differences.

3. Results

3.1. Effects on Body Weight and Colorectal Length

Figure 1A shows the changes in the body weight in each group. The net weight gain of the mice in the HFD group was 46.8% higher than that in the control group ($p < 0.05$), successfully establishing the high-fat-diet-induced obesity mouse model. The weight gain in the mice in the HFD + NPs + BC group was restored to a level similar to that of the control group ($p > 0.05$). However, the weight gain in the mice in the HFD + NPs&BC group was significant (an increase of 40.4%) compared to that in the control group.

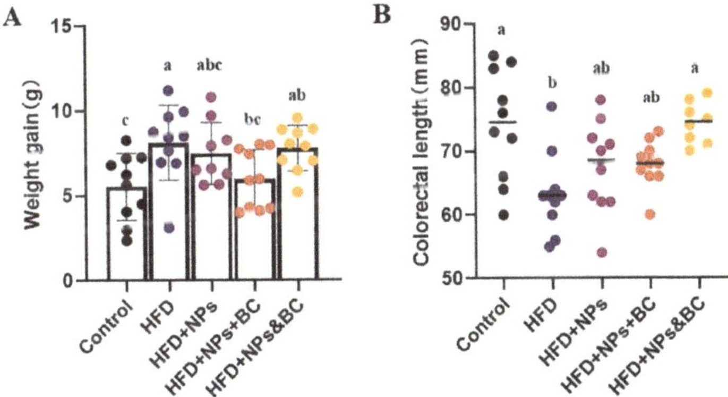

Figure 1. The effects on the body weight and colorectal length of mice in different groups. (**A**) The net weight gain before and after treatment in each group. (**B**) The colorectal length in different treatments. The different letters represent significant differences among groups ($p < 0.05$).

Colorectal length is an important pathological indicator for analyzing colorectal inflammation. As shown in Figure 1B, the colorectal lengths of the mice in each group were also measured. Compared with the control group, the decreased colorectal lengths of the mice in the HFD group and the HFD + NPs group suggest that mice fed with a high-fat diet and TiO_2 NPs could have reduced colorectal length. However, the colorectal lengths of the mice in the HFD + NPs&BC group were significantly more than those in the HFD group and the HFD + NPs group and recovered to the level of the control group after the mice received intervention with *B. coagulans*, demonstrating that *B. coagulans* can reduce the colorectal contraction induced by a high-fat diet and TiO_2 NPs.

3.2. Effects on Inflammatory Response and Oxidative Stress

As shown in Figure 2, tThe IL-1β, IL-6, and TNF-α levels were detected in serum and colon tissues (Figure 2). Our results showed that IL-1β levels in the serum significantly increased in the HFD + NPs group compared with those in the control and HFD groups, while *B. coagulans* administration significantly suppressed IL-1β production in the HFD + NPs + BC group and the HFD + NPs&BC group ($p < 0.05$) (Figure 2A). Similarly, the inhibiting effect of *B. coagulans* on IL-1β expression was observed in colon tissues (Figure 2B). Serum IL-6 levels did not differ significantly among the groups (Figure 2C), but differences in colon tissues were seen among the groups (Figure 2D). The levels of IL-6 significantly increased in the HFD group and the HFD + NPs group compared with the control group. The levels of IL-6 were, respectively, 26.0% and 41.3% lower in the HFD + NPs + BC group and the HFD + NPs&BC group than those in the HFD + NPs group. In addition, as shown in Figure 2F, the TNF-α levels in colon tissues were significantly higher in the HFD and HFD + NPs groups than in the control group. However, after *B. coagulans* treatment, the levels of TNF-α in the HFD + NPs&BC group were significantly lower than those in the HFD + NPs group ($p < 0.05$).

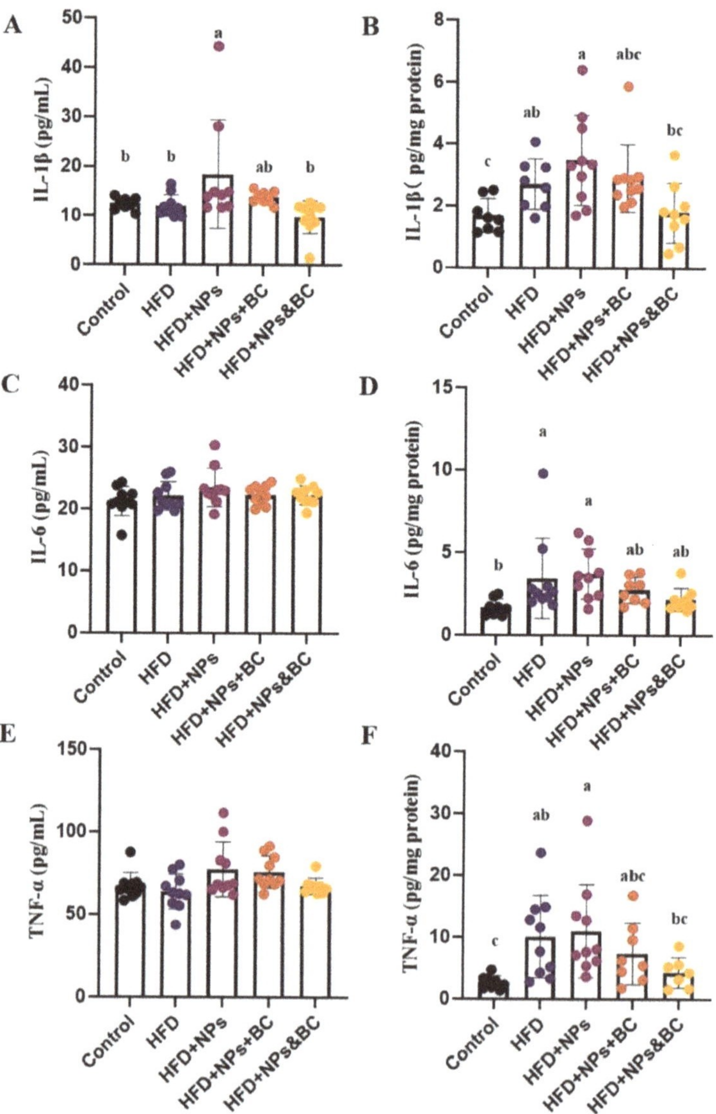

Figure 2. The effects of pro-inflammatory cytokine levels on serum and colon tissues in mice. (**A**) The IL-1β levels in serum, (**B**) the IL-1β levels in colon tissues, (**C**) the IL-6 levels in serum, (**D**) the IL-6 levels in colon tissues, (**E**) the TNF-α levels in serum, and (**F**) the TNF-α levels in colon tissues. The different letters represent significant differences among groups ($p < 0.05$).

As a typical antioxidant metalloenzyme, SOD is usually considered to be the primary enzyme to defend against ROS-mediated damage, helping balance oxidation and antioxidant activities [24–26]. Our results indicated that the sensitivity of SOD in serum is lower than that in colon tissues, but the SOD activities in the HFD group and the HFD + NPs group still decrease (Figure 3A). The SOD activities in colon tissues significantly reduced in the HFD group and the HFD + NPs group compared with those in the control group, whereas *B. coagulans* intervention increased the SOD activities to a level similar to that of the control group, as shown by the SOD activities of the HFD + NPs + BC group and the HFD + NPs + BC group. These results show that the antioxidant capacity of the intestinal tract was seriously impaired by a high-fat diet and the addition of TiO_2 NPs but the intervention with *B. coagulans* could enhance antioxidant capacity and reduce oxidative damage.

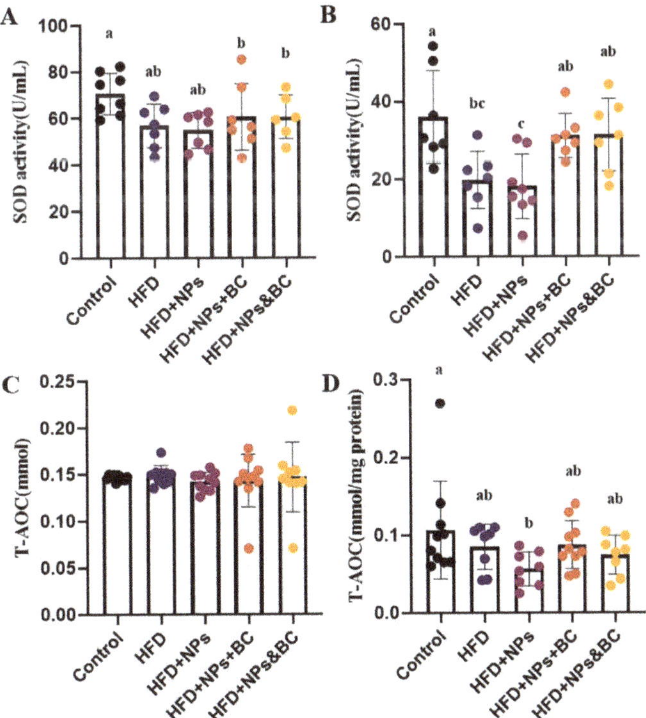

Figure 3. The effects on the antioxidant ability in serum and colon tissues of mice. (**A**) The SOD activities in serum, (**B**) the SOD activities in colon tissues, (**C**) the T-Aoc levels in serum, and (**D**) the T-Aoc levels in colon tissues. The different letters represent significant differences among groups ($p < 0.05$).

T-Aoc is one of the indicators for evaluating the total antioxidant level, reflecting the non-enzymatic antioxidant defense system [27]. As shown in Figure 3C, there was no significant difference in the serum T-Aoc levels between the groups. The T-Aoc levels in the colon tissues (Figure 3D) of the HFD + NPs group significantly decreased compared with those of the control group. However, after supplementation with *B. coagulans*, the T-Aoc levels showed only a slight upward trend, indicating the limited capacity of *B. coagulans* to inhibit the reduction of the total antioxidant level caused by TiO_2 NPs.

3.3. Effects on the Diversity of Gut Microbiota

After 16S rDNA gene sequencing and quality filtering, a total of 3470 amplicon sequence variants (ASVs) were identified in the five groups. According to the classification of ASVs in each group and the identification of taxonomics (Table 2), in general, the lower the taxon and the more the ASVs seem to be detected. In the control group and the HFD group, the ASVs were relatively abundant, but they decreased significantly in the HFD + NPs group. In the HFD + NPs + BC group, the amount of ASVs had a slight rebound. However, in the HFD + NPs&BC group, the amount of ASVs increased significantly. Generally speaking, *B. coagulans* supplementation did not resist the simultaneous effects of obesity and TiO$_2$ NPs but attenuated the reduction in the intestinal microbial abundance after cessation of TiO$_2$ NPs intake.

Table 2. Classification of ASVs in each group and the identification of taxonomics.

Group	Domain	Phylum	Class	Order	Family	Genus	Species
Control	1	22	60	123	171	244	93
HFD	1	23	54	117	173	275	111
HFD + NPs	1	13	17	37	55	138	81
HFD + NPs + BC	1	13	19	41	56	145	80
HFD + NPs&BC	1	13	18	43	72	163	109

Figure 4A–C presents the alpha diversity reflected by Chao1, Shannon, and Simpson indexes. It can be seen that the levels of alpha diversity did not differ significantly among groups, except for a slight increase in the alpha diversity in the HFD + NPs&BC group.

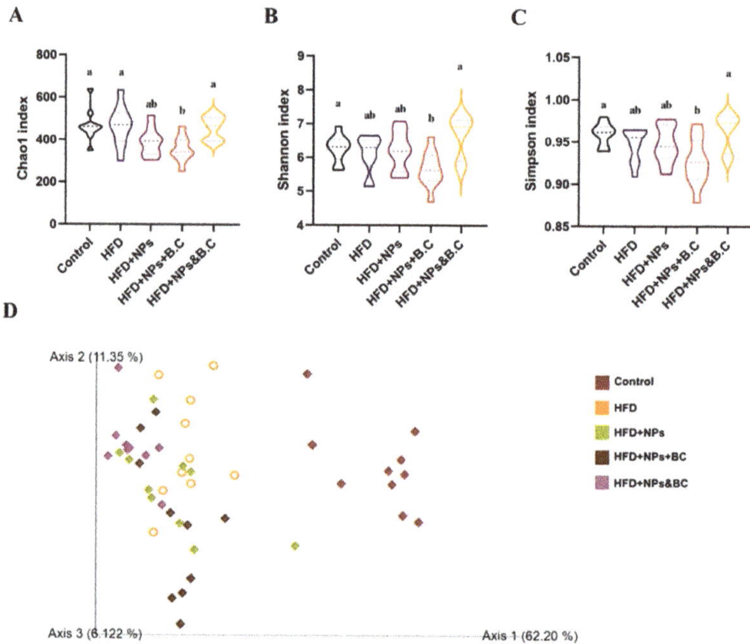

Figure 4. Alpha and beta diversity of gut microbiota. The (**A**) Chao1 index, (**B**) Shannon index, (**C**) Simpson index, and (**D**) PCA score plot of each group. The different letters represent significant differences among groups (*p* < 0.05).

To further identify the effect of *B. coagulans* administration on the composition of gut microbiota in mice treated with TiO$_2$ NPs, a principal component analysis (PCA) was performed. As seen in Figure 4D, there was a clear separation between the control group and other experimental groups, indicating that the gut microbiota changed greatly after the mice were fed a high-fat diet.

3.4. Effects on the Composition of Gut Microbiota

Figure 5 displays the differences in the distribution of bacterial groups at different taxonomic levels. It is obvious that the lower the taxonomic level, the more prominent the regulatory effect of *B. coagulans* on intestinal microbes. To this end, we further screened and analyzed microorganisms at lower taxonomic levels.

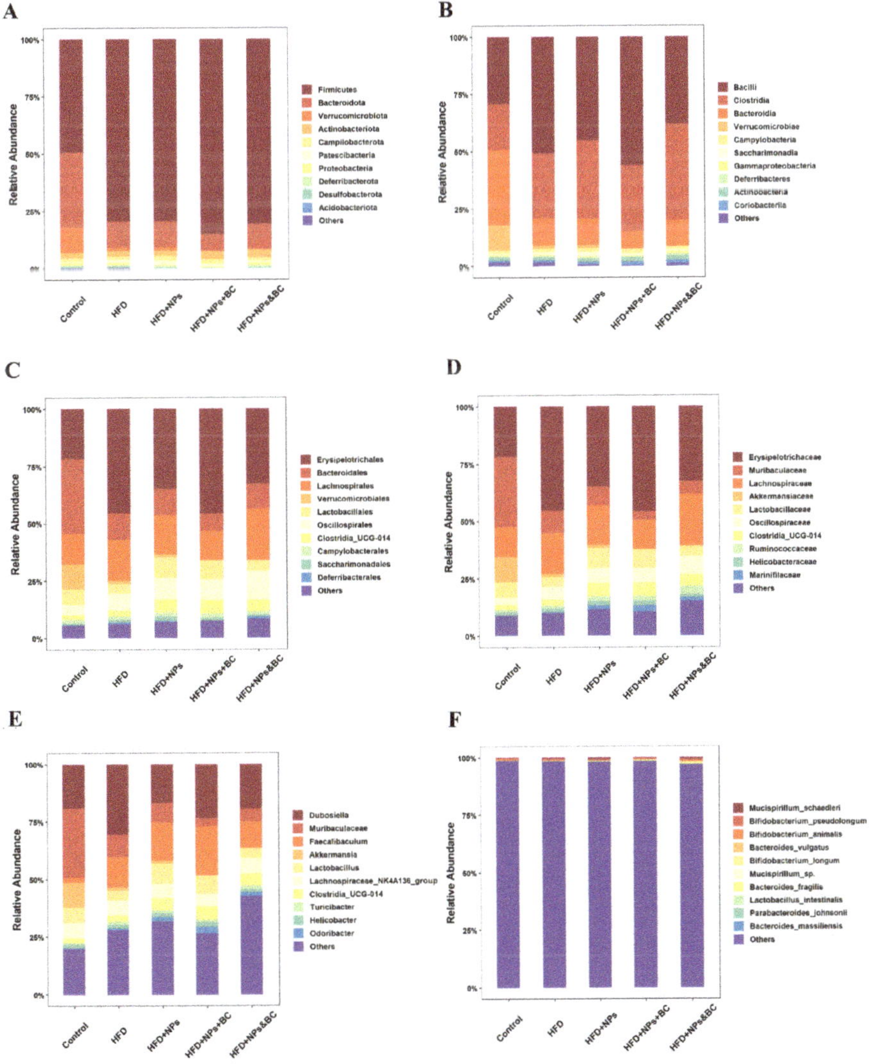

Figure 5. Distribution of bacterial groups at different taxonomic levels. (**A**) Phylum level, (**B**) class level, (**C**) order level, (**D**) family level, (**E**) genus level, and (**F**) species level.

At the phylum level, the dominant intestinal bacteria in mice in the five groups were Firmicutes and Bacteroidetes. The proportion of the two dominant bacteria in the control group was approximately 82.21%, and the proportions in the HFD, HFD + NPs, HFD + NPs + BC, and HFD + NPS&BC groups were 91.12%, 91.02%, 92.54%, and 91.77%, respectively (Figure 6A–C). The proportion of the dominant bacteria Firmicutes significantly increased in the other four groups compared with the control group. The ratio of Firmicutes and Bacteroidetes (F/B) altered when the relative quantity of Firmicutes and Bacteroidetes fluctuated. Compared with the HFD + NPs group, the F/B ratio in the HFD + NPS&BC group showed a slight downward trend.

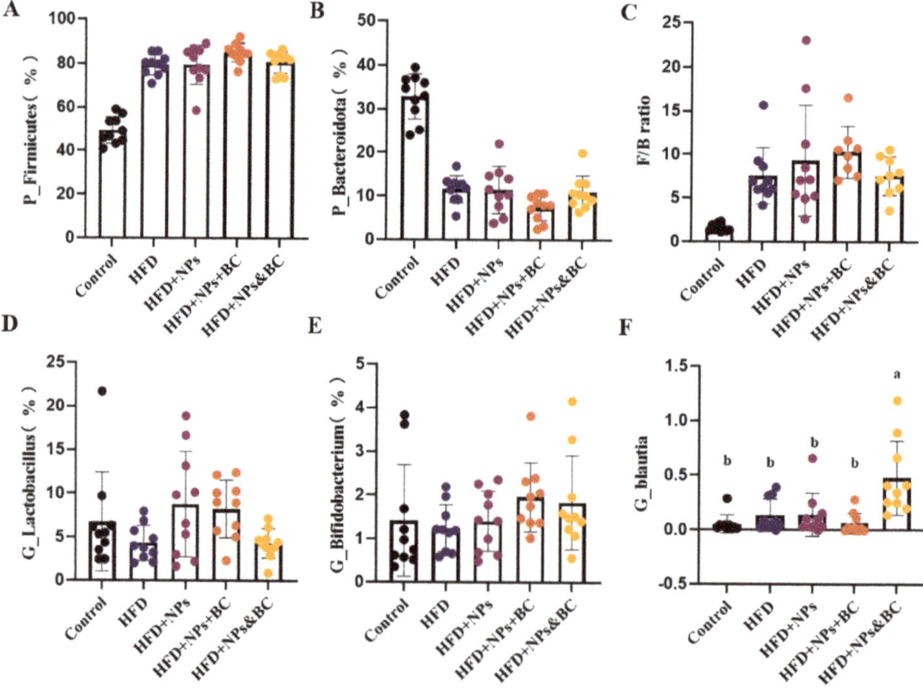

Figure 6. The relative abundance of microbiota at the phylum level (**A–C**) and the genus level (**D–F**). The different letters represent significant differences among groups ($p < 0.05$).

At the genus level, we screened out some characteristic bacteria (Figure 6D–F). Lactobacillus showed a downward trend after *B. coagulans* intervention, especially in the HFD + NPS&BC group. We also found a slight upward trend in the relative abundance of Bifidobacterium following *B. coagulans* intervention in the HFD + NPs + BC group and the HFD + NPS&BC group. Moreover, the relative abundance of Blautia in the HFD + NPS&BC group significantly increased, but there was no significant change in the HFD + NPs + BC group, suggesting that Blautia may be the key strain that cooperates with *B. coagulans*.

3.5. Effects on the Production of Short-Chain Fatty Acids

Short-chain fatty acids (SCFAs) are important microbial degradation metabolites and play a significant role in host health. The SCFA content in mouse feces was measured after eight weeks of different treatments. As shown in Figure 7, in the feces, there were significant differences ($p < 0.05$) in the three types of SCFAs: butyric acid, valeric acid, and isovaleric acid. Butyrate contributes a lot to colonic homeostasis and is the main energy source for colon cells [28]. In the HFD + NPs group, an inflammatory intestinal mucosa may have impaired butyrate metabolism, predisposing colon cells toward butyrate deficiency. Our results indicate that *B. coagulans* may improve cecal butyrate levels by repairing intestinal mucosal inflammation (Figure 7C). After *B. coagulans* intervention, there was a significant increase in butyrate levels in the HFD + NPs group compared to the control group. However, there were no significant changes in acetic acid and propionic acid after *B. coagulans* intervention ($p > 0.05$).

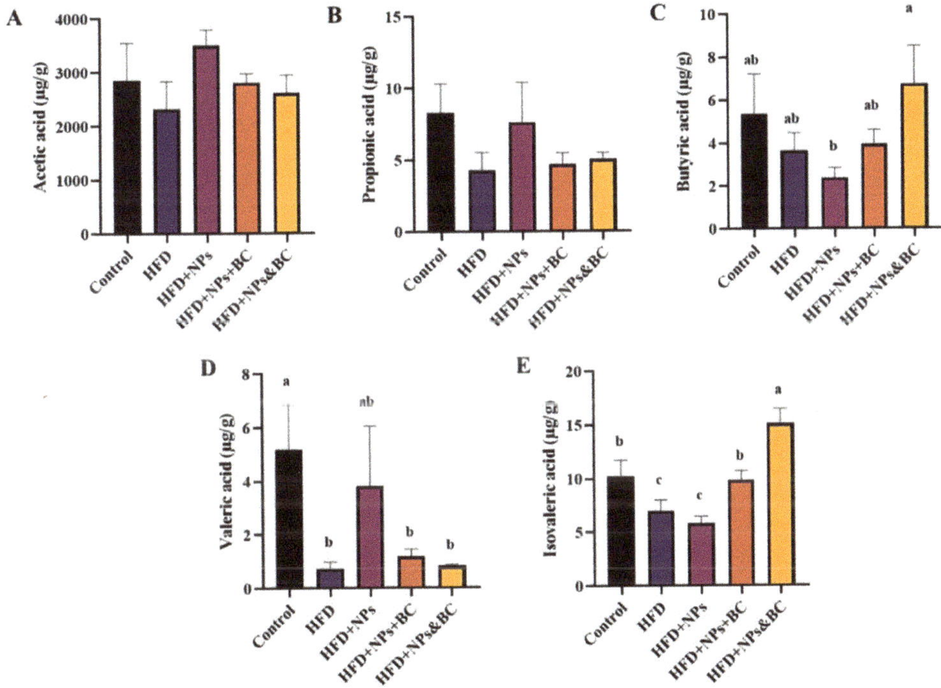

Figure 7. The levels of (**A**) acetic acid, (**B**) propionic acid, (**C**) butyric acid, (**D**) valeric acid, and (**E**) isovaleric acid. The different letters represent significant differences among groups ($p < 0.05$).

3.6. Effects on Fecal Metabolites

As shown in Figure 8A, the control group and the HFD + NPs group were completely separated from the other groups in the Sparse PLS discriminant analysis (sPLS–DA) score plots in the negative-ion mode. In addition, the HFD group was completely separated from the HFD + NPs group. This showed that the effect of TiO₂ NPs on the fecal metabolites under a high-fat diet is more profound. Further analysis of the two groups subjected to *B. coagulans* intervention pointed out that there was no obvious separation between the HFD group and the HFD + NPs + BC or the HFD + NPs&BC group, indicating that *B. coagulans* could not resist the effects of a high-fat diet in mice but the damage caused by TiO₂ NPs can be repaired by modulating fecal metabolites. Figure 8B displays the sPLS–DA scores in the positive-ion mode, which is basically consistent with the conclusions obtained in the negative-ion mode, but the distance between the HFD + NPs + BC group and the HFD + NPs group is relatively short, suggesting that the simultaneous intake of TiO₂ NPs and *B. coagulans* has a less regulatory effect on fecal metabolites. However, the HFD + NPs&BC group remained close to the HFD group, showing that the intervention strategy of the HFD + NPs&BC group was more effective.

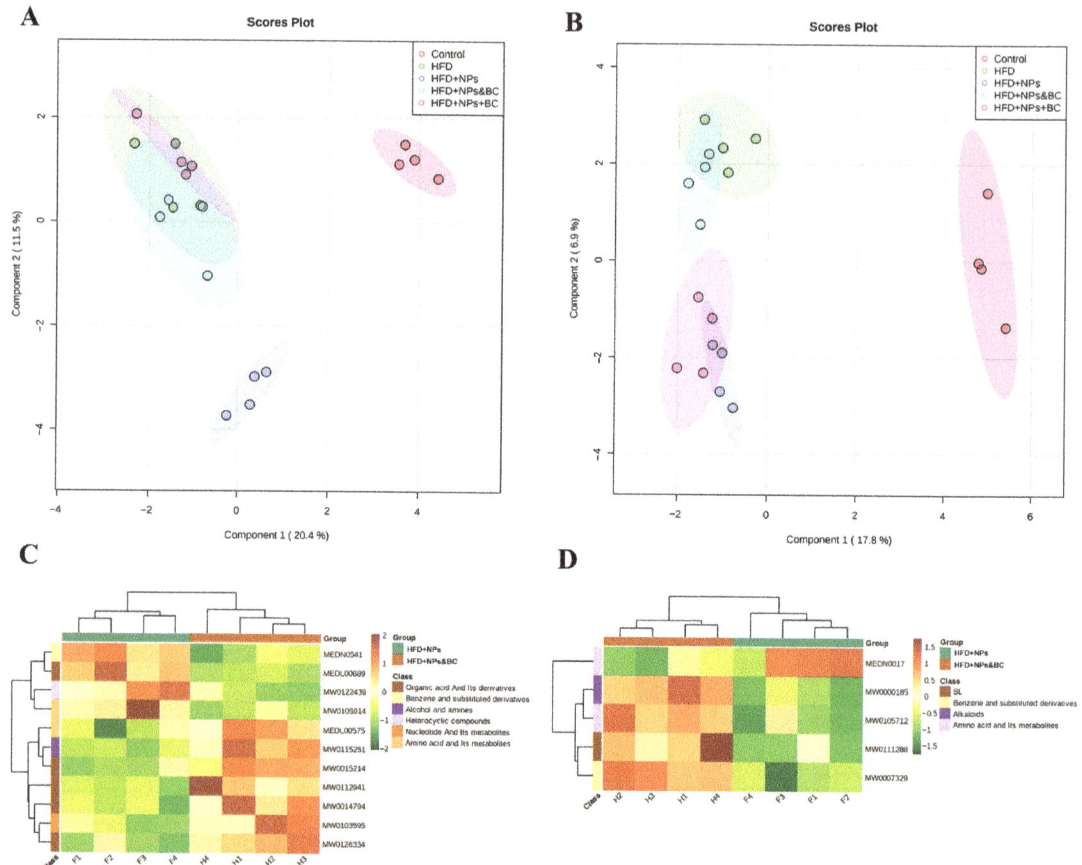

Figure 8. sPLS–DA score plots for each group in (**A**) the negative-ion mode and (**B**) the positive-ion mode. KEGG pathway analysis of differential metabolites for each group in (**C**) the negative-ion mode and (**D**) the positive-ion mode.

Furthermore, to explore the repair mechanism of *B. coagulans* in intestinal damage induced by TiO$_2$ NPs, we performed a KEGG cluster analysis of differential metabolites between the HFD group and the HFD + NPs&BC group (Figure 8C,D). The KEGG metabolic pathways containing at least five differential metabolites were selected, and the relative contents of all differential metabolites in these pathways were analyzed by cluster analysis to further study the metabolite changes in the HFD + NPs group and the HFD + NPs&BC group. The results showed that, in the fecal metabolites of the HFD + NPs group, the metabolic pathways of benzene and substituted derivatives, heterocyclic compounds, and sphingolipid were up-regulated, and the metabolic pathways of amino acid and its metabolites were down-regulated. In the HFD + NPs&BC group, the metabolic pathways of organic acid and its derivatives, nucleotide and its metabolites, amino acid and its metabolites, and alkaloids were up-regulated.

4. Discussion

There is increasing use of TiO$_2$ in daily life as a food additive and an antibacterial agent, for example, in food packaging, leading to concerns about its potential toxicity. More than 30% of the TiO$_2$ used as a food additive is at the nanoscale [3]. Obese people may consume more TiO$_2$ NPs than healthy people. Oral exposure of mice to TiO$_2$ NPs could lead to the development and exacerbation of inflammatory bowel disease in the mice by altering their intestinal barrier function and other pathways [29]. Previous studies have pointed out that oral exposure of obese mice to TiO$_2$ NPs can cause more adverse effects, such as stronger intestinal oxidative stress, increased levels of pro-inflammatory cytokines, and disturbance of intestinal microbes. However, there are few reports on intervention programs to reduce the intestinal damage caused by TiO$_2$ NPs in obese individuals. Therefore, it is of great importance to explore the impact of TiO$_2$ NPs on intestinal health from multiple perspectives and methods to alleviate this negative impact. Here, *B. coagulans* used in two intervention methods was investigated for its repair effect on TiO$_2$-NP-induced damage in mice fed a high-fat diet.

According to previous research, we established an obese mouse model and added 0.2% TiO$_2$ NPs to the high-fat diet to simulate the addition of TiO$_2$ NPs as a food additive in a daily high-fat diet. The results of mice weight showed that the simultaneous action of *B. coagulans* and TiO$_2$ NPs had a more positive regulatory effect on the body weight of mice. In addition, the colorectal lengths suggested that *B. coagulans* can prevent colorectal contraction caused by a high-fat diet and TiO$_2$ NPs. However, the results of blood glucose and blood lipids exhibited that *B. coagulans* did not reduce the levels of fasting blood glucose and blood lipids in mice (Figure S1). Therefore, we preliminarily concluded that the administration of *B. coagulans* can restore the body weight and colorectal length of mice to the normal level but does not address the imbalance of glucose and lipid metabolism caused by obesity.

When TiO$_2$ NPs pass through the digestive system and finally reach the intestine, they may enrich the inflammatory cells in the intestinal mucosa, which secrete a large number of inflammatory mediators, cytokines, toxins, and oxygen-free radicals [30]. Cytokines play an important role in the intestinal immune system [31,32]. As a part of the intestinal tract, the intestinal immune system can regulate the intestinal barrier function [33,34]. It is reported that TiO$_2$ NPs can aggravate intestinal inflammation by activating innate and adaptive immune responses [35]. Pro-inflammatory cytokines, such as IL-1β, IL-6, and TNF-α, serve to stimulate inflammatory responses and promote the occurrence of intestinal diseases [36,37]. In this study, the increased levels of IL-1β, IL-6, and TNF-α in colon tissues in the HFD group and the HFD + NPs group indicate the higher inflammation levels resulting from the exposure to TiO$_2$ NPs and the high-fat diet, which is consistent with previous studies [38–42]. However, *B. coagulans* intervention significantly reduced the inflammatory response, as shown by the reduced levels of pro-inflammatory cytokines in the HFD + NPs + BC group and the HFD + NPs&BC group. Inflammation and oxidative stress are closely related to pathophysiological processes, and one can be easily provoked by the other [43]. Inflammation is usually the initial response to tissue damage, with conditions leading to tissue damage that may be promoted by oxidative stress [44–47]. In fact, oxidative stress results from an imbalance between the production of pro-oxidants and their neutralization by antioxidants [48]. Here, the antioxidant ability was further evaluated, where the elevated SOD activities in colon tissues in the HFD + NPs + BC group and the HFD + NPs&BC group reaffirmed the protective effect of *B. coagulans* against intestinal damage. Moreover, the greater fluctuation of pro-inflammatory cytokines and oxidative stress levels in colon tissues t could be because intestinal tissue is more sensitive to inflammation than serum.

In addition, TiO$_2$ NPs have a certain antibacterial activity [49,50] and can promote microbial dysregulation [51], so gut microbes may be highly sensitive to them. Thus, we subsequently studied the changes in gut microbes. We found that *B. coagulans* could also ameliorate the reduction in intestinal microbial species resulting from exposure to TiO$_2$ NPs (Table 2) and modulate the composition of gut microbiota (Figures 5 and 6). Previously, our research group found that the relative abundance of Lactobacillus in the guts of mice exposed to TiO$_2$ NPs increased in a dose-dependent manner [10], which corresponded to other research, and the researchers suggested that the increase in Lactobacillus may be a key factor in the influence of TiO_2 NPs on gut microbiota [35]. Lactobacillus_gasseri was able to cause gangrene, leading to increased ROS production [52]. Notably, a reduction in Lactobacillus was observed in *B. coagulans* treatment. Blautia has been reported as a potential probiotic with beneficial metabolic activity for host health, protecting against inflammation and promoting SCFA activity [53]. The significant increase in the levels of Blautia in the HFD + NPs&BC group also supported the beneficial effect of *B. coagulans*. At the genus level, in addition to the top ten abundant microorganisms, we summed up the relative abundances of the remaining microorganisms (Figure S2). The significantly increased abundance of the HFD + NPs&BC group may be the reason for the previous increase in alpha diversity in this group.

Alterations in the gut microbiota may deliver signals through the gut and generate bacterial metabolites that change metabolism at different levels [35]. We performed fecal metabolomic analysis to understand the metabolism of gut microbes. Metabolites obtained in different modes were visualized using sPLS–DA. The results showed that the effectiveness of *B. coagulans* was influenced by the intervention methods and that the administration of TiO_2 NPs followed by *B. coagulans* (HFD + NPs&BC group) was more effective than the administration of TiO_2 NPs and *B. coagulans* simultaneously (HFD + NPs + BC group). In other words, for individuals whose intestinal health has already been impaired by TiO_2 NPs, it is more effective to use *B. coagulans* after stopping their exposure to TiO_2 NPs. To further explore the reasons for the better intervention effect in the HFD + NPs&BC group, we clustered the differential metabolites (Tables S4 and S5) between HFD + NPs and HFD + NPs&BC groups to further study the metabolic pathways. The metabolic pathways showed significant differences. Sphingolipid pathways were up-regulated, and the metabolic pathway of amino acid and its metabolites was down-regulated in the HFD + NPs group. However, amino acid and its metabolites were up-regulated in the HFD + NPs&BC group. Sphingolipid plays an important role in gastrointestinal immune homeostasis. Metabolites in this pathway are crucial in inflammatory signaling pathways, and some related metabolites can affect the integrity and function of the intestinal barrier and promote inflammation [54]. The amino acid metabolism pathway is the premise of various metabolic pathways, and amino acid metabolism disorders are related to the occurrence of diseases such as liver cirrhosis [55]. It can be seen that *B. coagulans* plays an important role in restoring the homeostasis of sphingolipids and amino acid metabolism.

According to our results, TiO_2 NPs affected gut health in mice on a high-fat diet, leading to, for example, oxidative stress, inflammatory responses, and disorders of gut microbiota and gut-associated metabolism. However, dietary intervention with probiotics was found to attenuate intestinal damage. *B. coagulans* could alleviate the intestinal inflammation and oxidative stress caused by TiO_2 NPs. In addition, the beneficial effects of *B. coagulans* were manifested as the modulation of gut microbiota and gut-associated metabolism. The regulation of gut microbiota was represented by decreased Lactobacillus and increased Bifidobacterium and Blautia. The opsonization of metabolic disorders was shown by the homeostasis of sphingolipids and amino acid metabolism. Notably, compared with the simultaneous intervention of TiO_2 NPs and *B. coagulans*, *B. coagulans* intervention following the exposure to TiO_2 NPs is more effective.

5. Conclusions

In this study, we focused on seeking preventive or therapeutic intervention programs to counteract the negative effects of TiO_2 NPs on the gut. Our results illustrated that *B. coagulans* could mitigate intestinal damage by decreasing the inflammatory response and oxidative stress levels and modulating intestinal microbial community structure and its metabolic profile, leading to a healthy tendency. Importantly, the current study pointed out that the intervention method used influenced *B. coagulans* effectiveness, as the administration of TiO_2 NPs followed by *B. coagulans* displayed a greater protective effect against intestinal injury. This research highlights the importance of dietary interventions in preventing or reducing the intestinal burden after dietary exposure to TiO_2 NPs, and more intervention modalities to restore intestinal health need to be investigated.

Supplementary Materials: The following supporting information can be downloaded at: https://www.mdpi.com/article/10.3390/foods11213368/s1. Table S1: Key parameters for GC-MS analysis; Table S2: The chromatographic column and mobile phase gradient program; Table S3: Mass spectrometry conditions in negative ion and positive ion modes; Table S4: Differential me-tabolites in the negative ion mode and its classification; Table S5: Differential metabolites in the positive ion mode and its classification; Figure S1: The levels of blood glucose, triglyceride and total cholesterol in different groups; Figure S2: The sum of the other microorganisms at the genus level.

Author Contributions: Conceptualization, H.Z.; data curation, C.Y., B.Z. and D.C.; formal analysis, H.Z.; methodology, Q.S. and C.Y.; project administration, H.Z.; resources, H.Z.; software, C.Y., B.Z. and D.C.; supervision, F.L. and H.Z.; visualization, Q.S. and C.Y.; writing—original draft, Q.S.; writing—review and editing, H.Z., All authors have read and agreed to the published version of the manuscript.

Funding: This work was supported by the Research and Development Plan for Key Fields in Guangdong Province (2019B020210001).

Institutional Review Board Statement: The animal study protocol was approved by the Institutional Review Board of JAK BIO company (JKX-2106-01).

Informed Consent Statement: Not applicable.

Data Availability Statement: The data presented in this study are available in this article.

Acknowledgments: Thanks a lot for Hongbin Wang's great contribution to data analysis and paper revision during the revision process.

Conflicts of Interest: The authors declare that they have no known competing financial interests or personal relationships that could have appeared to influence the work reported in this paper.

References

1. Baranowska-Wójcik, E.; Szwajgier, D.; Winiarska-Mieczan, A. A review of research on the impact of E171/TiO_2 NPs on the digestive tract. *J. Trace Elements Med. Biol.* **2022**, *72*. [CrossRef] [PubMed]
2. Baranowska-Wójcik, E. Factors Conditioning the Potential Effects TiO_2 NPs Exposure on Human Microbiota: A Mini-Review. *Biol. Trace Element Res.* **2021**, *199*, 4458–4465. [CrossRef] [PubMed]
3. Weir, A.; Westerhoff, P.; Fabricius, L.; Hristovski, K.; von Goetz, N. Titanium Dioxide Nanoparticles in Food and Personal Care Products. *Environ. Sci. Technol.* **2012**, *46*, 2242–2250. [CrossRef]
4. McClements, D.J.; Xiao, H.; Demokritou, P. Physicochemical and colloidal aspects of food matrix effects on gastrointestinal fate of ingested inorganic nanoparticles. *Adv. Colloid Interface Sci.* **2017**, *246*, 165–180. [CrossRef] [PubMed]
5. Heringa, M.B.; Geraets, L.; Van Eijkeren, J.C.H.; Vandebriel, R.J.; De Jong, W.H.; Oomen, A.G. Risk assessment of titanium dioxide nanoparticles via oral exposure, including toxicokinetic considerations. *Nanotoxicology* **2016**, *10*, 1515–1525. [CrossRef] [PubMed]
6. Younes, M.; Aquilina, G.; Castle, L.; Engel, K.-H.; Fowler, P.; Fernandez, M.J.F.; Fürst, P.; Gundert-Remy, U.; Gürtler, R.; Husoy, T.; et al. Safety assessment of titanium dioxide (E171) as a food additive. *EFSA J.* **2021**, *19*, e06585. [CrossRef]
7. Limage, R.; Tako, E.; Kolba, N.; Guo, Z.; García-Rodríguez, A.; Marques, C.N.H.; Mahler, G.J. TiO_2 Nanoparticles and Commensal Bacteria Alter Mucus Layer Thickness and Composition in a Gastrointestinal Tract Model. *Small* **2020**, *16*, e2000601. [CrossRef] [PubMed]
8. Powell, J.J.; Ainley, C.C.; Harvey, R.S.J.; Mason, I.M.; Kendall, M.D.; Sankey, E.A.; Dhillon, A.P.H.; Thompson, R.P. Characterisation of inorganic microparticles in pigment cells of human gut associated lymphoid tissue. *Gut* **1996**, *38*, 390–395. [CrossRef] [PubMed]
9. Brun, E.; Barreau, F.; Veronesi, G.; Fayard, B.; Sorieul, S.; Chanéac, C.; Carapito, C.; Rabilloud, T.; Mabondzo, A.; Herlin-Boime, N.; et al. Titanium dioxide nanoparticle impact and translocation through ex vivo, in vivo and in vitro gut epithelia. *Part. Fibre Toxicol.* **2014**, *11*, 13. [CrossRef]
10. Yang, C.; Tan, Y.; Li, F.; Wang, H.; Lin, Y.; Lu, F.; Zhao, H. Intestinal Microecology of Mice Exposed to TiO_2 Nanoparticles and Bisphenol A. *Foods* **2022**, *11*, 1696. [CrossRef] [PubMed]
11. Hassanein, K.M.A.; El-Amir, Y.O. Protective effects of thymoquinone and avenanthramides on titanium dioxide nanoparticles induced toxicity in Sprague-Dawley rats. *Pathol.-Res. Pract.* **2017**, *213*, 13–22. [CrossRef] [PubMed]
12. Jafari, A.; Rasmi, Y.; Hajaghazadeh, M.; Karimipour, M. Hepatoprotective effect of thymol against subchronic toxicity of titanium dioxide nanoparticles: Biochemical and histological evidences. *Environ. Toxicol. Pharmacol.* **2018**, *58*, 29–36. [CrossRef] [PubMed]

13. Ruiz, P.A.; Morón, B.; Becker, H.M.; Lang, S.; Atrott, K.; Spalinger, M.R.; Scharl, M.; Wojtal, K.A.; Fischbeck-Terhalle, A.; Frey-Wagner, I.; et al. Titanium dioxide nanoparticles exacerbate DSS-induced colitis: Role of the NLRP3 inflammasome. *Gut* **2017**, *66*, 1216–1224. [CrossRef] [PubMed]

14. Cao, X.; Han, Y.; Gu, M.; Du, H.; Song, M.; Zhu, X.; Ma, G.; Pan, C.; Wang, W.; Zhao, E.; et al. Foodborne Titanium Dioxide Nanoparticles Induce Stronger Adverse Effects in Obese Mice than Non-Obese Mice: Gut Microbiota Dysbiosis, Colonic Inflammation, and Proteome Alterations. *Small* **2020**, *16*, e2001858. [CrossRef] [PubMed]

15. Sonnenburg, J.L.; Bäckhed, F. Diet–microbiota interactions as moderators of human metabolism. *Nature* **2016**, *535*, 56–64. [CrossRef]

16. Natividad, J.M.; Lamas, B.; Pham, H.P.; Michel, M.-L.; Rainteau, D.; Bridonneau, C.; da Costa, G.; Van Hylckama Vlieg, J.; Sovran, B.; Chamignon, C.; et al. Bilophila wadsworthia aggravates high fat diet induced metabolic dysfunctions in mice. *Nat. Commun.* **2018**, *9*, 2802. [CrossRef]

17. Konuray, G.; Erginkaya, Z. Potential Use of Bacillus coagulans in the Food Industry. *Foods* **2018**, *7*, 92. [CrossRef]

18. Majeed, M.; Majeed, S.; Arumugam, S.; Ali, F.; Beede, K. Comparative evaluation for thermostability and gastrointestinal survival of probiotic *Bacillus coagulans* MTCC 5856. *Biosci. Biotechnol. Biochem.* **2020**, *85*, 962–971. [CrossRef]

19. Shinde, T.; Perera, A.P.; Vemuri, R.; Gondalia, S.V.; Karpe, A.V.; Beale, D.J.; Shastri, S.; Southam, B.; Eri, R.; Stanley, R. Synbiotic Supplementation Containing Whole Plant Sugar Cane Fibre and Probiotic Spores Potentiates Protective Synergistic Effects in Mouse Model of IBD. *Nutrients* **2019**, *11*, 818. [CrossRef]

20. Majeed, M.; Nagabhushanam, K.; Arumugam, S.; Majeed, S.; Ali, F. Bacillus coagulans MTCC 5856 for the management of major depression with irritable bowel syndrome: A randomised, double-blind, placebo controlled, multi-centre, pilot clinical study. *Food Nutr. Res.* **2018**, *62*, 1218. [CrossRef]

21. Majeed, M.; Nagabhushanam, K.; Natarajan, S.; Sivakumar, A.; Ali, F.; Pande, A.; Majeed, S.; Karri, S.K. Bacillus coagulans MTCC 5856 supplementation in the management of diarrhea predominant Irritable Bowel Syndrome: A double blind ran-domized placebo controlled pilot clinical study. *Nutr. J.* **2016**, *15*, 21. [CrossRef] [PubMed]

22. Majeed, M.; Nagabhushanam, K.; Natarajan, S.; Arumugam, S.; Pande, A.; Majeed, S.; Ali, F. A Double-Blind, Place-bo-Controlled, Parallel Study Evaluating the Safety of Bacillus coagulans MTCC 5856 in Healthy Individuals. *J. Clin. Toxicol.* **2016**, *6*, 283. [CrossRef]

23. Majeed, M.; Nagabhushanam, K.; Natarajan, S.; Sivakumar, A.; Ruiter, T.E.-d.; Booij-Veurink, J.; de Vries, Y.P.; Ali, F. Eval-uation of genetic and phenotypic consistency of Bacillus coagulans MTCC 5856: A commercial probiotic strain. *World J. Microbiol. Biotechnol.* **2016**, *32*, 60. [CrossRef] [PubMed]

24. Wang, C.; Nie, G.; Yang, F.; Chen, J.; Zhuang, Y.; Dai, X.; Liao, Z.; Yang, Z.; Cao, H.; Xing, C.; et al. Molybdenum and cadmium co-induce oxidative stress and apoptosis through mitochondria-mediated pathway in duck renal tubular epithelial cells. *J. Hazard. Mater.* **2019**, *383*, 121157. [CrossRef] [PubMed]

25. Pandey, S.; Parvez, S.; Sayeed, I.; Haque, R.; Bin-Hafeez, B.; Raisuddin, S. Biomarkers of oxidative stress: A comparative study of river Yamuna fish Wallago attu (Bl. & Schn.). *Sci. Total Environ.* **2003**, *309*, 105–115. [CrossRef]

26. Gao, M.; Liu, Y.; Song, Z. Effects of polyethylene microplastic on the phytotoxicity of di-n-butyl phthalate in lettuce (*Lactuca sativa* L. var. ramosa Hort). *Chemosphere* **2019**, *237*, 124482. [CrossRef]

27. Momeni, H.R.; Eskandari, N. Effect of curcumin on kidney histopathological changes, lipid peroxidation and total antioxidant capacity of serum in sodium arsenite-treated mice. *Exp. Toxicol. Pathol.* **2017**, *69*, 93–97. [CrossRef]

28. Mrozinska, S.; Kapusta, P.; Gosiewski, T.; Sroka-Oleksiak, A.; Ludwig-Słomczyńska, A.H.; Matejko, B.; Kiec-Wilk, B.; Bulanda, M.; Malecki, M.T.; Wolkow, P.P.; et al. The Gut Microbiota Profile According to Glycemic Control in Type 1 Diabetes Patients Treated with Personal Insulin Pumps. *Microorganisms* **2021**, *9*, 155. [CrossRef]

29. Barreau, F.; Tisseyre, C.; Ménard, S.; Ferrand, A.; Carriere, M. Titanium dioxide particles from the diet: Involvement in the genesis of inflammatory bowel diseases and colorectal cancer. *Part. Fibre Toxicol.* **2021**, *18*, 26. [CrossRef]

30. Liang, Y.; Li, C.; Liu, B.; Zhang, Q.; Yuan, X.; Zhang, Y.; Ling, J.; Zhao, L. Protective effect of extracorporeal membrane oxygenation on intestinal mucosal injury after cardiopulmonary resuscitation in pigs. *Exp. Ther. Med.* **2019**, *18*, 4347–4355. [CrossRef]

31. Polinska, B.; Matowicka-Karna, J.; Kemona, H. The cytokines in inflammatory bowel disease. *Postep. Hig. I Med. Dosw.* **2009**, *63*, 389–394. [PubMed]

32. Moldoveanu, A.C.; Diculescu, M.; Braticevici, C.F. Cytokines in inflammatory bowel disease. *Rom. J. Intern. Med. = Rev. Roum. De Med. Interne* **2015**, *53*, 118–127. [CrossRef]

33. Xu, C.-L.; Sun, R.; Qiao, X.-J.; Xu, C.-C.; Shang, X.-Y.; Niu, W.-N. Protective effect of glutamine on intestinal injury and bacterial community in rats exposed to hypobaric hypoxia environment. *World J. Gastroenterol.* **2014**, *20*, 4662–4674. [CrossRef] [PubMed]

34. Gong, Y.; Li, H.; Li, Y. Effects of *Bacillus subtilis* on Epithelial Tight Junctions of Mice with Inflammatory Bowel Disease. *J. Interf. Cytokine Res.* **2016**, *36*, 75–85. [CrossRef] [PubMed]

35. Chen, Z.; Han, S.; Zhou, D.; Zhou, S.; Jia, G. Effects of oral exposure to titanium dioxide nanoparticles on gut microbiota and gut-associated metabolism *in vivo*. *Nanoscale* **2019**, *11*, 22398–22412. [CrossRef]

36. Moldoveanu, B.; Otmishi, P.; Jani, P.; Walker, J.; Sarmiento, X.; Guardiola, J.; Saad, M.; Yu, J. Inflammatory mechanisms in the lung. *J. Inflamm. Res.* **2009**, *2*, 1–11.

37. Abbasi-Oshaghi, E.; Mirzaei, F.; Pourjafar, M. NLRP3 inflammasome, oxidative stress, and apoptosis induced in the intestine and liver of rats treated with titanium dioxide nanoparticles: In vivo and in vitro study. *Int. J. Nanomed.* **2019**, *14*, 1919–1936. [CrossRef]

38. Mu, W.; Wang, Y.; Huang, C.; Fu, Y.; Li, J.; Wang, H.; Jia, X.; Ba, Q. Effect of Long-Term Intake of Dietary Titanium Dioxide Nanoparticles on Intestine Inflammation in Mice. *J. Agric. Food Chem.* **2019**, *67*, 9382–9389. [CrossRef]

39. Cui, Y.; Liu, H.; Zhou, M.; Duan, Y.; Li, N.; Gong, X.; Hu, R.; Hong, M.; Hong, F. Signaling pathway of inflammatory responses in the mouse liver caused by TiO$_2$ nanoparticles. *J. Biomed. Mater. Res. Part A* **2011**, *96*, 221–229. [CrossRef]

40. Kongseng, S.; Yoovathaworn, K.; Wongprasert, K.; Chunhabundit, R.; Sukwong, P.; Pissuwan, D. Cytotoxic and inflammatory responses of TiO$_2$ nanoparticles on human peripheral blood mononuclear cells. *J. Appl. Toxicol.* **2016**, *36*, 1364–1373. [CrossRef]

41. Duan, Y.; Zeng, L.; Zheng, C.; Song, B.; Li, F.; Kong, X.; Xu, K. Inflammatory Links Between High Fat Diets and Diseases. *Front. Immunol.* **2018**, *9*, 2649. [CrossRef]

42. Cani, P.D.; Bibiloni, R.; Knauf, C.; Waget, A.; Neyrinck, A.M.; Delzenne, N.M.; Burcelin, R. Changes in Gut Microbiota Control Metabolic Endotoxemia-Induced Inflammation in High-Fat Diet-Induced Obesity and Diabetes in Mice. *Diabetes* **2008**, *57*, 1470–1481. [CrossRef]

43. Shi, Q.; Tang, J.; Wang, L.; Liu, R.; Giesy, J.P. Combined cytotoxicity of polystyrene nanoplastics and phthalate esters on human lung epithelial A549 cells and its mechanism. *Ecotoxicol. Environ. Saf.* **2021**, *213*, 112041. [CrossRef]

44. Thomson, A.; Hemphill, D.; Jeejeebhoy, K. Oxidative Stress and Antioxidants in Intestinal Disease. *Dig. Dis.* **1998**, *16*, 152–158. [CrossRef]

45. Liu, P.; Wang, Y.; Yang, G.; Zhang, Q.; Meng, L.; Xin, Y.; Jiang, X. The role of short-chain fatty acids in intestinal barrier function, inflammation, oxidative stress, and colonic carcinogenesis. *Pharmacol. Res.* **2021**, *165*, 105420. [CrossRef]

46. Qiao, R.; Sheng, C.; Lu, Y.; Zhang, Y.; Ren, H.; Lemos, B. Microplastics induce intestinal inflammation, oxidative stress, and disorders of metabolome and microbiome in zebrafish. *Sci. Total Environ.* **2019**, *662*, 246–253. [CrossRef]

47. Bondia-Pons, I.; Ryan, L.; Martinez, J.A. Oxidative stress and inflammation interactions in human obesity. *J. Physiol. Biochem.* **2012**, *68*, 701–711. [CrossRef]

48. Ahamed, M.; Akhtar, M.J.; Khan, M.M.; Alrokayan, S.; Alhadlaq, H. Oxidative stress mediated cytotoxicity and apoptosis response of bismuth oxide (Bi$_2$O$_3$) nanoparticles in human breast cancer (MCF-7) cells. *Chemosphere* **2019**, *216*, 823–831. [CrossRef]

49. Sohm, B.; Immel, F.; Bauda, P.; Pagnout, C. Insight into the primary mode of action of TiO$_2$ nanoparticles on *Escherichia coli* in the dark. *Proteomics* **2015**, *15*, 98–113. [CrossRef]

50. Kong, H.; Song, J.; Jang, J. Photocatalytic Antibacterial Capabilities of TiO$_2$−Biocidal Polymer Nanocomposites Synthesized by a Surface-Initiated Photopolymerization. *Environ. Sci. Technol.* **2010**, *44*, 5672–5676. [CrossRef]

51. Schwarzfischer, M.; Rogler, G. The Intestinal Barrier—Shielding the Body from Nano- and Microparticles in Our Diet. *Metabolites* **2022**, *12*, 223. [CrossRef]

52. Tleyjeh, I.M.; Routh, J.; Qutub, M.O.; Lischer, G.; Liang, K.V.; Baddour, L.M. Lactobacillus gasseri causing Fournier's gangrene. *Scand. J. Infect. Dis.* **2004**, *36*, 501–503. [CrossRef]

53. Liu, X.; Mao, B.; Gu, J.; Wu, J.; Cui, S.; Wang, G.; Zhao, J.; Zhang, H.; Chen, W. *Blautia*—A new functional genus with potential probiotic properties? *Gut Microbes* **2021**, *13*, 1–21. [CrossRef]

54. Rohrhofer, J.; Zwirzitz, B.; Selberherr, E.; Untersmayr, E. The Impact of Dietary Sphingolipids on Intestinal Microbiota and Gastrointestinal Immune Homeostasis. *Front. Immunol.* **2021**, *12*. [CrossRef]

55. Wei, X.; Jiang, S.; Zhao, X.; Li, H.; Lin, W.; Li, B.; Lu, J.; Sun, Y.; Yuan, J. Community-Metabolome Correlations of Gut Microbiota from Child-Turcotte-Pugh of A and B Patients. *Front. Microbiol.* **2016**, *7*, 1856. [CrossRef]

Article

Intestinal Microecology of Mice Exposed to TiO$_2$ Nanoparticles and Bisphenol A

Chen Yang [1], Youlan Tan [1], Fengzhu Li [1], Hongbin Wang [1,2], Ying Lin [3], Fuping Lu [1,2] and Huabing Zhao [1,2,*]

1   College of Biotechnology, Tianjin University of Science and Technology, 9 TEDA 13th Street, Tianjin 300457, China; chenyariapril@163.com (C.Y.); 17806284231@163.com (Y.T.); lifengzhu829@163.com (F.L.); whb@tust.edu.cn (H.W.); lfp@tust.edu.cn (F.L.)
2   Key Laboratory of Industrial Fermentation Microbiology, Ministry of Education, Tianjin University of Science and Technology, Tianjin 300450, China
3   School of Biological Science and Engineering, South China University of Technology, Guangzhou 510006, China; feylin@scut.edu.cn
*   Correspondence: zhaohuabing@tust.edu.cn

Abstract: Exposure to titanium dioxide nanoparticles (TiO$_2$ NPs) and bisphenol A (BPA) is ubiquitous, especially through dietary and other environmental pathways. In the present study, adult C57BL/6J mice were exposed to TiO$_2$ NPs (100 mg/kg), BPA (0, 5, and 50 mg/kg), or their binary mixtures for 13 weeks. The 16S rDNA amplification sequence analysis revealed that co-exposure to TiO$_2$ NPs and BPA altered the intestinal microbiota; however, this alteration was mainly caused by TiO$_2$ NPs. Faecal metabolomics analysis revealed that 28 metabolites and 3 metabolic pathways were altered in the co-exposed group. This study is the first to reveal the combined effects of TiO$_2$ NPs and BPA on the mammalian gut microbial community and metabolism dynamics, which is of great value to human health. The coexistence of TiO$_2$ NPs and BPA in the gut poses a potential health risk due to their interaction with the gut microbiota.

Keywords: TiO$_2$ NPs; BPA; 16S rDNA; faecal metabolomic

Citation: Yang, C.; Tan, Y.; Li, F.; Wang, H.; Lin, Y.; Lu, F.; Zhao, H. Intestinal Microecology of Mice Exposed to TiO$_2$ Nanoparticles and Bisphenol A. *Foods* **2022**, *11*, 1696. https://doi.org/10.3390/foods11121696

Academic Editors: Lavanya Reddivari, Mary Anne Amalaradjou and Evandro Leite de Souza

Received: 22 April 2022
Accepted: 5 June 2022
Published: 9 June 2022

Publisher's Note: MDPI stays neutral with regard to jurisdictional claims in published maps and institutional affiliations.

1. Introduction

The balance between the species within each microbial community is central to microbial homeostasis. The gut microbiome serves various functions, including energy harvesting, the neutralisation of carcinogens and drugs, and the modulation of intestinal mobility and immune responses [1–3]. Microbial dysbiosis can aggravate various diseases, including diabetes, obesity, Crohn's disease, and inflammatory bowel disease [1,4–6]. In addition, the gut microbiota is highly sensitive to exogenous stressors, including nanoparticles (NPs) and persistent organic pollutants (POPs) [7–10].

TiO$_2$ NPs, a widely used nanomaterial, are used in various areas, such as cosmetics, food, and paints. TiO$_2$ is generally incorporated into many daily necessities, (e.g., candies, cheese, sauces, skim milk, and ice cream) as a food colouring agent due to its brightness and high refractive index. In addition, it is used as a flavour enhancer in off-white foods, including dry vegetables, nuts, seeds, soups, wasabi, beer, and wine. Therefore, ingestion may be an important mode of TiO$_2$ exposure, and the amount of TiO$_2$ exposure varies with dietary intake. Dietary exposure to TiO$_2$ was estimated to be 2–3 mg/kg/day for children and approximately 1 mg/kg/day for other age groups in the United Kingdom (UK) [11]. Notably, for those who prefer sweets, the consumption of TiO$_2$ can be several times higher.

Bisphenol A (BPA) is one of the most widely used industrial compounds in the world [12] and is extensively used in daily commodities, such as in plastic containers, food package materials, and food can linings. BPA molecules are linked by ester bonds that are subjected to hydrolysis when exposed to high temperatures or acidic/basic substances, leading to the release of BPA into the environment, posing a potential risk to living beings [13]. Studies have demonstrated that BPA can leach from the epoxy resins of metallic food cans [14–16]. Furthermore, as a representative POP and environmental obesogens, BPA is distinguished by its global dispersion, numerous routes of exposure, multiple vector species, and strong bioaccumulation [17,18].

Several studies have shown that once TiO_2 NPs enter circulation, they cause damage to multiple tissues, such as those in the kidney, liver, brain [19], intestine [20], and lung [21]. The intestine, the longest residence organ for TiO_2 NPs following their ingestion, is the most sensitive. Some studies have demonstrated that the long-term oral use of foods containing TiO_2 NPs may increase the risk of colorectal cancer [7]. In vitro studies have shown that TiO_2 NPs have antimicrobial activity and can promote the production of reactive oxygen species [22,23]. Many studies on the gut have found that they can reduce intestinal microbial diversity and short-chain fatty acid production [20,24,25]. In a manner that is similar to that of TiO_2 NPs, BPA can alter the composition of various intestinal microbiota [8]. Studies have shown that diseases such as type 2 diabetes (T2D) and obesity are closely linked to imbalances in intestinal microecology and to the accumulation of the environmental "carcinogen" BPA [26]. Furthermore, it has been documented that interaction between TiO_2 NPs and BPA alters their innate bioavailability and toxicity [27].

There are few studies on the effects of combined exposure to TiO_2 NPs and BPA on the gut microbiota. These studies are limited to zebrafish and have found that co-exposure to BPA and TiO_2 NPs changed the gut microbial community in a dose-dependent manner [28]. However, as mice are phylogenetically closer to humans and possess physiological and biochemical indicators and regulatory mechanisms that are similar to those in humans, research on mice is of greater significance. Therefore, we investigated the health effects of TiO_2 NPs and BPA on mice.

In this study, we subchronically co-exposed C57BL/6J mice to TiO_2 NPs and BPA. The effects of co-exposure on the structure and diversity of the mouse intestinal microflora were studied using high-throughput sequencing. The metabolic components of the mouse intestine were analysed using chromatography–mass spectrometer (GC-MS)-based on untargeted metabolomics. Finally, we studied the accumulation of contaminants and their main products and possible transformation processes in the intestine and flora.

2. Materials and Methods

2.1. Chemicals

TiO_2 NPs and BPA were purchased from Sigma-Aldrich (St. Louis, MO, USA). The purities of TiO_2 NPs and BPA were greater than 99.7% and 99%, respectively. BPA was initially dissolved in dimethyl sulfoxide (DMSO; purity > 99%) and olive oil. The final DMSO concentration of the solution was 0.1% (v/v).

TiO_2 NPs were dispersed in phosphate-buffered saline (PBS) according to the exposure dose. The suspension was sonicated in an ultrasonic cell disruptor for 20 min for homogenisation. The size and shape of the TiO_2 NPs were characterised using transmission electron microscopy (TEM, Talos G2 200X, Hillsboro, OR, USA). The size distribution and average diameter (d_{TEM}) of the particles were analysed using ImageJ software (Version 1.8.0).

2.2. Animals and Treatment

Healthy three-week-old C57BL/6J male mice were provided by SPF (Beijing, China) Biotechnology Co., Ltd. Animal ethics approval was approved by the Animal Ethics Committee of JAK BIO company (JKX-2106-01). The mice were housed in a humidity-controlled room with a 12-h light/dark cycle and free access to water and food (AIN93G). Following a week of acclimatisation, the mice were gavaged with either TiO_2 NPs (100 mg/kg/day) or BPA (0, 5, and 50 mg/kg/day) alone or in combination. Table 1 lists the specific groupings. After grouping, canagliflozin was administered once daily by oral gavage for 13 weeks. The animals were weighed weekly so that their gavage volumes could be adjusted. BPA was administered as a repeated gavage of 5 mg BPA/kg body weight and 50 mg BPA/kg body weight, a dose corresponding to the lowest observable adverse effect level (LOAEL) and no observed adverse effect level (NOAEL) [29]. The exposure dose of 100 mg/kg/day for the TiO_2 NPs is equivalent to the number of sweets consumed by teenagers [30]. The mice were sacrificed by cervical dislocation after 13 weeks of exposure.

Table 1. Animal grouping and treatments.

Groups, $n = 6$	Treatment	Gavage Dose and Duration (13 Weeks)
Control	Olive Oil + DMSO (Vehicle)	10 mL/kg/day
BPA0Ti100	TiO_2 NPs + Olive Oil + DMSO	0 mg/kg/day + 100 mg/kg/day
BPA5	BPA + Olive Oil + DMSO	5 mg/kg/day
BPA5Ti100	BPA + TiO_2 NPs + Olive Oil + DMSO	5 mg/kg/day + 100 mg/kg/day
BPA50	BPA + Olive Oil + DMSO	50 mg/kg/day
BPA50Ti100	BPA + TiO_2 NPs + Olive Oil + DMSO	50 mg/kg/day + 100 mg/kg/day

n = number of animals used in each group.

To compare the titanium (Ti) content in faeces under the combined effect of TiO_2 NPs and BPA and under the effect of single TiO_2 NPs, a single TiO_2 NPs exposure model was established. The groupings are presented in Table S1.

2.3. Quantification of Faecal Titanium in Mice

The Ti concentration was measured in the faecal samples that were collected from the mice before sacrifice. Using a microwave assisted reaction system, the faecal samples (100–300 mg) were weighed and put in a Teflon microwave digestion vessel along with 8 mL of 68% nitric acid and 2 mL of 30% hydrogen peroxide. After cooling, the digestion solution was transferred into a new tube and diluted to 50 mL for detection. Inductively coupled plasma mass spectrometry (ICP-MS, Thermo-Fisher Scientific Inc., Bremen, Germany) was used to determine the Ti content of the samples. Indium (20 ng/mL) was used as the internal standard. The detection limit was determined to be 0.076 ng/mL.

2.4. Detection of Inflammatory Cytokines and Oxidative Stress Biomarkers

The expression levels of interleukin 6 (IL-6) and interleukin 10 (IL-10) were determined using enzyme-linked immunosorbent assay (ELISA) kits for mice (Shanghai Enzyme-linked Biotechnology Co., Ltd., Shanghai, China).

Furthermore, the oxidative stress levels were assessed by measuring the in-serum levels of superoxide dismutase (SOD) and reactive oxygen species (ROS). While the SOD levels were measured using commercial kits (Nanjing Jiancheng Bioengineering Institute, Nanjing, China), for ROS measurement, ELISA kits were used.

2.5. 16S rDNA Sequencing and Gut Microbiota Analysis

Before the animals were sacrificed, thirty-six faecal samples were collected from the six groups. The samples were placed in 1.5 mL tubes, snap-frozen on dry ice, and kept at $-80\,°C$. A faecal genomic DNA extraction kit was used to extract the faecal genomic DNA according to the manufacturer's instructions (Tiangen Biotech Co., Ltd., Beijing, China). The Illumina HiSeq platform was used to analyse the 16S ribosomal RNA genes (16S rDNA) in the faecal samples (Shenzhen HuaDa Gene Co., Ltd., Shenzhen, China). Reactions were performed in triplicate, and the genomic DNA was amplified using bacterial 16S RNA gene (V3-V4 regions)-specific primers: 341F (5′-ACTCCTACGGGAGGCAGCAG-3′) and 806R (5′-GGACTACHVGGGTWTCTAAT-3′). The QIIME (Version 1.8.0) and R (Version 4.0.5) software packages were used to analyse the data obtained.

2.6. Measurement of Short-Chain Fatty Acids (SCFAs)

The SCFAs in the caecal contents, including acetic acid (AA), propionic acid (PA), isobutyric acid (IBA), butyric acid (BA), isovaleric acid (IVA), and hexanoic acid (HA), were assayed through gas chromatography–mass spectrometry (GC-MS, Agilent, Santa Clara,

CA, USA). Caecal contents (100 mg) were suspended in 100 μL of a 20% phosphoric acid solution and adequately homogenised for 2 min using a vortex mixer. The suspension was centrifuged at $18,000 \times g$ for 10 min. The supernatant resulting from the centrifugation was extracted with 500 μL of ethyl acetate followed by centrifugation at $14,000 \times g$ for 10 min. A 500 μM amount of 4-Methylvaleric acid was added as the internal standard. The parameters for GC-MS analysis are presented in Table S2 in the Supplementary Information. The Agilent Mass Hunter software was used to process the data.

2.7. Metabolomic Analysis

(1) The homogenisation of faecal matter and sample preparation

Approximately 40 mg of each sample was weighed and added to 0.5 mL of pre-cooled MeOH/H$_2$O (8/2, *v/v*) buffer containing 0.6 μg/mL ribitol as an internal standard. The faeces were homogenised using a tissue homogeniser and then centrifuged at 13,500 rpm for 15 min at 4 °C. The supernatant (300 μL) was collected and dried under a nitrogen stream. Subsequently, 50 μL of freshly prepared methoxyamine hydrochloride (20 mg/mL pyridine) was added to the dried sample, and the mixture was incubated at 60 °C for 30 min. Next, 30 μL of N-Methyl-N-(trimethylsilyl) trifluoroacetamide with 1% trimethylchlorosilane (MSTFA + 1%TMCS) and 30 μL pyridine was added, and the sample was further incubated at 60 °C for 30 min.

(2) Analysis of non-targeted metabolomics

Detection was performed using a GC-MS instrument (5977 B, Agilent, USA). The parameters for GC/MS analysis are shown in Table S3 in the Supplementary Information, and Mass Profiler Professional (MPP) was used for data normalisation and statistical analysis. A total of 189 metabolites were identified and integrated from the GC/MS analysis of the faecal samples.

2.8. Histology

The animals were sacrificed after 13 weeks. Colons were fixed in 4% paraformaldehyde for 24 h and then processed for paraffin embedding, sectioning, and haematoxylin and eosin (H&E) staining, as previously described [25].

2.9. Statistical Analysis

SPSS 26.0 software (IBM SPSS Inc., Chicago, IL, USA) was used to analyse the data and to provide a randomization schedule. One-way analysis of variance (ANOVA) was used to assess the differences in the mean values of various parameters, and the Waller–Duncan test was used as a post doc test. After normalising the sample (sum normalisation) and scaling the data (auto-scaling) using MetaboAnalyst 5.0 [31], PLS-DA, heatmap, and KEGG pathway analyses were performed. The Spearman correlations between the gut metabolites and microbiota at the genus level were determined, and heatmaps were generated using the psych package (https://CRAN.R-project.org/package=psych (accessed on 20 September 2021)) in R (Version 4.0.5). The level of statistical significance for all tests was set at $p < 0.05$.

3. Results

3.1. Physicochemical Properties of TiO$_2$ NPs

The majority of the TiO$_2$ NPs used in this study were cuboidal and anatase crystals with a purity of 99.90%. As shown in Figure 1, the size of most of the TiO$_2$ NPs measured by TEM was 10–30 nm.

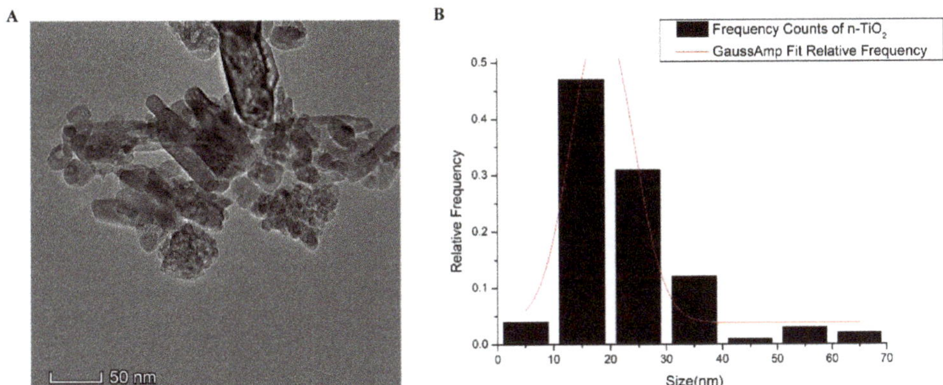

Figure 1. (**A**) Morphology of TiO$_2$ NPs observed by TEM; (**B**) size distribution of TiO$_2$ NPs measured by TEM.

3.2. Effects of Coexposure of TiO$_2$ NPs and BPA on the Diversity of Gut Microbiota

After 16S rDNA gene sequencing and quality filtering, 279.05 million paired reads were obtained, corresponding to a mean of 7.75 $\pm$ 2.5 thousand paired reads per sample. A total of 1412 operational taxonomic units (OTUs) were obtained from the six groups. The species accumulation curve (Figure S1) for all of the samples supports the adequacy of the sampling efforts.

To evaluate differences in the composition of the gut microbiota, $\alpha-$ and $\beta-$diversity analyses were performed. Compared to the BPA-only exposure groups, the Chao1 (Figure 2A) and Shannon (Figure 2B) indexes were all reduced in the combined group, suggesting that the co-exposure of BPA and TiO$_2$ NPs can significantly reduce the diversity of the gut microbiota and that TiO$_2$ NPs may play a dominant role in the combined effect. The β-diversity analysis was performed by dividing the groups into two groups according to BPA-only exposure and co-exposure to TiO$_2$ NPs and BPA, and it revealed a clear separation between the two groups (Figure 2C). These results suggest that the diversity of mouse intestinal flora may be strongly influenced by the combined effects of TiO$_2$ NPs and BPA.

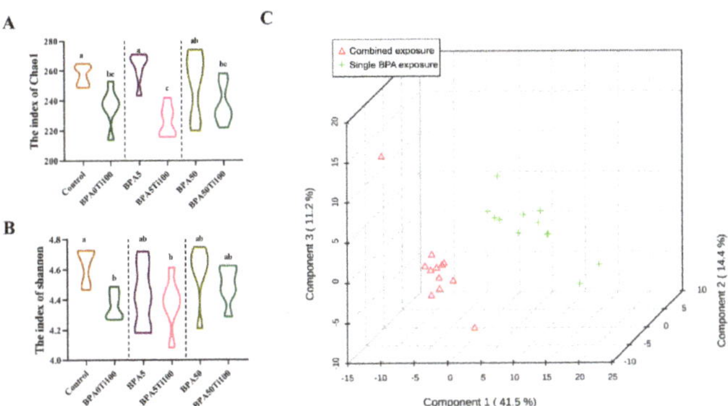

Figure 2. α- and β-diversity of intestinal microbiota. (**A**) Chao1, (**B**) Shannon, and (**C**) PCA score plots of the combined exposure groups (BPA5Ti100 and BPA50Ti100) and the single BPA exposure groups (BPA5 and BPA50). The same letters represent no significant differences among groups ($p > 0.05$).

3.3. Effects of Coexposure of TiO$_2$ NPs and BPA on the Composition of Gut Microbiota

Bacteroidetes, Firmicutes, Proteobacteria, Verrucomicrobia, and TM7 were dominant at the phylum level (Figure 3A,C−H). Compared to the control group, the single BPA exposure groups did not experience significant alterations in the dominant bacteria at the phylum level, whereas the BPA0Ti100 group demonstrated a greater impact on Firmicutes and Bacteroidetes, with a 39% decrease in Firmicutes and a 27% increase in Bacteroidetes ($p < 0.05$) (Figure 3C,D). However, the effect of the combined exposure on the relative abundance of Bacteroidetes and Firmicutes gradually declined as the dose of BPA increased. As a consequence, the effect of combined exposure to BPA and TiO$_2$ NPs on the abundance of major species in the intestine was not significant: a plausible explanation for this phenomenon is that BPA antagonised the effect of the TiO$_2$ NPs.

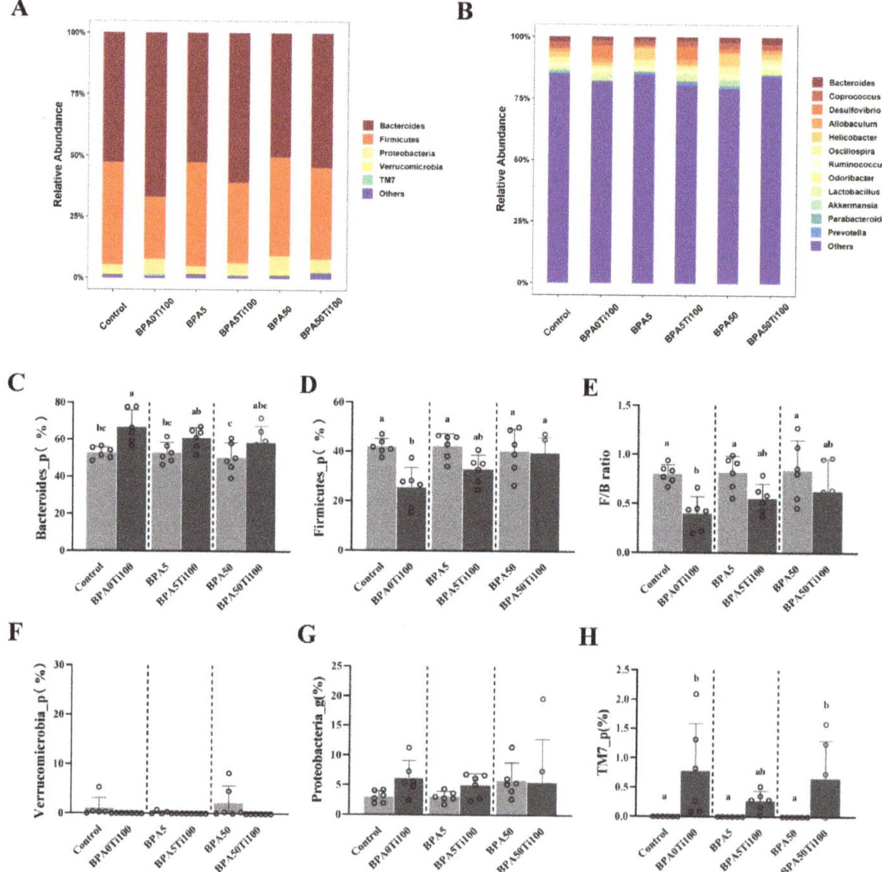

Figure 3. Compositional changes in the microbiota at the (**A**) phylum level and (**B**) genus level. (**C–H**) The relative abundance of faecal microbiota at the phylum level. Data represent the means of the relative abundance of bacteria taxa ± SEM. The same letters represent no significant differences among groups ($p > 0.05$).

The relative abundance of Bacteroidetes did not differ significantly among the co-exposure groups. However, as the BPA concentration increased, the relative abundance of Firmicutes in the co-exposure groups also increased, with a significant difference observed between the BPA0Ti100 and BPA50Ti100 groups ($p < 0.01$). The ratio of Firmicutes and Bacteroidetes (F/B) altered when the relative quantity of Firmicutes fluctuated. Previous studies have reported that the ratio of Firmicutes to Bacteroidetes is strongly associated with various metabolic diseases such as obesity [32]. The F/B ratio of the BPA0Ti100 group was approximately 50% lower than that of the control group. However, as the concentration of BPA increased, the F/B ratio in the co-exposure groups increased, indicating that BPA antagonised the effect of TiO_2 NPs on the F/B ratio.

In addition, we observed an abnormal increase in the relative abundance of the inflammatory bowel disease (IBD)-related bacteria TM7 in the group supplemented with TiO_2 NPs [33]. A previous study indicated that IBD can be aggravated by TiO_2 NPs, which corresponds to our results [34].

At the genus level, the combination of TiO_2 NPs and BPA affected the composition of the gut microbial community, altering the populations of *Lactobacillus*, *Oscillospira*, and *Odoribacter* (Figure 3B and Figure S2). The relative abundance of *Lactobacillus* was almost zero in the BPA-only group but significantly increased in the combined group receiving both TiO_2 NPs and BPA. The abundance of *Oscillospira* was found to be significantly lower in the BPA50Ti100 group than it was in the BPA50 group ($p < 0.05$). The abundance of *Odoribacter* decreased by 52.7% in the BPA5Ti100 group compared to in the BPA5 group. The BPA50Ti100 group showed a 47.4% increase over the BPA50 group. In conclusion, it can be stated that TiO_2 NPs strongly influence the relative abundance of inflammation-related intestinal microbes.

3.4. Effects of Coexposure of TiO_2 NPs and BPA on the Production of Short-Chain Fatty Acid

SCFAs are significant microbial degradation metabolites that play vital roles in host health. Acetic acid, propionic acid, and butyric acid are the main components of SCFAs, accounting for 90–95% of the total amount of SCFAs.

Compared to the control group, the low-dose BPA-only group showed a decrease in acetic acid, propionic acid, and butyric acid (Figure 4A–C). On the other hand, the high-dose BPA-only group showed the opposite result. Compared to the corresponding BPA-only group, the levels of acetic acid, propionic acid, and butyric acid increased in the BPA5Ti100 group, whereas there was a decrease in the levels of propionic acid and butyric acid in the BPA50Ti100 group, which may represent the biological effect of BPA being antagonised by TiO_2 NPs.

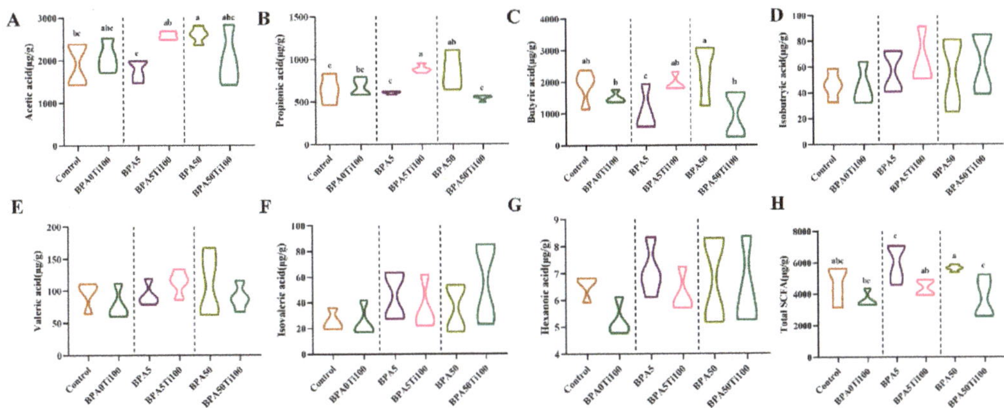

Figure 4. The levels of (**A**) Acetic acid; (**B**) Propionic acid; (**C**) Butyric acid; (**D**) Isobutyric acid; (**E**) Valeric acid; (**F**) Isovaleric acid; (**G**) Hexanoic acid and (**H**) Total SCFA. The same letters represent no significant differences among groups ($p > 0.05$).

In terms of total SCFAs, our results revealed that the combined effect of TiO_2 NPs and BPA considerably lowered SCFA levels in the caecum compared to each BPA-only group (Figure 4H). The TiO_2 NPs effectively reduced the total number of SCFAs, which was mainly due to the reduction in propionic acid and butyric acid. The total SCFAs remained somewhat steady in the composite group compared to the group BPA0Ti100 group, which was most likely because the TiO$_2$ NPs antagonised the biological impact of BPA.

3.5. Effects of Coexposure of TiO$_2$ NPs and BPA on the Inflammatory Response

Our data indicate that the combination of TiO$_2$ NPs and BPA causes intestinal microbial dysregulation in mice, in which TiO$_2$ NPs play a critical role in microbial dysbiosis. Previously, TiO$_2$ NP-induced alterations in the gut microbiota have been reported to be associated with IBD [7]. Therefore, we hypothesised that the gut microbial dysbiosis caused by TiO$_2$ NPs and BPA might be related to IBD. Figure 5A–F show the changes in the colon morphology of mice induced by oral exposure to TiO$_2$ NPs and BPA for 13 weeks. At various doses of BPA-only exposure, no significant differences were observed in the H&E staining results compared to the control group, while the BPA50Ti100 group showed a moderate distortion of the crypt structure and moderate goblet cell loss.

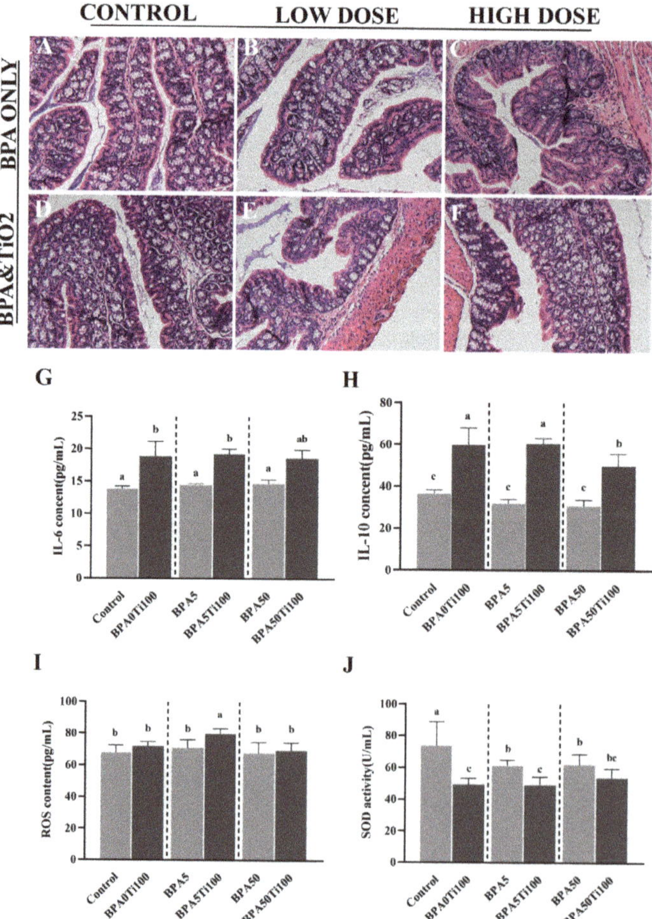

Figure 5. Inflammatory response of TiO$_2$ NPs and BPA. (**A–F**) Histological characterisation (H&E staining) of the colonic tissue of mice (200×); (**G**) in-serum IL-6 levels; (**H**) in-serum IL-10 levels; (**I**) in-serum ROS levels; (**J**) in-serum SOD levels. The same letters represent no significant differences among groups ($p > 0.05$).

Additionally, we measured the concentrations of the pro-inflammatory cytokine IL-6 (Figure 5G) and anti-inflammatory cytokine IL-10 in the serum (Figure 5H). The IL-6 and IL-10 levels were elevated in all of the combined exposure groups, with significant differences being observed in the IL-6 and IL-10 levels in the BPA0Ti100 and BPA5Ti100 groups compared to those observed in the BPA-only group ($p < 0.05$). Considering that the progression and pathogenesis of IBD are associated with oxidative stress and irregularly elevated ROS levels [35], we examined the ROS (Figure 5I) and SOD levels (Figure 5J). We found that the ROS content in the serum was significantly higher in the BPA5Ti100 group than it was in the other groups ($p < 0.05$). SOD plays a vital role in reducing ROS production. In this study, the SOD levels in the BPA0Ti100 group were found to be similar to those in the BPA5Ti100 and BPA50Ti100 groups. Therefore, it is reasonable to infer that TiO$_2$ NPs are an important contributor to SOD reduction.

3.6. Effects of Coexposure of TiO$_2$ NPs and BPA on the Gut-Associated Metabolism

The metabolomic analysis yielded 189 different metabolic features defined by their retention time and mass/charge. One sample was removed because of a technical failure. The interference effect of exposure to TiO$_2$ NPs and BPA (BPA0Ti100, BPA5Ti100, and BPA50Ti100) on the metabolites was elucidated through a heatmap based on the metabolite peak area ratios, which were calculated by dividing the metabolite peak areas by the internal standard peak areas (Figure 6A). After adjusting for multiple comparisons, 28 metabolites were found to be significantly altered.

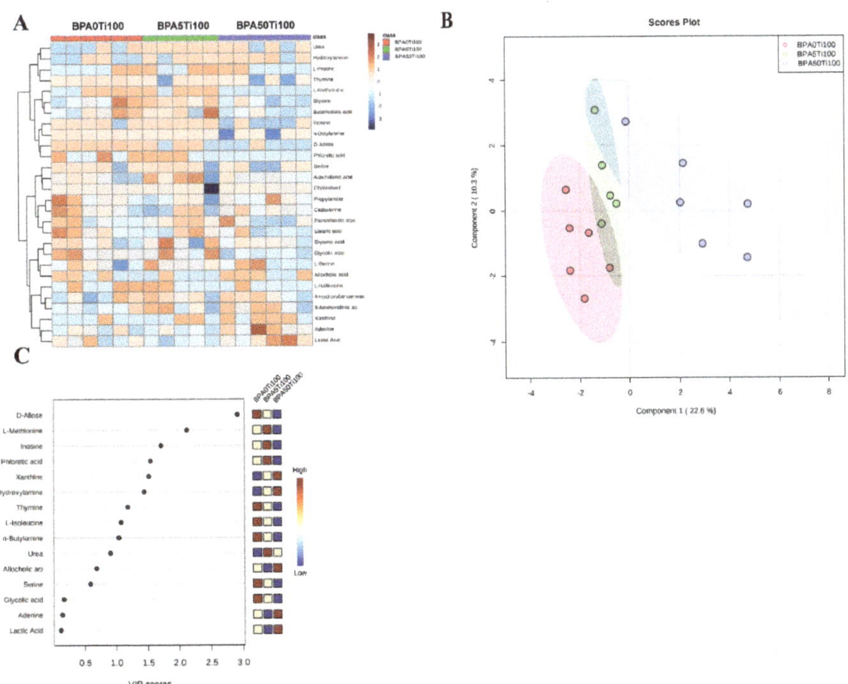

Figure 6. Important discriminatory metabolites were identified by clustering, correlation, and multivariate analysis between the BPA0Ti100, BPA5Ti100, and BPA50Ti100 groups. (**A**) Hierarchical clustering analysis (HCA) for the metabolites in the three groups based on their z-normalised abundances. (**B**) PLS-DA 2D score plot displaying the grouped discrimination of the Ti, 5 Ti, and 50 Ti groups by the first two P.C.s. (**C**) Variable importance in projection (VIP) scores of the important discriminatory metabolites obtained from the PLS-DA models.

The PLS-DA 2D scores clearly showed the differences between the BPA0Ti100, BPA5Ti100, and BPA50Ti100 groups based on the first two principal components, component1 (22.6%) and component2 (10.3%), suggesting BPA5Ti100 and BPA50Ti100 group-specific metabolomic abundance and signatures (Figure 6B). The PLS-DA VIP score plot showed that several specific metabolites such as D-Allose, L-Methionine, inosine, phloretic acid, xanthine, and hydroxylamine were able to differentiate between the mice from the BPA0Ti100, BPA5Ti100, and BPA50Ti100 groups (Figure 6C). These metabolites are intimately connected to amino acids, carbohydrates, and purine metabolism. Our data strongly suggest that in the combined groups, the faecal metabolites varied with the BPA dose.

Figure 7 shows the KEGG pathway analysis of differential metabolites between the samples in the BPA0Ti100 group and BPA50Ti100-treated group. The *X*-axis represents the pathway impact obtained by the out-degree centrality algorithm. The size of the point is related to the pathway impact. The *Y*-axis represents the negative logarithm of the *p*-value ($-\log(p)$) obtained by the pathway enrichment analysis. The yellow and red colour change of the point is positively related to the $-\log(p)$. The names of the pathways are labelled in the graph with $-\log(p) > 1$ or with a pathway impact > 0.1.

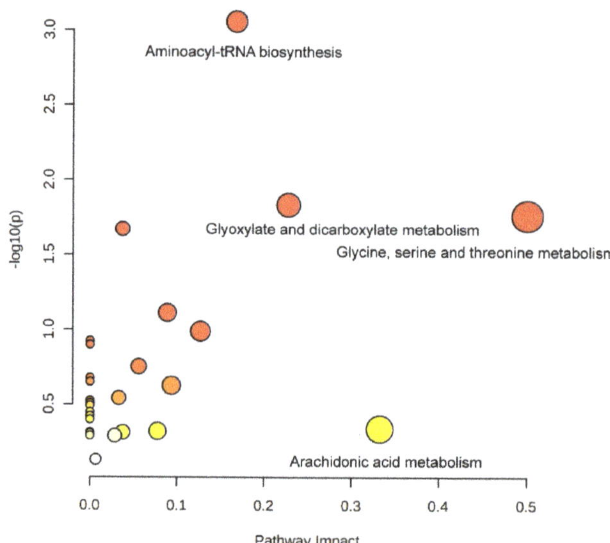

Figure 7. KEGG pathway analysis of differential metabolites between the samples in the BPA0Ti100 and BPA5Ti100 groups. X-Axis represents the Pathway Impact obtained by the out-degree centrality algorithm. The size of the point is related to the Pathway Impact. Y-Axis represents the negative logarithm of the p-value ($-\log(p)$) obtained by the pathway enrichment analysis. The yellow-red color change of the point is positively related to the $-\log(p)$. The names of pathways are labeled in the graph with $-\log(p) > 1$ or pathway impact > 0.1.

The KEGG database was used to undertake pathway analysis of differential metabolites, revealing the effects of co-exposure of two pollutants. When we compared the BPA0Ti100 and BPA5Ti100 groups, no significant changes in the pathway were observed. The BPA5Ti100 group was compared to the BPA50Ti100 group, and it was found that the pathways remained the same as those in the BPA0Ti100 and BPA50Ti100 groups (Figure 7). Among them, aminoacyl-tRNA biosynthesis; glyoxylate and dicarboxylate metabolism; glycine, serine, and threonine metabolism were significantly altered in the group that was co-exposed to BPA and the TiO_2 NPs. The results showed that pathways related to amino acid metabolism, genetic information processing, and carbohydrate metabolism were significantly enriched.

3.7. The Relationship between Faecal Metabolites and Gut Microbiota

Correlations between microbes and metabolites arose due to either the catabolism/anabolism of the metabolites by microbes or via the stimulation/inhibition of microbial growth by metabolites. We computed Spearman's correlation coefficients for the metabolites and microbiome variables to study the potential dependencies between the microbiome composition and faecal metabolism (Figure 8). The results showed that most of the altered bacteria were significantly correlated with a series of differential metabolites (r > ±0.5, $p < 0.05$). These bacteria mainly belonged to the Firmicutes (*Ruminococcus, Oscillospira, Lactobacillus*), Proteobacteria (*Helicobacter, Desulfovibrio*), and Bacteroidetes (*Odoribacter*) phyla. The significantly correlated metabolites were assigned to lipid metabolism (stearic acid), amino acid metabolism (l-serine, glyceric acid, glycine, l-isoleucine, l-methionine, 5-Aminovaleric acid, cadaverine), carbohydrate metabolism (glyceric acid, d-allose, and lactic acid), nucleotide metabolism (inosine, thymine, adenine, xanthine), biosynthesis of other secondary metabolites (glycolic acid, allocholic acid), and metabolites without metabolic pathways (palmitelaidic acid, phloretic acid, and 4-Hydroxybenzeneacetic acid). Our study demonstrated that the co-exposure of mice to TiO$_2$ NPs and BPA disturbed both the gut microbiome and faecal metabolome, and the altered gut microbiota affected metabolism.

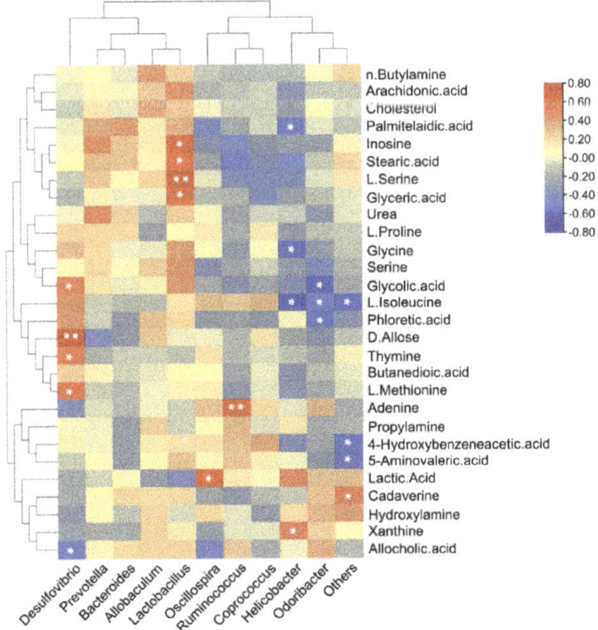

Figure 8. Spearman correlation heatmap analysis between gut microbiota and significant changes in faecal metabolites. *p*-values computed by Spearman correlation tests (* $p < 0.05$, ** $p < 0.01$).

4. Discussion

BPA is an endocrine disruptor that has the potential to mimic oestrogen and can thereby exert harmful health effects [36]. The extensive use of polycarbonate plastics and epoxy resins in consumer items has led to widespread BPA exposure in humans, with BPA identified in the urine of 92.6% of the American population [37]. Previous studies have shown that a single exposure to TiO$_2$ NPs or BPA disrupts the metabolic balance, intestinal microbiota, intestinal oxidative stress, and inflammatory reactions [38–41]. When TiO$_2$ NPs are utilised as medication carriers in the human body or when people consume the food containing E171 (36% of which is constituted by TiO$_2$ NPs) [42], interactions between TiO$_2$

NPs and BPA may occur [43]. Previous studies have noted that TiO_2 NPs could enhance the bioavailability and toxicity of co-existing toxicants in the aquatic phase. The simultaneous presence of BPA and TiO_2 NPs causes neurotoxicity [27], reproductive toxicity [44], and disturbances in the intestinal microecology of zebrafish [28]. However, the effects of the simultaneous oral exposure of mammals to BPA and TiO_2 NPs remain unclear. In this work, we evaluated the subchronic and combined impacts of TiO_2 NPs and BPA on the intestinal microbiota dynamics in C57BL/6J mouse models and analysed faecal metabolites to decipher the association between the two. The study revealed that combined exposure to TiO_2 NPs and BPA altered the composition of the gut microbial community, disturbed the synthesis of metabolites such as SCFAs, and induced intestinal damage and inflammation.

Toxicological research on the effects of TiO_2 NPs and BPA on the gut microbiota is limited. An environmental toxicology study found that combined exposure to BPA and TiO_2 NPs shifted the intestinal microbiota in zebrafish in an antagonistic manner when the BPA concentration was low but in a synergistic manner at a higher BPA concentration [28]. We did not observe this phenomenon in this study. This could be attributed to differences in the doses of BPA and TiO_2 NPs, routes of administration, and model animals. However, we did detect some major alterations in the gut microbiota. Compared to the combined group, the changes in the microbiota in the BPA-only group were minor. The alterations in the microbiota were significant, even in the BPA0Ti100 group. Evidently, the TiO_2 NPs and BPA worked together to influence the gut microbial community, and the TiO_2 NPs were mostly responsible for the alterations in the gut bacteria that were observed in the combined exposure group. The relative abundance of *Lactobacillus* was significantly higher in the co-exposure group compared to in the BPA-only group, with significant differences being observed between the BPA5 and corresponding BPA5Ti100 groups ($p < 0.05$). An in vitro study showed that the coexistence of *Lactobacillus* with nanoparticles weakened the nano-effects by controlling enzyme production [45]. Additionally, *Lactobacillus_gasseri* has been discovered to be a cause of Fournier's gangrene and can produce ROS such as H_2O_2 [46]. It is possible that the greater abundance of *Lactobacillus* in the BPA5Ti100 group led to an increase in the ROS levels in this study. In the combined TiO_2 NPs and BPA exposure groups, both *Oscillospira* and *Odoribacter* were reduced to various degrees. The BPA5Ti100 group had a much lower abundance of *Odoribacter* than the other groups did.

Similarly, research on diethylhexyl phthalate and BPA-only exposure revealed a decrease in *Odoribacter* [39,47]. Both *Oscillospira* and *Odoribacter* are producers of butyrate in the intestine [48,49]. Notably, changes in the relative abundance were not dose-dependent regarding the BPA content in the combined exposure group. This could be due to two causes. On the one hand, the toxic effects of BPA as an endocrine disruptor are more potent at low doses than at high doses within a certain range [50]. This effect even persisted when the TiO_2 NPs were used. BPA and TiO_2 NPs may be mutually synergistic in the intestine. As shown in the previous results for faecal Ti content (Figure S3), the Ti content in mice in the BPA5Ti100 group was slightly lower than it was in the BPA50Ti100 group. In other words, the amount of Ti that remained in the mice was slightly increased. The interaction between the TiO_2 NPs and BPA in the gut is thought to result in increased toxicity at low BPA concentrations.

However, one limitation of this study was the use of 16S sequencing. Some bacteria could not be identified to the species level on the basis of 16S rDNA gene sequence analysis, due to minimal sequence diversity. Further investigations combined with other examination approaches should be conducted to give a clearer picture of the microbial community alternation.

Changes in the gut microbiota can send signals through the gut and produce bacterial metabolites that alter metabolism at different levels. Therefore, we performed a metabolomic analysis of the obtained mouse faeces to further explore the metabolic changes associated with the intestine [40]. Bioinformatics analysis of the faecal metabolome revealed that the BPA5Ti100 group had the greatest changes in metabolomic characteristics when compared to the BPA0Ti100 group, whereas only a few metabolites showed significant changes in the BPA50Ti100 group. This may also be due to the low dose impact indicated earlier.

In the intestine, proteins are initially broken down into peptides or amino acids by proteases that are released by the intestinal flora. Amino acid metabolism is tightly tied to gut bacteria and is altered when the intestinal flora is disrupted. Several amino acids, including L-methionine and L-isoleucine, were shown to have varied abundances in distinct metabolites. Differences in thymine, inosine, adenine, and xanthine expression suggest intestinal purine and pyrimidine metabolism issues. Allocholic acid is a bile acid that aids fat and sterol excretion, absorption, and transport. Lactic acid can be produced in the intestine by certain microbes, (e.g., *streptococci*). Although lactic acid makers can ferment carbohydrates to produce lactic acid, lactic acid accumulation lowers the intestinal pH, irritates the intestine, and increases inflammation in disease situations [51].

The metabolites were mapped to the KEGG metabolic pathway database to better understand the groups to which the differential metabolites belong. According to the KEGG enrichment study, the metabolic pathways generated by the combined action of BPA and TiO_2 NPs differed significantly. The significantly altered metabolites were mostly implicated in the anabolic pathways for aminoacyl-tRNA synthesis; glyoxylate and dicarboxylic acid metabolism as well as in the glycine, tryptophan, and serine metabolic pathways. The differential metabolism of these amino acids and aromatic amino acids can reduce the risk of T2DM and obesity in individuals and are favourably related to insulin resistance and hyperglycaemia [52,53]. The aminoacyl tRNA synthesis pathway involves a variety of amino acid metabolic pathways, including glycine, tryptophan, and serine metabolism, which may affect the translation process by influencing the tRNA function of these amino acids. Chen et al. reported that the intake of TiO_2 NPs affected the aminyl tRNA synthesis pathway in rat intestines [40]. Therefore, we hypothesised that this abnormality in the metabolism under the combined effect of the two contaminants was caused by TiO_2 NPs. In terms of energy supply, glyoxylate and dicarboxylic acid metabolism are a TCA cycle back-supplementation pathway that provides the TCA cycle with an efficient energy-producing function. Abnormal glyoxylate and dicarboxylic acid metabolism indicate an abnormal energy supply to the intestinal flora. Spearman's correlation analysis revealed a significant association between the genus-level gut microbiota and faecal metabolites. These findings imply that disruptions in the gut microbiota may result in significant changes in lipid, amino acid, carbohydrate, and nucleotide metabolism.

According to our results, the synergism between the TiO_2 NPs and BPA was the main cause of the altered intestinal microbiota and faecal metabolites. The TiO_2 NPs caused the greatest disruption to the intestinal bacteria, but when combined with BPA, they also exhibited aberrant changes in intestinal microbes such as *Odoribacter*. The alterations mentioned above were most likely caused by the unique adsorption behaviour of the organic pollutants BPA and TiO_2 NPs, which influence the release of complex kinetics and alter their bioavailability, bioaccumulation, and toxicity under combined exposure.

5. Conclusions

In this study, we investigated the effects of subchronic exposure to TiO_2 NPs and BPA on the intestinal microbiota of mice. We discovered noteworthy correlations between the intestinal metabolites and the gut microbiota in mice that had been co-exposed to TiO_2 NPs and BPA, and a range of altered metabolites was detected. However, the interaction between BPA and TiO_2 NPs is complex and cannot be simply described as being synergistic or antagonistic. More studies are needed to elucidate the mechanism of action between the two. Furthermore, the health risks that are associated with dietary exposure to TiO_2 NPs and BPA should be noted.

Supplementary Materials: The following are available online at https://www.mdpi.com/article/10.3390/foods11121696/s1. Figure S1: Rarefaction curve representing species richness; Figure S2: Changes at the genus level in both BPA-only exposure groups and in the groups receiving combined exposure to TiO2 NPs and BPA; Figure S3: Faecal titanium content is combined in the TiO_2 NPs and BPA exposure groups and in the single-TiO_2 NPs exposure groups; Table S1: Animal groupings and treatments; Table S2: Key parameters for GC-MS analysis (SCFAs); Table S3: Key parameters for GC-MS analysis (non-targeted metabolomics).

Author Contributions: Conceptualization, H.W.; data curation, C.Y.; formal analysis, C.Y. and H.Z.; methodology, C.Y., Y.T., F.L. (Fengzhu Li) and H.W.; project administration, Y.L.; resources, F.L. (Fuping Li); software, C.Y.; supervision, Y.L., F.L. (Fuping Li) and H.Z.; visualization, C.Y.; writing—original draft, C.Y.; writing—review and editing, H.Z. All authors have read and agreed to the published version of the manuscript.

Funding: This work was supported by the Research and Development Plan for Key Fields in Guangdong Province (2019B020210001).

Institutional Review Board Statement: The animal study protocol was approved by the Institutional Review Board of JAK BIO company (JKX-2106-01).

Data Availability Statement: The data presented in this study are available in this article.

Conflicts of Interest: The authors declare no conflict of interest.

References

1. Clark, A.; Mach, N. Exercise-Induced Stress Behavior, Gut-Microbiota-Brain Axis and Diet: A Systematic Review for Athletes. *J. Int. Soc. Sports Nutr.* **2016**, *13*, 43. [CrossRef] [PubMed]
2. Ling, Z.; Cheng, Y.; Yan, X.; Shao, L.; Liu, X.; Zhou, D.; Zhang, L.; Yu, K.; Zhao, L. Alterations of the Fecal Microbiota in Chinese Patients With Multiple Sclerosis. *Front. Immunol.* **2020**, *11*, 590783. [CrossRef] [PubMed]
3. Simon, H.; Vartanian, V.; Wong, M.H.; Nakabeppu, Y.; Sharma, P.; Lloyd, R.S.; Sampath, H. OGG1 Deficiency Alters the Intestinal Microbiome and Increases Intestinal Inflammation in a Mouse Model. *PLoS ONE* **2020**, *15*, e0227501. [CrossRef] [PubMed]
4. Bäckhed, F.; Ding, H.; Wang, T.; Hooper, L.V.; Koh, G.Y.; Nagy, A.; Semenkovich, C.F.; Gordon, J.I. The Gut Microbiota as an Environmental Factor That Regulates Fat Storage. *Proc. Natl. Acad. Sci. USA* **2004**, *101*, 15718–15723. [CrossRef]
5. Svolos, V.; Gkikas, K.; Gerasimidis, K. Diet and Gut Microbiota Manipulation for the Management of Crohn's Disease and Ulcerative Colitis. *Proc. Nutr. Soc.* **2021**, *80*, 409–423. [CrossRef]
6. Turnbaugh, P.J.; Ley, R.E.; Mahowald, M.A.; Magrini, V.; Mardis, E.R.; Gordon, J.I. An Obesity-Associated Gut Microbiome with Increased Capacity for Energy Harvest. *Nature* **2006**, *444*, 1027–1031. [CrossRef]
7. Barreau, F.; Tisseyre, C.; Ménard, S.; Ferrand, A.; Carriere, M. Titanium Dioxide Particles from the Diet: Involvement in the Genesis of Inflammatory Bowel Diseases and Colorectal Cancer. *Part. Fibre Toxicol.* **2021**, *18*, 26. [CrossRef]
8. Ni, Y.; Hu, L.; Yang, S.; Ni, L.; Ma, L.; Zhao, Y.; Zheng, A.; Jin, Y.; Fu, Z. Bisphenol A Impairs Cognitive Function and 5-HT Metabolism in Adult Male Mice by Modulating the Microbiota-Gut-Brain Axis. *Chemosphere* **2021**, *282*, 130952. [CrossRef]
9. Shih, M.-K.; Tain, Y.-L.; Chen, Y.-W.; Hsu, W.-H.; Yeh, Y.-T.; Chang, S.K.C.; Liao, J.-X.; Hou, C.-Y. Resveratrol Butyrate Esters Inhibit Obesity Caused by Perinatal Exposure to Bisphenol A in Female Offspring Rats. *Molecules* **2021**, *26*, 4010. [CrossRef]
10. Yu, R.; Ahmed, T.; Jiang, H.; Zhou, G.; Zhang, M.; Lv, L.; Li, B. Impact of Zinc Oxide Nanoparticles on the Composition of Gut Microbiota in Healthy and Autism Spectrum Disorder Children. *Materials* **2021**, *14*, 5488. [CrossRef]
11. Winkler, H.C.; Notter, T.; Meyer, U.; Naegeli, H. Critical Review of the Safety Assessment of Titanium Dioxide Additives in Food. *J. Nanobiotechnology* **2018**, *16*, 51. [CrossRef] [PubMed]
12. Vandenberg, L.N.; Maffini, M.V.; Sonnenschein, C.; Rubin, B.S.; Soto, A.M. Bisphenol-A and the Great Divide: A Review of Controversies in the Field of Endocrine Disruption. *Endocr. Rev.* **2009**, *30*, 75–95. [CrossRef] [PubMed]
13. Welshons, W.V.; Nagel, S.C.; vom Saal, F.S. Large Effects from Small Exposures. III. Endocrine Mechanisms Mediating Effects of Bisphenol A at Levels of Human Exposure. *Endocrinology* **2006**, *147* (Suppl. 6), S56–S69. [CrossRef] [PubMed]
14. Vandenberg, L.N.; Hauser, R.; Marcus, M.; Olea, N.; Welshons, W.V. Human Exposure to Bisphenol A (BPA). *Reprod. Toxicol.* **2007**, *24*, 139–177. [CrossRef] [PubMed]
15. Brotons, J.A.; Olea-Serrano, M.F.; Villalobos, M.; Pedraza, V.; Olea, N. Xenoestrogens Released from Lacquer Coatings in Food Cans. *Environ. Health Perspect.* **1995**, *103*, 608–612. [CrossRef] [PubMed]
16. Yoshida, T.; Horie, M.; Hoshino, Y.; Nakazawa, H.; Horie, M.; Nakazawa, H. Determination of Bisphenol A in Canned Vegetables and Fruit by High Performance Liquid Chromatography. *Food Addit. Contam.* **2001**, *18*, 69–75. [CrossRef] [PubMed]
17. Stacy, S.L.; Eliot, M.; Calafat, A.M.; Chen, A.; Lanphear, B.P.; Hauser, R.; Papandonatos, G.D.; Sathyanarayana, S.; Ye, X.; Yolton, K.; et al. Patterns, Variability, and Predictors of Urinary Bisphenol A Concentrations during Childhood. *Environ. Sci. Technol.* **2016**, *50*, 5981–5990. [CrossRef]

18. Lama, S.; Vanacore, D.; Diano, N.; Nicolucci, C.; Errico, S.; Dallio, M.; Federico, A.; Loguercio, C.; Stiuso, P. Ameliorative Effect of Silybin on Bisphenol A Induced Oxidative Stress, Cell Proliferation and Steroid Hormones Oxidation in HepG2 Cell Cultures. *Sci. Rep.* **2019**, *9*, 3228. [CrossRef]

19. Bakour, M.; Hammas, N.; Laaroussi, H.; Ousaaid, D.; Fatemi, H.E.; Aboulghazi, A.; Soulo, N.; Lyoussi, B. Moroccan Bee Bread Improves Biochemical and Histological Changes of the Brain, Liver, and Kidneys Induced by Titanium Dioxide Nanoparticles. *Biomed. Res. Int.* **2021**, *2021*, 6632128. [CrossRef]

20. Pinget, G.; Tan, J.; Janac, B.; Kaakoush, N.O.; Angelatos, A.S.; O'Sullivan, J.; Koay, Y.C.; Sierro, F.; Davis, J.; Divakarla, S.K.; et al. Impact of the Food Additive Titanium Dioxide (E171) on Gut Microbiota-Host Interaction. *Front. Nutr.* **2019**, *6*, 57. [CrossRef]

21. Jensen, D.M.; Løhr, M.; Sheykhzade, M.; Lykkesfeldt, J.; Wils, R.S.; Loft, S.; Møller, P. Telomere Length and Genotoxicity in the Lung of Rats Following Intragastric Exposure to Food-Grade Titanium Dioxide and Vegetable Carbon Particles. *Mutagenesis* **2019**, *34*, 203–214. [CrossRef] [PubMed]

22. Kong, L.; Barber, T.; Aldinger, J.; Bowman, L.; Leonard, S.; Zhao, J.; Ding, M. ROS Generation Is Involved in Titanium Dioxide Nanoparticle-Induced AP-1 Activation through P38 MAPK and ERK Pathways in JB6 Cells. *Environ. Toxicol.* **2021**, *37*, 237–244. [CrossRef]

23. Harun, A.M.; Noor, N.F.M.; Zaid, A.; Yusoff, M.E.; Shaari, R.; Affandi, N.D.N.; Fadil, F.; Rahman, M.A.A.; Alam, M.K. The Antimicrobial Properties of Nanotitania Extract and Its Role in Inhibiting the Growth of Klebsiella Pneumonia and Haemophilus Influenza. *Antibiotics* **2021**, *10*, 961. [CrossRef] [PubMed]

24. Mao, Z.; Li, Y.; Dong, T.; Zhang, L.; Zhang, Y.; Li, S.; Hu, H.; Sun, C.; Xia, Y. Exposure to Titanium Dioxide Nanoparticles During Pregnancy Changed Maternal Gut Microbiota and Increased Blood Glucose of Rat. *Nanoscale Res. Lett.* **2019**, *14*, 26. [CrossRef] [PubMed]

25. Cao, X.; Han, Y.; Gu, M.; Du, H.; Song, M.; Zhu, X.; Ma, G.; Pan, C.; Wang, W.; Zhao, E.; et al. Foodborne Titanium Dioxide Nanoparticles Induce Stronger Adverse Effects in Obese Mice than Non-Obese Mice: Gut Microbiota Dysbiosis, Colonic Inflammation, and Proteome Alterations. *Small* **2020**, *16*, e2001858. [CrossRef]

26. Malaisé, Y.; Menard, S.; Cartier, C.; Gaultier, E.; Lasserre, F.; Lencina, C.; Harkat, C.; Geoffre, N.; Lakhal, L.; Castan, I.; et al. Gut Dysbiosis and Impairment of Immune System Homeostasis in Perinatally Exposed Mice to Bisphenol A Precede Obese Phenotype Development. *Sci. Rep.* **2017**, *7*, 14472. [CrossRef]

27. Fu, J.; Guo, Y.; Yang, L.; Han, J.; Zhou, B. Nano-TiO$_2$ Enhanced Bioaccumulation and Developmental Neurotoxicity of Bisphenol a in Zebrafish Larvae. *Environ. Res.* **2020**, *187*, 109682. [CrossRef]

28. Chen, L.; Guo, Y.; Hu, C.; Lam, P.K.S.; Lam, J.C.W.; Zhou, B. Dysbiosis of Gut Microbiota by Chronic Coexposure to Titanium Dioxide Nanoparticles and Bisphenol A: Implications for Host Health in Zebrafish. *Environ. Pollut.* **2018**, *234*, 307–317. [CrossRef]

29. Shelby, M.D. NTP-CERHR Monograph on the Potential Human Reproductive and Developmental Effects of Bisphenol A. *NTP CERHR MON* **2008**. v, vii–ix, 1-64 passim.

30. Khan, S.T.; Saleem, S.; Ahamed, M.; Ahmad, J. 1Survival of Probiotic Bacteria in the Presence of Food Grade Nanoparticles from Chocolates: An in Vitro and in Vivo Study. *Appl. Microbiol. Biotechnol.* **2019**, *103*, 6689–6700. [CrossRef]

31. Chong, J.; Xia, J. MetaboAnalystR: An R Package for Flexible and Reproducible Analysis of Metabolomics Data. *Bioinformatics* **2018**, *34*, 4313–4314. [CrossRef] [PubMed]

32. Masumoto, S.; Terao, A.; Yamamoto, Y.; Mukai, T.; Miura, T.; Shoji, T. Non-Absorbable Apple Procyanidins Prevent Obesity Associated with Gut Microbial and Metabolomic Changes. *Sci. Rep.* **2016**, *6*, 31208. [CrossRef] [PubMed]

33. Kuehbacher, T.; Rehman, A.; Lepage, P.; Hellmig, S.; Fölsch, U.R.; Schreiber, S.; Ott, S.J. Intestinal TM7 Bacterial Phylogenies in Active Inflammatory Bowel Disease. *J. Med. Microbiol.* **2008**, *57 Pt 12*, 1569–1576. [CrossRef] [PubMed]

34. Ruiz, P.A.; Morón, B.; Becker, H.M.; Lang, S.; Atrott, K.; Spalinger, M.R.; Scharl, M.; Wojtal, K.A.; Fischbeck-Terhalle, A.; Frey-Wagner, I.; et al. Titanium Dioxide Nanoparticles Exacerbate DSS-Induced Colitis: Role of the NLRP3 Inflammasome. *Gut* **2017**, *66*, 1216–1224. [CrossRef]

35. Zhao, L.; Du, X.; Tian, J.; Kang, X.; Li, Y.; Dai, W.; Li, D.; Zhang, S.; Li, C. Berberine-Loaded Carboxylmethyl Chitosan Nanoparticles Ameliorate DSS-Induced Colitis and Remodel Gut Microbiota in Mice. *Front. Pharmacol.* **2021**, *12*, 644387. [CrossRef]

36. Martinez, A.M.; Cheong, A.; Ying, J.; Xue, J.; Kannan, K.; Leung, Y.-K.; Thomas, M.A.; Ho, S.-M. Effects of High-Butterfat Diet on Embryo Implantation in Female Rats Exposed to Bisphenol A. *Biol. Reprod.* **2015**, *93*, 147. [CrossRef]

37. Calafat, A.M.; Ye, X.; Wong, L.-Y.; Reidy, J.A.; Needham, L.L. Exposure of the U.S. Population to Bisphenol A and 4-Tertiary-Octylphenol: 2003–2004. *Environ. Health Perspect.* **2008**, *116*, 39–44. [CrossRef]

38. Gálvez-Ontiveros, Y.; Páez, S.; Monteagudo, C.; Rivas, A. Endocrine Disruptors in Food: Impact on Gut Microbiota and Metabolic Diseases. *Nutrients* **2020**, *12*, E1158. [CrossRef]

39. Reddivari, L.; Veeramachaneni, D.N.R.; Walters, W.A.; Lozupone, C.; Palmer, J.; Hewage, M.K.K.; Bhatnagar, R.; Amir, A.; Kennett, M.J.; Knight, R.; et al. Perinatal Bisphenol A Exposure Induces Chronic Inflammation in Rabbit Offspring via Modulation of Gut Bacteria and Their Metabolites. *mSystems* **2017**, *2*, e00093-17. [CrossRef]

40. Chen, Z.; Han, S.; Zhou, D.; Zhou, S.; Jia, G. 1Effects of Oral Exposure to Titanium Dioxide Nanoparticles on Gut Microbiota and Gut-Associated Metabolism in Vivo. *Nanoscale* **2019**, *11*, 22398–22412. [CrossRef]

41. Li, J.; Yang, S.; Lei, R.; Gu, W.; Qin, Y.; Ma, S.; Chen, K.; Chang, Y.; Bai, X.; Xia, S.; et al. 1Oral Administration of Rutile and Anatase TiO(2) Nanoparticles Shifts Mouse Gut Microbiota Structure. *Nanoscale* **2018**, *10*, 7736–7745. [CrossRef] [PubMed]

42. Weir, A.; Westerhoff, P.; Fabricius, L.; Hristovski, K.; von Goetz, N. Titanium Dioxide Nanoparticles in Food and Personal Care Products. *Environ. Sci. Technol.* **2012**, *46*, 2242–2250. [CrossRef] [PubMed]

43. Liu, K.; Lin, X.; Zhao, J. Toxic Effects of the Interaction of Titanium Dioxide Nanoparticles with Chemicals or Physical Factors. *Int. J. Nanomed.* **2013**, *8*, 2509–2520. [CrossRef]

44. Fang, Q.; Shi, Q.; Guo, Y.; Hua, J.; Wang, X.; Zhou, B. Enhanced Bioconcentration of Bisphenol A in the Presence of Nano-TiO_2 Can Lead to Adverse Reproductive Outcomes in Zebrafish. *Environ. Sci. Technol.* **2016**, *50*, 1005–1013. [CrossRef] [PubMed]

45. García-Rodríguez, A.; Moreno-Olivas, F.; Marcos, R.; Tako, E.; Marques, C.N.H.; Mahler, G.J. The Role of Metal Oxide Nanoparticles, Escherichia Coli, and Lactobacillus Rhamnosus on Small Intestinal Enzyme Activity. *Environ. Sci. Nano* **2020**, *7*, 3940–3964. [CrossRef] [PubMed]

46. Tleyjeh, I.M.; Routh, J.; Qutub, M.O.; Lischer, G.; Liang, K.V.; Baddour, L.M. Lactobacillus Gasseri Causing Fournier's Gangrene. *Scand. J. Infect. Dis.* **2004**, *36*, 501–503. [CrossRef] [PubMed]

47. Lei, M.; Menon, R.; Manteiga, S.; Alden, N.; Hunt, C.; Alaniz, R.C.; Lee, K.; Jayaraman, A. Environmental Chemical Diethylhexyl Phthalate Alters Intestinal Microbiota Community Structure and Metabolite Profile in Mice. *mSystems* **2019**, *4*, e00724-19. [CrossRef]

48. Yang, J.; Wei, H.; Zhou, Y.; Szeto, C.-H.; Li, C.; Lin, Y.; Coker, O.O.; Lau, H.C.H.; Chan, A.W.; Sung, J.J.; et al. High-Fat Diet Promotes Colorectal Tumorigenesis through Modulating Gut Microbiota and Metabolites. *Gastroenterology* **2021**, *162*, 135–149. [CrossRef]

49. Bendová, B.; Piálek, J.; Ďureje, Ľ.; Schmiedová, L.; Čížková, D.; Martin, J.-F.; Kreisinger, J. How Being Synanthropic Affects the Gut Bacteriome and Mycobiome: Comparison of Two Mouse Species with Contrasting Ecologies. *BMC Microbiol.* **2020**, *20*, 194. [CrossRef]

50. Vandenberg, L.N.; Colborn, T.; Hayes, T.B.; Heindel, J.J.; Jacobs, D.R.; Lee, D.-H.; Shioda, T.; Soto, A.M.; vom Saal, F.S.; Welshons, W.V.; et al. Hormones and Endocrine-Disrupting Chemicals: Low-Dose Effects and Nonmonotonic Dose Responses. *Endocr. Rev.* **2012**, *33*, 378–455. [CrossRef]

51. Zhang, Y.-L.; Cai, L.-T.; Qi, J.-Y.; Lin, Y.-Z.; Dai, Y.-C.; Jiao, N.; Chen, Y.-L.; Zheng, L.; Wang, B.-B.; Zhu, L.-X.; et al. Gut Microbiota Contributes to the Distinction between Two Traditional Chinese Medicine Syndromes of Ulcerative Colitis. *World J. Gastroenterol.* **2019**, *25*, 3242–3255. [CrossRef] [PubMed]

52. Zhou, Y.; Men, L.; Pi, Z.; Wei, M.; Song, F.; Zhao, C.; Liu, Z. Fecal Metabolomics of Type 2 Diabetic Rats and Treatment with Gardenia Jasminoides Ellis Based on Mass Spectrometry Technique. *J. Agric. Food Chem.* **2018**, *66*, 1591–1599. [CrossRef] [PubMed]

53. Zhu, Y.; Cong, W.; Shen, L.; Wei, H.; Wang, Y.; Wang, L.; Ruan, K.; Wu, F.; Feng, Y. Fecal Metabonomic Study of a Polysaccharide, MDG-1 from Ophiopogon Japonicus on Diabetic Mice Based on Gas Chromatography/Time-of-Flight Mass Spectrometry (GC TOF/MS). *Mol. Biosyst.* **2014**, *10*, 304–312. [CrossRef] [PubMed]

Article

Quality Properties of Dry-Aged Beef (Hanwoo Cattle) Crust on Pork Patties

Jeong-Ah Lee [1], Hack-Youn Kim [2,*] and Kuk-Hwan Seol [1,*]

[1] National Institute of Animal Science, Rural Development Administration, Wanju 55365, North Jeolla, Korea; 2970703@naver.com
[2] Department of Animal Resources Science, Kongju National University, Yesan 32439, Chungnam, Korea
* Correspondence: kimhy@kongju.ac.kr (H.-Y.K.); seolkh@korea.kr (K.-H.S.)

Abstract: This study evaluated the effects of crust derived from dry-aged beef (Hanwoo cattle) on the quality of pork patties. Pork patty samples were prepared with different amounts of crust (0—control, 1, 2, and 3%). The protein, fat, and ash contents in the crust samples were significantly higher than those in the control sample ($p < 0.05$). The CIE b* value of uncooked pork patties with crust added was significantly lower than that of the control patties ($p < 0.05$). The pH and CIE L* values of uncooked patty batter samples decreased with increasing concentrations of crust ($p < 0.05$). However, the viscosity increased proportionally with an increase in crust ($p < 0.05$). Samples containing 3% crust showed significantly higher uncooked and cooked CIE a*, water-holding capacity, cooking yield, and shear force than the control sample ($p < 0.05$). Moreover, samples containing 2% and 3% crust showed significantly lower diameter and thickness reductions than those of the control sample ($p < 0.05$). The sensory evaluation conferred by the crust was significantly higher than that of the control sample ($p < 0.05$). Overall, our results suggest that pork patties supplemented with 3% crust have improved properties.

Keywords: crust; dry-aged; natural flavor enhancer; patties; Hanwoo cattle

Citation: Lee, J.-A.; Kim, H.-Y.; Seol, K.-H. Quality Properties of Dry-Aged Beef (Hanwoo Cattle) Crust on Pork Patties. *Foods* **2022**, *11*, 2191. https://doi.org/10.3390/foods11152191

Academic Editors: Wenjie Sui and Fuping Lu

Received: 1 July 2022
Accepted: 22 July 2022
Published: 23 July 2022

Publisher's Note: MDPI stays neutral with regard to jurisdictional claims in published maps and institutional affiliations.

1. Introduction

In Korea, there has been an increase in consumer preference for Hanwoo dry-aged beef, a high-quality meat product with unique flavor and tenderness [1]. Dry aging meat occurs in an aerobic environment under specific humidity and temperature conditions for several weeks. This dry-aged meat develops characteristics that are different from those of unaged meat. Continuous proteolysis tenderizes the meat and produces a softer texture [2], while the byproducts of proteolysis and lipolysis, free amino acids and free fatty acids, respectively, add a distinctive flavor [3,4]. However, because the surface of dry-aged meat is exposed to air (under aerobic conditions), moisture is continuously evaporated and results in a hardened surface known as the crust [5,6]. The crust formed accounts for approximately 35% of dry-aged meat and is trimmed after processing as it cannot be used for roasting. As such, discarded crust is the primary reason for the low yield and high price of dry-aged meat [6].

However, non-edible crust can provide economic value. Since it has a concentrated "dry-aging flavor" owing to moisture evaporation, it can be used as a flavor enhancer [7]. Crust can also be used as a natural and bioactive food additive because it exhibits higher antioxidant and antihypertensive properties than raw, wet-aged, and edible dry-aged meat [8]. Interestingly, Park et al. [9] reported that the addition of crust to beef patties enhanced their taste, flavor, and tenderness, suggesting the potential use of crust as an additive for meat products.

Ground meat products, such as patties and sausages, are the most common form of processed meat products consumed worldwide [10,11]. Since the introduction of hamburgers in Korea in 1973, the market has expanded to include various types of ground meat products [12]. Patties are manufactured by shaping ground beef and pork combined with various ingredients that enhance sensory properties, such as flavor, juiciness, and texture [13]. Some meat products, such as patties and sausages, are popular, but consumer preferences are shifting toward health and wellness concerns. In response, researchers have generated ground meat products with improved quality via the addition of high-quality functional ingredients tailored to consumer needs [14–18]. Most of these functional ingredients are derived from natural sources, of which plant-based materials form the majority, and commonly unconsumed meat byproducts. Since the crust obtained from dry-aged meat is extracted from edible meat, research on its application in various products, such as products derived from cattle, must be performed to determine its suitability as an additive.

Therefore, in this study, we analyzed the physicochemical properties of the addition of crust derived from dry-aged beef (Hanwoo cattle) to pork patties.

2. Materials and Methods

2.1. Hanwoo Crust and Pork Pattie Preparation

The dry-aging and crust separation process of beef loin (Hanwoo) was as follows: six pieces of beef loin (M. longissimus dorsi) were obtained from six carcasses two days after being slaughtered (Hanwoo cattle, Korean quality grade 2, Tobawoo, Sejong, Korea). Samples were dry aged in a refrigerator (aging temperature, 4 ± 1 °C; humidity, $65 \pm 5\%$; air velocity, 5 ± 3 m/s) for four weeks. After dry aging, the crust was collected from the outermost edge (0.5 ± 0.2 cm) of the dry-aged beef loin, then lyophilized in a freeze dryer (FDU-1110, Eyela, Tokyo, Japan) at -70 °C for 48 h. Lastly, the lyophilized crust derived from Hanwoo beef loin was pulverized using a mixer (MQ5135, Braun, Kronberg im Taunus, Germany) and used in this study.

Pork patties were formulated with various amounts of crust (Table 1). To prepare the meat batter samples, pork hind leg meat and back fat were purchased from a local market (I home meat, Seoul, Korea). Pork hind leg meat and back fat were cut and ground using a grinder (PA-82, Mainca, Barcelona, Spain) attached to a 3 mm plate. Next, the ground meat (70%) and back fat (15%) were mixed with iced water (15%), garlic powder (0.5%), sugar (1%), nitrite pickling salt (1.2%; nitrite content: 6000 ppm), onion powder (0.5%), and crust at different concentrations (0, 1, 2, and 3%). Each mixture was used to prepare pork patties (100 g each) using a pressure device. Pork patties were thermally processed in a chamber (10.10ESI/SK, Alto Shaam, Menomonee Falls, WI, USA) at 80 °C for 30 min.

Table 1. Formulation of the pork patties used in this study.

Ingredients (%)		Crust (%)			
		0 (Control)	1	2	3
Main	Pork lean meat	70	70	70	70
	Pork back fat	15	15	15	15
	Ice	15	15	15	15
Additives	NPS [1]	1.2	1.2	1.2	1.2
	Sugar	1	1	1	1
	Garlic powder	0.5	0.5	0.5	0.5
	Onion powder	0.5	0.5	0.5	0.5
	Crust	0	1	2	3

[1] NPS: nitrite pickling salt.

2.2. Determination of the Proximate Composition

The proximate composition of the cooked patties was determined according to the AOAC method [19]. The moisture, protein, fat, and ash contents were measured using 105 °C oven drying, Kjeldahl, Soxhlet, and 550 °C dry ashing methods, respectively.

2.3. Determination of the pH Values

To measure the pH of the cooked pork patties, 5 g of each sample was placed in a conical tube with 20 mL of distilled water. The samples were homogenized using an Ultra-Turrax homogenizer (HMZ-20DN, Poolim Tech, Seongnam, Korea) at 8000 rpm for 1 min. The pH values of the mixtures of samples were determined using a glass electrode pH meter (Model 340; Mettler-Toledo, Schwerzenbach, Switzerland).

2.4. Determination of the Color of Samples

The surface area color of the cooked patties was measured using a colorimeter (CR-10. Minolta, Tokyo, Japan) for CIE L* (lightness), CIE a* (redness), and CIE b* (yellowness) values. A white standard plate was used as a reference (CIE L*: +97.83; CIE a*: −0.43; CIE b*: +1.98).

2.5. Determination of the Water-Holding Capacity (WHC)

The WHC was determined according to a method adapted from [20]. Briefly, 5 g of each sample was placed inside a conical tube containing cotton and a filter paper. Samples were centrifuged (Supra R22, Hanil, Gimpo, Korea) at $1092 \times g$, 4 °C for 10 min. After centrifugation, the WHC (as per the weight of the water-drained sample) was calculated using the following formula

$$\text{WHC } (\%) = \frac{A - B}{B} \times 100 \tag{1}$$

A: Sample mass before centrifugation (g) × water content of sample (%)) ÷ 100.

B: Sample mass before centrifugation (g) - sample mass after centrifugation (g).

2.6. Determination of the Cooking Yield

The cooking yield of patty samples was determined by calculating the mass before and after cooking, as follows

$$\text{Cooking yield } (\%) = \frac{\text{Sample mass after cooking (g)}}{\text{Sample mass before cooking (g)}} \times 100 \tag{2}$$

2.7. Determination of Viscosity

The apparent viscosity of the patty batter samples was measured using a viscometer (Merlin VR, Rheosys, Hamilton Township, NJ, USA) equipped with a 30-mm cone and a 25-mm co-axial cylinder. Measurements were conducted for 60 s at 20 °C at a head speed of 20 rpm. The measured apparent viscosity values were expressed in Pa·s.

2.8. Determination of the Diameter and Thickness Reduction Ratio

The diameter and thickness reduction ratios were determined by measuring the diameter and thickness of the patties before and after cooking using Vernier calipers (CD-15APX; Mitutoyo Co., Kawasaki-shi, Japan). Values are expressed as percentages, according to the following formulas

$$\text{Diameter reduction ratio } (\%) = \frac{\text{Diameter before cooking (mm)} - \text{Diameter after cooking (mm)}}{\text{Diameter before cooking (mm)}} \times 100 \tag{3}$$

$$\text{Thickness reduction ratio } (\%) = \frac{\text{Diameter before cooking (mm)} - \text{Diameter after cooking (mm)}}{\text{Diameter before cooking (mm)}} \times 100 \tag{4}$$

2.9. Determination of the Shear Force

The shear force was determined for each sample by attaching them to 2.0 cm^3 sample blocks and evaluated using a Texture Analyzer with an attached v-blade (head speed, 2.0 mm/s; distance, 22.0 mm; downforce, 5 g; TA 1, Ametek, Largo, FL, USA), and the measurements were expressed in Newtons (N).

2.10. Sensory Evaluation

The sensory evaluation of this study was approved by the Ethics Committee of Kongju National University (Authority IRB No: KNU 2020-15), and sensory evaluation was performed by ten sensory panelists. The panelists were trained with commercially available pork patties for 7 d (1 h session per day) to ensure familiarization with the sensory properties of pork patties. The sensory evaluations were color, flavor, tenderness, juiciness, and overall acceptability, and the samples were scored using a 10-point descriptive scale.

2.11. Statistical Analysis

All the experimental data were collected from at least three independent trials. The data were statistically analyzed using analysis of variance for all variables, followed by Duncan's multiple range test ($p < 0.05$), and the general linear model in SAS software (version 9.3, SAS Institute, Cary, NC, USA). Values are expressed as the mean $\pm$ SD.

3. Results and Discussion

3.1. Effect of Lyophilized Crust Derived from Dry-Aged Beef (Hanwoo) Supplementation on Proximate Composition

The proximate composition of cooked pork patties, according to the amount of crust derived from dry-aged beef (Hanwoo), is shown in Table 2. Crust addition significantly lowered the moisture content compared with the control samples ($p < 0.05$).

Table 2. Proximate composition of the pork patties formulated with various levels of crust derived from dry-aged beef (Hanwoo).

Traits	Crust (%)			
	0 (Control)	1	2	3
Moisture (%)	62.38 ± 0.67 [a]	59.09 ± 1.66 [b]	58.17 ± 0.11 [b]	57.67 ± 0.23 [b]
Protein (%)	21.20 ± 0.25 [c]	23.02 ± 0.15 [b]	23.33 ± 0.51 [ab]	23.80 ± 0.12 [a]
Fat (%)	14.67 ± 0.58 [c]	16.50 ± 0.71 [b]	17.67 ± 0.58 [b]	19.33 ± 0.58 [a]
Ash (%)	1.46 ± 0.25 [b]	1.79 ± 0.10 [a]	1.74 ± 0.09 [a]	1.78 ± 0.04 [a]

All values are expressed as mean ± SD. [a–c] Means on the same row with different letters are significantly different ($p < 0.05$).

In contrast, protein and fat contents significantly increased with increasing crust content compared with the control samples ($p < 0.05$). Campbell et al. [21] and Lee et al. [22] reported that dry-aged meat has higher protein and ash contents, which are major components of muscle tissue, because a large amount of moisture evaporates during the aging process. As such, our results are consistent with existing research demonstrating that increasing the amount of crust increases the protein and ash contents [23]. Crust has high protein and fat content as it originates from the surface of Hanwoo beef, and it contains almost no moisture owing to dry aging. Similarly, when lyophilized protein additives are added to meat products, protein, fat, and ash contents increase as the moisture content decreases [24]. Therefore, it is believed that meat products with high protein, fat, and ash content can be obtained via the addition of crust.

3.2. Effect of Lyophilized Crust Derived from Dry-Aged Beef (Hanwoo) Supplementation on pH and Color

Table 3 shows the pH and CIE color of uncooked and cooked pork patties with crust derived from dry-aged beef (Hanwoo). The pH of the uncooked pork patties significantly decreased as the amount of crust increased ($p < 0.05$). The pH of the cooked patties also tended to decrease with increasing crust content. In addition, pork patties containing 3% crust showed a significantly lower pH than the control sample ($p < 0.05$). Therefore, we assumed that the pH of pork patties is affected by this crust addition because the pH of meat products can vary with the pH range of additive(s) [25]. Because the pH of the crust was 5.67, it is not surprising that higher amounts of crust led to greater

decreases in the pH. Interestingly, the pH values were higher in cooked versus uncooked samples, which was consistent with studies reporting that the pH increases as basic active species (e.g., imidazolium present in the amino acid histidine) are exposed to the external environment in the context of heat-induced protein denaturation [26].

Table 3. pH and color of pork patties formulated with various levels of crust derived from dry-aged beef (Hanwoo).

Traits			Crust (%)			
			0 (Control)	1	2	3
pH		Uncooked	6.14 ± 0.01 [a]	6.13 ± 0.01 [b]	6.09 ± 0.01 [c]	6.08 ± 0.01 [d]
		Cooked	6.31 ± 0.01 [a]	6.31 ± 0.01 [a]	6.27 ± 0.01 [b]	6.25 ± 0.01 [c]
Color	Uncooked	CIE L*	62.70 ± 1.27 [a]	60.65 ± 1.63 [ab]	59.50 ± 0.28 [b]	57.65 ± 0.49 [b]
		CIE a*	3.80 ± 0.01 [b]	4.0 ± 0.14 [b]	4.35 ± 0.07 [b]	5.40 ± 0.71 [a]
		CIE b*	14.20 ± 0.01 [a]	13.60 ± 0.14 [b]	11.60 ± 0.14 [c]	11.25 ± 0.21 [c]
	Cooked	CIE L*	67.53 ± 0.51 [a]	63.43 ± 0.42 [b]	62.75 ± 0.07 [bc]	62.50 ± 0.00 [c]
		CIE a*	4.00 ± 0.28 [b]	4.15 ± 0.21 [b]	4.75 ± 0.21 [b]	5.85 ± 0.49 [a]
		CIE b*	13.45 ± 0.35	12.40 ± 0.14	12.20 ± 0.42	11.50 ± 1.70

All values are expressed as mean $\pm$ SD. [a–d] Means on the same row with different letters are significantly different ($p < 0.05$).

With respect to color, the lightness of the uncooked and cooked samples tended to decrease as the crust content increased. In addition, lightness before cooking significantly decreased as the amount of crust increased ($p < 0.05$). The redness of uncooked and cooked patties tended to increase with increasing crust content, and the highest value was observed in patties containing 3% crust ($p < 0.05$). The yellowness of the uncooked samples significantly decreased with increasing crust added compared with the control sample ($p < 0.05$). The results in this work are similar to those reported in a study by Kim et al. [27]: lightness and yellowness decreased as redness increased due to the increase in the myoglobin content (red-colored component) present in the meat additive. As aging time increases, myoglobin in meat is oxidized to metmyoglobin, which is a discolored brown. This is a result of the relative increase in myoglobin content as moisture evaporates [28]. Nevertheless, the crust color is similar to that of typical meat products. Overall, we showed that the change in color was affected by the lightness (37.82), redness (12.87), and yellowness (2.08) of the Hanwoo crust. Finally, the yellowness of cooked samples is likely to be less affected in pork patties because the proteins contained in the crust are heat-denatured by the Maillard reaction [29].

3.3. Effect of Lyophilized Crust Derived from Dry-Aged Beef (Hanwoo) Supplementation on Water-Holding Capacity (WHC) and Cooking Yield

The WHC and cooking yield of meat products depend on the cross-linked state of salt-soluble proteins, water molecules, and fat globules [30], which are affected by the structure, amount, protein content, and physical properties [31]. Figure 1 shows the WHC of uncooked pork patty batter and cooking yield of pork patties with different crust levels derived from dry-aged beef (Hanwoo).

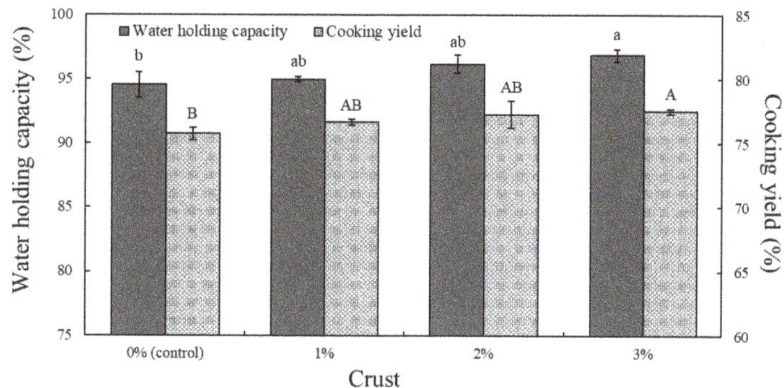

Figure 1. Water-holding capacity (WHC) and cooking yield of pork patties formulated with various levels of crust derived from dry-aged beef (Hanwoo). [a,b,A,B] represent statistically significant differences comparing the same color bar graphs ($p < 0.05$).

The WHC and cooking yield tended to increase with an increasing amount of crust; in particular, significantly higher values were obtained in pork patties containing 3% crust than in the control samples ($p < 0.05$). The WHC change is probably due to the rehydration of the lyophilized crust, which increased with the amount of crust added [32]. In a similar study, Choi et al. [33] added lyophilized mealworms to pork patties and noted that the WHC increased because of reduced cooking loss. The high cooking yield of samples containing crust is attributed to stronger bonds between protein molecules and water molecules [34], resulting in a lower cooking loss. The lyophilized crust with a high rehydration potential strengthened the bond with water molecules in the patties and improved the cooking yield and WHC, which is characteristic of highly rehydratable additives [24].

3.4. Effect of Lyophilized Crust Derived from Dry-Aged Beef (Hanwoo) Supplementation on Viscosity

The viscosity of uncooked pork patty batter containing crust derived from dry-aged beef (Hanwoo) is shown in Figure 2. The apparent viscosity increased with an increasing amount of crust, and all samples containing crust showed higher apparent viscosities compared with the control samples. This viscosity increase was attributed to the enhanced WHC of the added crust. Similar to these results, June et al. [35] demonstrated that the addition of lyophilized protein-type walnuts to a non-Newtonian fluid led to an increase in viscosity.

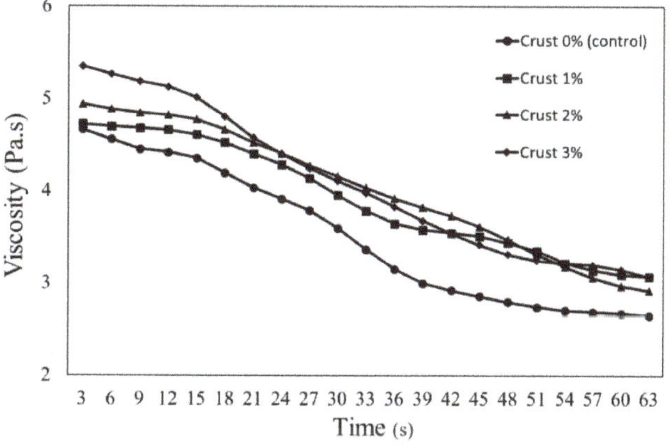

Figure 2. Apparent viscosity of pork patties formulated with various levels of crust derived from dry-aged beef (Hanwoo).

The viscosity of uncooked ground meat is affected by physical properties, such as WHC, protein shrinkage, and solubility, and it exhibits a flow curve characteristic of a non-Newtonian fluid [36]. The apparent viscosity of emulsified patties also decreased with increasing rotation time, and this behavior is characteristic of thixotropic fluids; protein molecules (e.g., in ground meat) form a more ordered arrangement with increasing rotation time, leading to a gradual decrease in viscosity [37].

3.5. Effect of Lyophilized Crust Derived from Dry-Aged Beef (Hanwoo) Supplementation on Diameter and Thickness Reduction Ratios

The diameter and thickness reduction ratios of the cooked pork patties with various crust levels are shown in Figure 3. The diameter and thickness reduction ratios were significantly lower in pork patties containing 2% and 3% crust than in the control samples ($p < 0.05$).

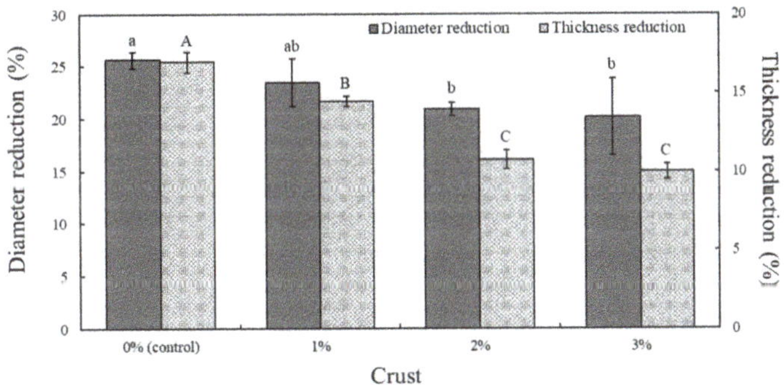

Figure 3. Diameter and thickness reduction ratios of pork patties formulated with various levels of crust derived from dry-aged beef (Hanwoo). [a,b,A–C] Represent statistically significant differences comparing the same color bar graphs ($p < 0.05$).

These results were consistent with previous findings that showed that non-meat proteins, such as wheat germ and soy proteins, added to patties resulted in lower cooking loss and shrinkage [38]. Our results were also similar with those of a study reporting a smaller reduction in the diameter and thickness of pork patties prepared with isolated soy protein that increased the bond strength between water and fat, thus minimizing size loss during cooking [12]. The diameter and thickness reduction ratios in pork patties are also related to organoleptic properties. Berry and Leddy [39] reported that patties with a higher cooking loss also showed poorer sensory properties. Therefore, the addition of crust enhanced the WHC and cooking yield of pork patties, thereby suppressing their size reduction during cooking. Notably, this is a positive factor concerning the hydration properties and palatability of meat products.

3.6. Effect of Lyophilized Crust Derived from Dry-Aged Beef (Hanwoo) Supplementation on Shear Force

Figure 4 shows the shear force measurements of pork patties containing crusts derived from dry-aged beef (Hanwoo). The shear force of pork patties with crust tended to increase, and samples that contained 2% or 3% crust showed significantly higher values ($p < 0.05$).

Lu and Chen [40] noted that improved binding ability in meat is enhanced both by protein–protein and protein–fat binding strength, which thereby increases the shear force. Similarly, we believe that the shear force of patties increased owing to the enhanced binding strength between proteins and fat in the pork patty batter mixture with added crust. These results are also in agreement with a study on hamburger patties supplemented with lyophilized tofu powder (mainly globulin protein) that increased patty hardness with increasing levels of powder [41]. Sample hardness containing isolated soy protein and dietary fiber was also shown to be higher in a study by Choi et al. [42]. In addition, this study also reported that isolated soy and wheat fiber proteins affect the water-holding capacity, emulsifying capacity, gel-forming capacity, and adhesion between particles. Furthermore, the textural properties of meat products are affected by the condition of the raw meat and the type of additive [43,44]. Likewise, we believe that the shear force of patties increased with increasing crust content, owing to both the increase in protein content and the decrease in moisture content in patty composition.

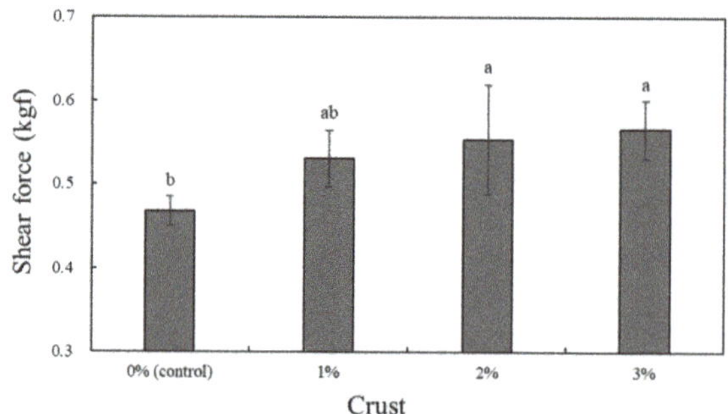

Figure 4. Shear force of pork patties formulated with various levels of crust derived from dry-aged beef (Hanwoo). [a,b] Represent statistically significant differences comparing the same color bar graphs ($p < 0.05$).

3.7. Sensory Evaluation

Sensory evaluation of pork patties containing crust derived from dry-aged beef (Hanwoo) is presented in Table 4.

Table 4. Sensory evaluation of pork patties formulated with various levels of crust derived from dry-aged beef (Hanwoo).

Traits	Crust (%)			
	0 (Control)	1	2	3
Color	8.00 ± 0.93 [b]	8.75 ± 0.93 [ab]	9.38 ± 0.74 [a]	9.38 ± 0.74 [a]
Flavor	8.13 ± 0.83 [b]	8.63 ± 0.52 [bc]	9.25 ± 0.46 [ab]	9.50 ± 1.07 [a]
Tenderness	9.00 ± 0.76	9.25 ± 0.46	9.13 ± 0.83	8.88 ± 0.64
Juiciness	8.38 ± 0.52 [b]	9.00 ± 0.53 [ab]	9.13 ± 0.64 [a]	9.00 ± 0.76 [ab]
Overall acceptability	8.00 ± 0.93 [b]	8.81 ± 0.53 [a]	9.44 ± 0.50 [a]	9.38 ± 0.92 [a]

All values are expressed as mean ± SD. [a–c] Means on the same row with different letters are significantly different ($p < 0.05$).

The sensory evaluation of color of 2% and 3% crust samples was significantly higher than that of the control sample ($p < 0.05$). The flavor tended to be higher as the added amount of crust increased, whereas tenderness showed no significant difference between all samples. The 2% crust sample received a significantly higher juiciness evaluation score than the control sample ($p < 0.05$). The 2% and 3% crust samples received significantly higher evaluations of overall acceptability than the control sample ($p < 0.05$). This evaluation was similar to a study by Park et al. [9], which reported that beef patties supplemented with crust obtained from dry-aged beef sirloin received higher scores for taste, flavor, juiciness, and overall acceptability compared to the control samples. As the meat protein and fat decompose during the dry-aging process, the content of inosine 5'-monophosphate disodium (IMP) increases, which is a flavor enhancer that elicits umami. Consequently, the enhanced flavor and umami were noted, and the products were considered more acceptable overall. Such differences in sensory properties depended on crust addition. Therefore, the added crust enhanced the flavor and played the role of an umami enhancer, and this added crust imparted a dry-aging flavor and enhanced organoleptic properties in these pork patties.

4. Conclusions

In this study, pork patties were prepared using added crust from dry-aged beef loin (Hanwoo cattle) as a flavor enhancer, and their qualities were analyzed. The experimental results demonstrated that increasing the amount of crust added to pork patties improved their physical properties, such as WHC, cooking yield, viscosity, and diameter and thickness reduction ratios. Moreover, the addition of crust to pork patties resulted in a better sensory evaluation. Therefore, the addition of 3% crust to pork patties should lead to the production of patties with better quality and enhanced flavor.

Author Contributions: Conceptualization, H.-Y.K.; methodology, H.-Y.K.; software, J.-A.L. and K.-H.S.; validation, K.-H.S. and H.-Y.K.; formal analysis, J.-A.L.; investigation, J.-A.L.; resources, H.-Y.K.; data curation, H.-Y.K. and K.-H.S.; writing—original draft preparation, J.-A.L.; writing—review and editing, J.-A.L., K.-H.S. and H.-Y.K.; supervision, H.-Y.K.; project administration, H.-Y.K.; funding acquisition, K.-H.S. All authors have read and agreed to the published version of the manuscript.

Funding: This research was supported by Basic Science Research Program through the National Research Foundation of Korea (KRF) funded by the Ministry of Education (2018R1D1A1B07049938) and "Cooperative Research Program for Agricultural Science & Technology Development (PJ01620103)" National Institute of Animal Science, Rural Development Administration.

Institutional Review Board Statement: Sensory evaluation of this study was approved by the Kongju National University's Ethics Committee (Authority No: KNU_IRB_2020-15).

Informed Consent Statement: Informed consent was obtained from all subjects involved in the study.

Data Availability Statement: All related data and methods are presented in this paper. Additional inquiries should be addressed to the corresponding author.

Conflicts of Interest: The authors declare no conflict of interest.

References

1. Cho, S.; Kang, S.M.; Ahn, D.; Kim, Y.; Lee, E.; Ba, H.V.; Kim, Y.; Seong, P.; Kim, J.; Park, B. Effect of dry aging condition on yield, microbial growth and storage stability of bottom round muscle from Hanwoo beef. *Ann. Anim. Resour. Sci.* **2018**, *29*, 106–114. [CrossRef]
2. Kim, J.H.; Cho, S.H.; Seong, P.N.; Hah, K.H.; Kim, H.K.; Park, B.Y.; Lee, J.M.; Kim, D.H.; Ahn, C.N. Effect of aging temperature and time on the meat quality of longissimus muscle from Hanwoo steer. *Korean J. Food Sci. Anim.* **2007**, *27*, 171–178. [CrossRef]
3. Smith, R.D.; Nicholson, K.L.; Nicholson, J.D.W.; Harris, K.B.; Miller, R.K.; Griffin, D.B.; Savell, J.W. Dry versus wet aging of beef: Retail cutting yields and consumer palatability evaluations of steaks from US Choice and US Select short loins. *Meat Sci.* **2008**, *79*, 631–639. [CrossRef]
4. Laville, E.; Sayd, T.; Morael, M.; Blinet, S.; Chambon, C.; Lepetit, J.; Renand, G.; Hocquette, J.G. Proteome changes during meat aging in tough and tender beef suggest the improtance of apoptosis and protein solubility for beef aging and tenderization. *J. Agric. Food Chem.* **2009**, *57*, 10755–10764. [CrossRef]
5. Dashdorj, D.; Tripathi, V.K.; Cho, S.; Kim, Y.; Hwang, I. Dry aging of beef; Review. *J. Anim. Sci. Technol.* **2016**, *58*, 20. [CrossRef] [PubMed]
6. Lee, H.J.; Choe, J.; Kim, M.; Kim, H.C.; Yoon, J.W.; Jo, C. Role of moisture evaporation in the taste attributes of dry- and wet-aged beef determined by chemical and electronic tongue analyses. *Meat Sci.* **2019**, *151*, 82–88. [CrossRef] [PubMed]
7. Gorraiz, C.; Beriain, M.J.; Chasco, J.; Insausti, K. Effect of aging time on volatile compounds, odor, and flavor of cooked beef from Pirenaica and Friesian bulls and heifers. *J. Food Sci.* **2002**, *67*, 916–922. [CrossRef]
8. Choe, J.; Park, B.; Lee, H.J.; Jo, C. Potential antioxidant and angiotensin i-converting enzyme inhibitory activity in crust of dry-aged beef. *Sci. Rep.* **2020**, *10*, 7883. [CrossRef]
9. Park, B.; Yong, H.I.; Choe, J.; Jo, C. Utilization of the crust from dry-aged beef to enhance flavor of beef patties. *Korean J. Food Sci. Anim.* **2018**, *38*, 1019–1028. [CrossRef]
10. Jung, I.C.; Park, H.S.; Choi, Y.J.; Park, S.S.; Kim, M.J.; Park, K.S. The effect of adding lotus root and leaf powder on the quality characteristics of cooked pork patties. *Korean J. Food Cook. Sci.* **2011**, *27*, 783–791. [CrossRef]
11. Choi, Y.J.; Park, K.S.; Jung, I.C. Quality characteristics of ground pork meat containing hot water extract from dandelion (*Taraxacum officinale*). *J. East Asian Soc. Diet. Life* **2015**, *25*, 651–659. [CrossRef]
12. Choi, Y.S.; Jeon, K.H.; Park, J.D.; Sung, J.M.; Seo, D.H.; Ku, S.K.; Oh, N.S.; Kin, Y.B. Comparison of pork patties quality characteristics with various binding agents. *Korean J. Food Cook. Sci.* **2015**, *31*, 588–595. [CrossRef]
13. Barbut, S.; Wood, J.; Marangoni, A. Potential use of organogels to replace animal fat in comminuted meat products. *Meat Sci.* **2016**, *122*, 155–162. [CrossRef] [PubMed]

14. Kim, H.E.; Chin, K.B. Antioxidant activities of brussels sprouts powder and its application to pork patties on the physicochemical properties and antioxidant activity during refrigerated storage. *J. Korean Soc. Food Sci. Nutr.* **2018**, *47*, 733–741. [CrossRef]

15. Kim, H.Y.; Kim, G.W.; Jeong, H.G. Development of *tteokgalbi* added with red pepper seed powder. *J. Korean Soc. Food Sci. Nutr.* **2016**, *45*, 255–260. [CrossRef]

16. Jeong, H.G.; Kim, H.Y. Development of *tteokgalbi* added pig skin gelatine powder. *J. Korean Soc. Food Sci. Nutr.* **2016**, *45*, 1147–1152. [CrossRef]

17. Joo, S.Y.; Choi, H.Y. Antioxidant activity and quality characteristics of pork patties added with saltwort (*Salicornia herbacea* L.) powder. *J. Korean Soc. Food Sci. Nutr.* **2014**, *43*, 1189–1196. [CrossRef]

18. Jung, H.O.; Lee, J.J. Quality and srorage characteristics of pork Teokgalbi with added rosemary (*Rosemarinus officinalis*) extract powder. *Korean J. Community Living Sci.* **2016**, *27*, 509–520. [CrossRef]

19. AOAC. *Official Methods of Analysis of AOAC International*, 19th ed.; AOAC International: Gaithersburg, MD, USA, 2012; p. 931.

20. Cabling, M.M.; Kang, H.S.; Lopez, B.M.; Jang, M.; Kim, H.S.; Nam, K.C.; Nam, J.G.; Choi, J.G.; Seo, K.S. Estimation of genetic associations between production and meat quality traits in duroc pigs. *Asian-Australas. J. Anim. Sci.* **2015**, *28*, 1061–1065. [CrossRef]

21. Campbell, R.E.; Hunt, M.C.; Levis, P.; Chambers, E. Dry-aging effects on palatability of beef longissimus muscle. *J. Food Sci.* **2001**, *66*, 196–199. [CrossRef]

22. Lee, C.W.; Lee, S.H.; Min, Y.; Lee, S.; Joo, C.; Jung, S. Quality improvement of strip loin from Hanwoo with low quality grade by dry aging. *Korean J. Food Nutr.* **2015**, *28*, 415–421. [CrossRef]

23. Abdullah, N.; Wahab, N.; Saruan, N.; Matias-Peralta, H.M.; Xavier, N.R.; Muhammad, N.; Talip, B.A.; Bakar, M.F.A. Effect of replacing coconut milk with almond milk in spicy coconut gravy on its sensorial, nutritional and physical properties. *Mater. Today-Proc.* **2018**, *5*, 21919–21925. [CrossRef]

24. Park, S.Y.; Seol, K.H.; Kim, H.Y. Effect of dry-aged beef crust levels on quality properties of brown sauce. *Food Sci. Anim. Resour.* **2020**, *40*, 699–709. [CrossRef] [PubMed]

25. Kim, I.S.; Jang, A.; Jin, S.K.; Lee, M.; Jo, C. Effect of marination with mixed salt and kiwi juice and cooking methods on the quality of pork loin-based processed meat product. *J. Korean Soc. Food Sci. Nutr.* **2008**, *37*, 217–222. [CrossRef]

26. Forrest, J.C.; Aberle, E.D.; Hedrick, H.B.; Judge, M.D.; Merkel, R.A. Principles of Meat Processing. In *Principles of Meat Science*; Schweigert, B.S., Ed.; WH Freeman and Co: New York, NY, USA, 1975; pp. 190–226.

27. Kim, M.; Choe, J.; Lee, H.J.; Yoon, Y.; Yoon, S.; Jo, C. Effects of aging and aging method on physicochemical and sensory traits of different beef cuts. *Food Sci. Anim. Resour.* **2019**, *39*, 54–64. [CrossRef]

28. Mancini, R.A.; Hunt, M.C. Current research in meat color. *Meat Sci.* **2005**, *71*, 100–121. [CrossRef]

29. Young, O.A.; West, J. Meat Color. In *Meat Science and Applications*; Hui, Y.H., Nio, W.K., Rogers, R.W., Young, O.A., Eds.; Marcel Dekker: New York, NY, USA, 2001; pp. 39–70.

30. Park, J.W. A Study on the Quality Property of Pork Patties Supplemented by Concentrated Soy Protein (CSP). Master's Thesis, Chung-Ang University, Seoul, Korea, 2011; pp. 4–15.

31. Mittal, G.S.; Usborne, W.R. Meat emulsion extenders. *Food Technol.* **1985**, *39*, 121–130.

32. Eshtiachi, M.N.; Stute, R.; Knorr, D. High-pressure and freesing pretreatment effects on drying, rehydraion, texture and color of green beans, carrots and potatoes. *J. Food Sci.* **1994**, *59*, 1168–1170. [CrossRef]

33. Choi, J.H.; Yong, H.I.; Ku, S.K.; Kim, T.K.; Choi, Y.S. The qyality characteristics of pork patties according to the replacement of mealworm (*Tenebrio molitor* L.). *Korean J. Food Cook. Sci.* **2019**, *35*, 441–449. [CrossRef]

34. Yang, H.D. Quality of Meat. In *Meat Science*, 1st ed.; Yang, H.D., Ed.; Sunjin Publishing: Gyeonggi-do, Korea, 2018; pp. 167–172.

35. June, J.H.; Yoon, J.Y.; Kim, H.S. A study on the development of 'hodojook'. *J. Korean Soc. Food Cult.* **1998**, *13*, 509–518.

36. Hamm, R. On the rheology of minced meat. *J. Texture Stud.* **1975**, *6*, 281–296. [CrossRef]

37. Kim, H.W.; Choi, J.H.; Choi, Y.S.; Han, D.J.; Kim, H.Y.; Lee, M.A.; Shim, S.Y.; Kim, C.J. Effect of wheat fiber and isolated soy protein on the quality characteristics of frankfurter type sausages. *Korean J. Food Sci. Anim. Resour.* **2009**, *29*, 475–481. [CrossRef]

38. Brown, L.M.; Zayas, J.F. Corn germ protein flour as an extender in broiled patties. *J. Food Sci.* **1990**, *55*, 888–892. [CrossRef]

39. Berry, B.W.; Leddy, K.F. Effects of freezing rate, frozen storage temperature and storage time on tenderness values of beef patties. *J. Food Sci.* **1989**, *54*, 291–296. [CrossRef]

40. Lu, G.H.; Chen, T.C. Application of egg white and plasma powders as muscle food binding agents. *J. Food Eng.* **1999**, *42*, 147–151. [CrossRef]

41. Choi, S.H.; Kim, D.S. Quality characteristics of hamburger patties adding with Tofu powder. *Culin. Sci. Hosp. Res.* **2014**, *20*, 28–40.

42. Choi, Y.S.; Lee, M.A.; Jeong, J.Y.; Choi, J.H.; Han, D.J.; Lee, E.S.; Kim, C.J. Effects of wheat fiber on the quality of meat batter. *Korean J. Food Sci. Anim.* **2007**, *27*, 22–28. [CrossRef]

43. Song, H.I.; Moon, K.I.; Moon, Y.H.; Jung, I.C. Quality and storage of hamburger during low temperature storage. *Korean J. Food Sci. Anim. Resour.* **2000**, *20*, 72–78.

44. Moon, Y.H.; Kang, S.J.; Hyon, J.S.; Kang, H.G.; Jung, I.C. Comparison of the palatability related with characteristics of beef carcass grade B2 and D. *J. Korean Soc. Food Sci. Nutr.* **2001**, *30*, 1152–1157.

Article

Steam Explosion Pretreatment for Improving Wheat Bran Extrusion Capacity

Lan Wang [1,2,*]**, Tairan Pang** [1,2]**, Feng Kong** [1,2] **and Hongzhang Chen** [1,2]

[1] State Key Laboratory of Biochemical Engineering, Beijing Key Laboratory of Biomass Refining Engineering, Institute of Process Engineering, Chinese Academy of Sciences, No. 1 Bei-Er-Jie, Zhongguancun, Haidian District, Beijing 100190, China

[2] University of Chinese Academy of Sciences, No. 19(A) Yuquan Road, Shijingshan District, Beijing 100049, China

* Correspondence: wanglan@ipe.ac.cn; Tel.: +86-010-8254-4978

Abstract: Extrusion improves the texture of wheat bran and enhances its product edibility, making it a promising processing method. However, the extrusion performance of wheat bran without any treatment is not satisfactory and limits the utilization of wheat bran in food processing. In this study, steam explosion pretreatment was used to treat wheat bran to investigate its promotion of wheat bran extrusion. The results showed that steam explosion could increase the extrusion ratio of wheat bran extrudate by 36%. Grinding the steam-exploded wheat bran extrudate yields wheat bran flour with smaller particle sizes and higher cell wall breakage. Fourier transform infrared spectroscopy and chemical composition results revealed that steam explosion degraded insoluble dietary fiber and disrupted the dense structure of the cell wall in wheat bran. The water-extracted arabinoxylan and soluble dietary fiber content of steam-exploded wheat bran were 13.95% and 7.47%, respectively, improved by 1567.42% and 241.75% compared to untreated samples. The total phenol and flavonoid contents, water solubility index, and cation exchange capacity of steam-exploded wheat bran extrudate were all superior to raw wheat bran extrudate. In summary, this study demonstrates that steam explosion improves the extrusion capacity of wheat bran and facilitates its utilization.

Keywords: steam explosion; wheat bran; dietary fiber; extrusion capacity

Citation: Wang, L.; Pang, T.; Kong, F.; Chen, H. Steam Explosion Pretreatment for Improving Wheat Bran Extrusion Capacity. *Foods* **2022**, *11*, 2850. https://doi.org/10.3390/foods11182850

Academic Editors: Fuping Lu and Wenjie Sui

Received: 8 August 2022
Accepted: 2 September 2022
Published: 15 September 2022

Publisher's Note: MDPI stays neutral with regard to jurisdictional claims in published maps and institutional affiliations.

1. Introduction

Extrusion is a typical thermo-mechanical processing technology that processes raw materials at high temperatures with a short mechanical shear [1,2]. The extruded product has an expanded structure and ripening characteristics that attract consumers [3]. Extrusion can disrupt the physical structure of macromolecules (such as proteins and starches) into small molecules, thus changing the properties of the food ingredients and thereby improving the taste of the food and facilitating its digestion and absorption by the body [4]. However, extruded products generally have a high-calorie content, limiting their development [5].

Products with high dietary fiber content and low energy density are more acceptable to consumers [6]. Dietary fiber consists of cellulose, lignin, and hemicellulose (highly substituted insoluble arabinoxylan) [6,7]; it is beneficial to human health and improves intestinal health as well as prevents some diseases such as constipation and colon cancer [8]. Wheat bran, a by-product of wheat flour processing, is rich in dietary fiber and phenolic compounds and has potential use as a food product, which can positively affect the nutritional value of foods [9–11].

However, the high dietary fiber content in whole wheat makes extrusion challenging [6]. This phenomenon is mainly due to the insoluble dietary fiber in wheat bran, which affects starch extrusion [12]. Increasing the wheat bran content in the extrusion formulation can reduce the extrusion ratio, thus reducing the textural properties of the product [6,13]. In addition, wheat bran can cause the extruder to run erratically, leading to the machine being clogged or even exploding. Wheat bran contains 44–50% dietary fiber, most of which is insoluble [14], resulting in inferior bran extrusion properties. Numerous studies have pointed out that soluble dietary fibers provide better extrusion performance and more favorable textures of products than insoluble dietary fibers [6]. Thus, it is necessary to find ways to avoid this negative effect of insoluble dietary fiber in wheat bran.

Steam explosion is a green and efficient method that uses highly permeable saturated vapor to pretreat the material and release the pressure instantaneously to achieve a cell wall disruption effect [15]. Steam explosion is principally applied to the pretreatment of lignocellulosic materials (e.g., wheat straw, maize straw) and can destroy the structure of insoluble dietary fibers such as cellulose and hemicellulose [16]. This study, therefore, investigates the steam explosion treatment as a method to improve the expansion properties of wheat bran extrudates. The expansion rate and grindability of wheat bran before and after steam explosion were characterized. Subsequently, the effects of steam explosion and extrusion were evaluated on the chemical composition of wheat bran. Finally, the improvement of wheat bran in functional properties (water holding capacity, swelling, ion exchange properties) after the two treatments was also studied. This study provides a novel strategy for processing wheat-bran-based extruded foods and is expected to broaden the path for the utilization of wheat bran products towards efficient utilization.

2. Materials and Methods

2.1. Materials

Wheat bran was provided by Langfang Mountain Food Co., Ltd. (Hebei, China). The compositions of wheat bran were 37.6% starch and 48.6% total dietary fiber on a dry-weight basis. Heat-stable α-amylase, protease, and amyloglucosidase were purchased from Shanghai Macklin Biochemical Co., Ltd. (Shanghai, China). All other chemicals used in the experiments were of analytical grade.

2.2. Sample Preparation

Wheat bran was rehydrated with distilled water for about 30 min to adjust the initial moisture content to 30% (w/w). Steam explosion pretreatment was performed in a 10 L self-designed batch vessel that consisted of three units: a steam generator, a reaction chamber, and two reception chambers (Weihai Automatic Control Reactor Co., Ltd., Weihai, China), each controlled by a ball valve between the units. [1,17,18] Wheat bran was loaded into the reaction chamber, treated under 0.8 MPa of water vapor, and held for 5 min. Then, the discharge valve was opened to allow the reaction chamber to unload the pressure quickly, and wheat bran entered the receiving chamber. Raw and steam-exploded wheat bran samples were extruded in an extruder (SYSLG30-IV, Shan Dong Saibainuo Machinery Co., Ltd., Dezhou, China) at a moisture content of 33%, with a maximum temperature of 140 °C and a feeding speed of 15 Hz. Wheat bran extrudate was dried at 60 °C for 12 h, then packed in polyethylene bags and stored at 4 °C until analyzed.

2.3. Characterization

2.3.1. Fourier Transform Infrared Spectroscopy

Fourier transform infrared spectroscopy (FTIR) was performed on the samples according to Zhao et al. [1] using an FTIR-8400S spectrometer (Shimadzu, Japan). Specifically, 1 mg of the dried sample was mixed with 100 mg of KBr and then pressed for 10 min at 8 tons to prepare the discs. Semi-quantitative analysis of the FTIR spectra, according to the method [2], and the band at 1519 cm^{-1} was used as a reference band to estimate the relative intensity of other bands [1].

2.3.2. Chemical Composition Analysis of Wheat Bran

Dietary fiber in wheat bran samples was analyzed using the American Association of Cereal Chemists (AACC) method 32-07. Total arabinoxylans and water-extractable arabinoxylans content were determined using the method of Hashimoto et al. [3].

2.3.3. Examination of Extrusion Ratio and Grinding Performance of Extruded Wheat Bran

The extrusion ratio of the extruded wheat bran was referenced in Zhu et al. [7] and Alam et al. [19], respectively. The die diameter used in the experiments was 3 mm. After grinding 100 g samples in an FW-400A grinder (Zhongxingweiye, Beijing, China) for 2 min, the particle size distribution was measured at room temperature using an NKT6100-B laser particle size meter (Naikete, Jinan, China). The particle size distributions and the particle sizes corresponding to the cumulative percentage of particle size distribution for samples reaching 10%, 50%, and 90% (D10, D50, and D90) were calculated by the software provided with the instrument [8]. The cell wall breakage ratio of samples was calculated using the method of Xiao et al. [9].

2.3.4. Total Phenolic Content and Total Flavonoid Content of Extruded Wheat Bran

Total phenolic content was measured by the Folin–Ciocalteu method using gallic acid as the standard, as described by Wu et al. [4]. Absorbance was measured using a UV–vis spectrophotometer (UV-5800PC, Shanghai Metash Instruments Co., Ltd., China) at 760 nm against a reagent blank. The total phenolic content was expressed as milligrams of gallic acid equivalents per gram of dry sample weight (mg of GAE/g) through the calibration curve of gallic acid.

The total flavonoid content was determined using an aluminum chloride colorimetric method described by Jia et al. [6]. The absorbance was measured at 510 nm using a UV–vis spectrophotometer. The total flavonoid content was expressed as milligrams of catechin equivalents per gram of dry sample weight (mg of CAE/g) using the calibration curve of (±)-catechin.

2.3.5. Expansion Capacity and Cation Exchange Capacity Detection of Extruded Wheat Bran

The expansion capacity of samples was determined using the method of Raghavendra et al., [10] water solubility index was measured as described by Jafari et al., [11] and cation exchange capacity was determined according to Ralet et al. [12]; a 0.25 g sample was taken in 10 mL of 0.1 mol/L HCl solution, placed at room temperature for 24 h, and then filtered to wash away the excess acid. The filtered residue was added to 100 mL 5% sodium chloride solution, and the solution was titrated with 0.01 mol/L sodium hydroxide solution. Instead of hydrochloric acid, distilled water was used to determine the sodium hydroxide consumed as a blank control. The cation exchange capacity was calculated by the following equation:

$$CEC = \frac{(V_1 - V_2) \times C}{m} \tag{1}$$

where CEC is the cation exchange capacity of the sample (mL/g), V_1 is the volume of sodium hydroxide solution consumed by the sample (mL), V_2 is the volume of sodium hydroxide solution consumed by the blank (mL), m is the mass of the sample (g), and C is the concentration of the sodium hydroxide solution (mol/L).

2.3.6. Statistical Analysis

The results are reported as mean ± standard deviation of at least three replicated determination results. Experimental data were processed by one-way analysis of variance using IBM SPSS Statistics 20 (IBM, Armonk, NY, USA) with Duncan's multiple range test ($p < 0.05$).

3. Results and Discussion

3.1. Functional Groups and Chemical Composition of Wheat Bran before and after Steam Explosion

FTIR spectra can effectively characterize the functional groups in raw materials and reflect the changes in the chemical structure of wheat bran before and after the steam explosion. Figure 1 illustrates the positions of the absorption peaks in the FTIR spectra of wheat bran before and after the steam explosion. The wavenumbers peaked at 3348 (OH), 2923 (CH), 1411 (CH$_2$), 1242 (CO), and 1041 cm^{-1} (CO), indicating the presence of cellulose, hemicellulose, and lignin in wheat bran. The absorbance peak at 1242 cm^{-1} is associated with the C–O stretching vibration of the acetyl group in the lignin, and the peak at 1411 cm^{-1} is attributed to the bending vibration of the C–H group of the aromatic ring in the polysaccharides [20]. The band at 1041 cm^{-1} is owed to the C–O stretching modes of the hydroxyl and ether groups in the cellulose. The peak at 2905 cm^{-1} is a characteristic band for the C–H stretching vibration of methyl and methylene in the cellulose and hemicellulose components [21]. The final peak at 3345 cm^{-1} is owed to the presence of O–H stretching vibrations and the hydrogen bond of the hydroxyl groups. The reduction observed in the relative intensity of the 1041 cm^{-1} band after steam explosion suggests that steam explosion promotes the degradation of dextran and xylan to glucose and xylose, respectively [1]. The relative intensity of the peak at 1242 cm^{-1} decreased, suggesting that steam explosion reduced the relative content of the guaiacyl lignin units [1]. The spectra presented peaks at 1650 cm^{-1}, corresponding to amide I in proteins. The result showed that steam explosion reduced the intensities of the amide I bands and might degrade protein [13]. The relative intensity of the 2923 cm^{-1} band, attributed to the C-H vibration of methyl and methylene [14], decreased from 1.2385 in raw wheat bran to 1.0782 in steam-exploded wheat bran (Table 1), indicating a reduction in the methyl and methylene components of steam-exploded wheat bran, further reflecting the reduced dietary fiber content [15,16]. The peak in this spectrum at 3348 cm^{-1} was associated with hydrogen and hydroxyl-bound O-H [22]. The decrease in the relative peak intensity of the steam-exploded wheat bran indicates a reduction in O-H bonds. The chemical structure of wheat bran was significantly altered after the steam explosion. Steam explosion pretreatment reduced the crystallinity of the cellulose in wheat bran and degraded the dietary fiber, thus promoting the expansion ratio and grindability of the extruded wheat bran [21].

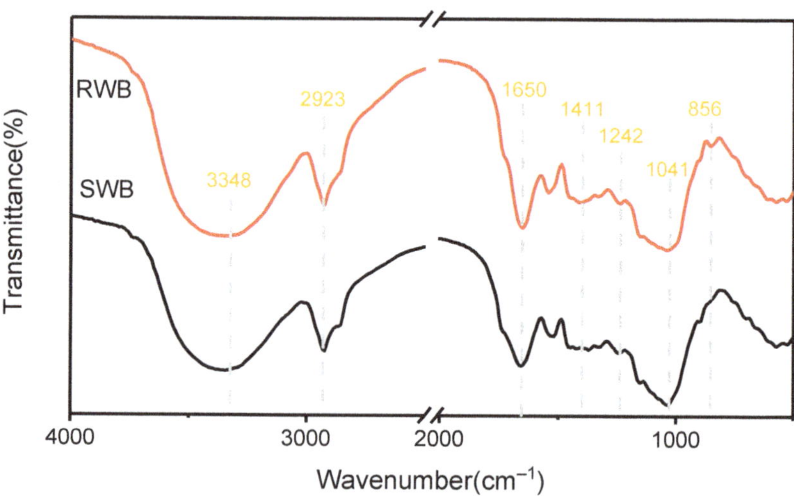

Figure 1. FTIR spectroscopy of raw wheat bran (RWB) and steam-exploded wheat bran (SWB).

Table 1. FTIR semi-quantitative analysis of RWB and SWB.

Band (cm^{-1})	Assignment	Relative Intensity	
		RWB	SWB
1041	COC stretching typical of glucan and xylan	2.3993	1.3404
1242	CO stretching vibration of acetyl group in the lignin	1.2049	1.0844
1519	Aromatic skeletal stretching	1.0000	1.0000
1650	CO stretching of amide I	1.6237	1.1483
2923	CH in cellulose and hemicellulose	1.2385	1.0782
3348	OH stretching and hydrogen bonds	1.8799	1.1769

The steam explosion and extrusion process can produce a series of chemical reactions that result in variations in the chemical and nutritional composition of the sample. The arabinoxylan, water-extractable arabinoxylan, insoluble dietary fiber, and soluble dietary fiber contents of samples are shown in Figure 2. It can be seen that the insoluble dietary fiber content of wheat bran decreased remarkably after the steam explosion (from 45.50% to 34.40%), while the percentage of soluble dietary fiber increased (from 3.09% to 7.47%), which is consistent with our previous study [17]. The high content of insoluble dietary fiber can cause a reduction in extrusion performance [23], while soluble dietary fiber provides a higher extrusion rate [24]. Therefore, the steam explosion pretreatment process, which degrades insoluble dietary fiber to soluble dietary fiber, dramatically improves wheat bran extrusion performance.

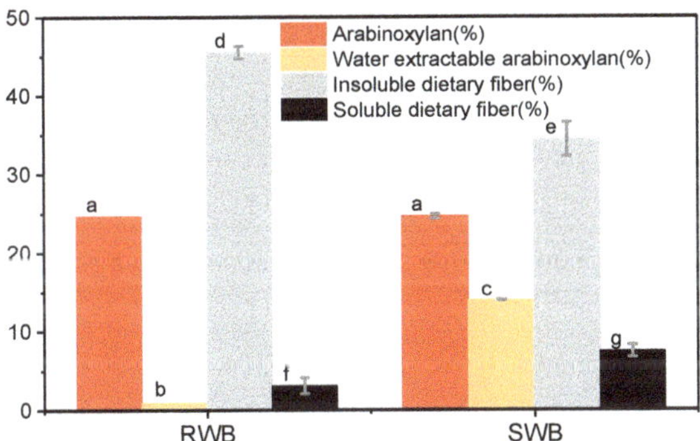

Figure 2. Chemical composition of raw wheat bran (RWB) and steam-exploded wheat bran (SWB); means that are significantly different ($p < 0.05$) are indicated by different letters.

In addition, arabinoxylan is an essential component of wheat bran cell wall polysaccharides [25,26]. The water-extractable arabinoxylan content in water can be utilized as an indicator to investigate the degradation effect of steam explosion treatments on the cell wall [27,28]. It was clear that the water-extractable arabinoxylan content of the steam-exploded sample was much higher than that of the non-steam-exploded sample (by about 1567.4%). On one hand, the high water-extractable arabinoxylan content of steam-exploded wheat bran confirms the destructive effect of steam explosion on the wheat bran cell wall; on the other hand, this may facilitate its extrusion performance [29].

3.2. Extrusion Performance and Grindability of Samples

The extrusion ratio is the most important index used to measure the performance of extruded products. The extrusion performance of steam-exploded and non-steam-exploded samples is shown in Figure 3a. The steam-exploded wheat bran exhibited better extrusion performance, with an extrusion ratio of 1.60, while the non-steam-exploded wheat bran was almost unexpanded (extrusion ratio of 1.16). Moreover, steam explosion pretreatment reduced problems such as the clogging, smoking, and spraying of wheat bran during the extrusion process, which made the extrusion of wheat bran accessible. The above results indicate that the untreated wheat bran has almost no extrusion capacity, while steam explosion gives better extrusion properties to wheat bran.

To further characterize the expansibility of steam-exploded wheat bran, the grindability of extruded wheat bran samples was investigated. The particle size distribution of the two samples is shown in Figure 3b and Table 2. The particle size distribution (D90) of the two samples was quite different, and the steam explosion pretreatment resulted in a lower particle size distribution of the extruded wheat bran flour under the same grinding conditions. The causes of this can be traced to the extruded wheat bran, which contains a large number of internal voids that lead to its loose structure, with reduced mechanical strength. The mechanical strength of samples with high extrusion ratios can be reduced more significantly and thus are more susceptible to being ground into fine flour [30]. Thus, the enhancement of wheat bran extrusion properties by steam explosion pretreatment was demonstrated based on the fact that steam-exploded wheat bran was more likely to be ground into fine flour. In addition, the cell wall breakage ratios of extruded samples without and after pretreatment were 25.05% and 47.00%, respectively. This indicates that the steam explosion pretreatment facilitates the cell wall breakage rate in wheat bran flour, which could improve the release level of intracellular substances and is more conducive to improving the nutritional value of the food and promoting the absorption of nutrients by the human body.

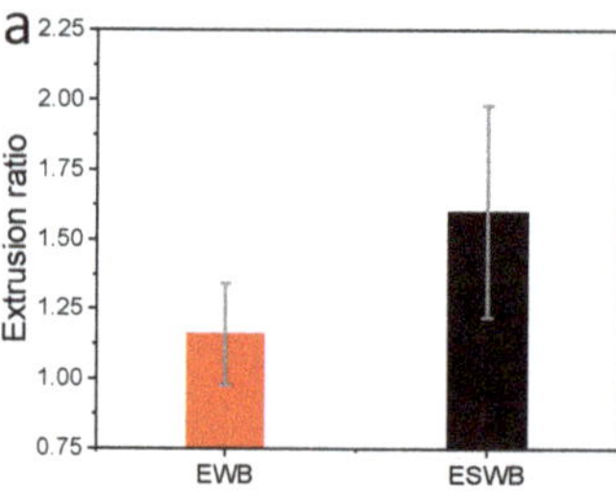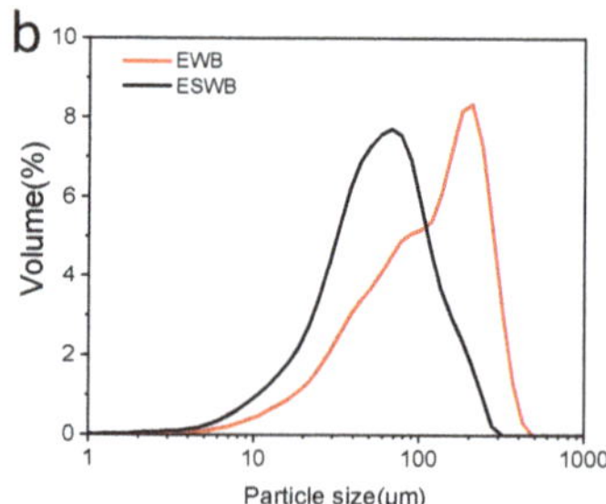

Figure 3. Macrostructural parameters (**a**) and particle size distributions (**b**) of extruded wheat bran (EWB) and extruded steam-exploded wheat bran (ESWB).

Table 2. Particle size distribution of EWB and ESWB flour.

Samples	D10 (μm)	D50 (μm)	D90 (μm)	Cell Wall Breakage Ratio (%)
EWB	26.97 ± 0.60	109.15 ± 3.00	239.07 ± 4.33	25.05 ± 0.62
ESWB	17.07 ± 1.27	52.48 ± 2.40	126.79 ± 4.48	47.00 ± 1.71

3.3. Functional Properties of Samples

The water solubility index indicates the decomposition degree of sugar and fiber substances in the sample [31]. Higher solubility indicates that the sample contains fewer macromolecules and more substances that the human body can absorb more easily. In Figure 4a, the water solubility indices of wheat bran with different treatments are shown. The water solubility index of the samples treated with steam explosion (0.22 and 0.18 g/g) was significantly higher than those of the untreated samples (0.11 and 0.10 g/g), which could be attributed to the soluble pentosan content increase due to the disruption of the wheat bran cell wall through the steam explosion [17] (Figure 4a). The extrusion process loosens the structure of wheat bran and makes it easy to expand. Therefore, the expansion properties of the extrusion-treated wheat bran samples were all stronger than those of the untreated samples (Figure 4b).

Compared with the non-steam-exploded sample, the total flavonoid content and total phenolic content of steam-exploded wheat bran were increased by 177.27% and 223.85%, respectively (Figure 4c,d). This phenomenon may be due to the steam explosion disrupting the dense structure of the cell wall of wheat bran and promoting the dissolution of the flavonoid and phenolic compounds [17]. In addition, cell wall friction damage during the steam explosion process resulted in the release of conjugated phenolic compounds that were bound to polysaccharides or proteins [32]. It indicates that steam explosion treatment facilitates nutrient extraction from wheat bran and allows for the efficient utilization of wheat bran [33]. The total phenolic and flavonoid contents of extruded wheat bran were not significantly changed (Figure 4c,d). This may be due to the intracellular substance dissolution promoted in wheat bran cells by the extrusion process; the high temperature in this process resulted in the inactivation of phenolics and flavonoids, which counteracted each other [34].

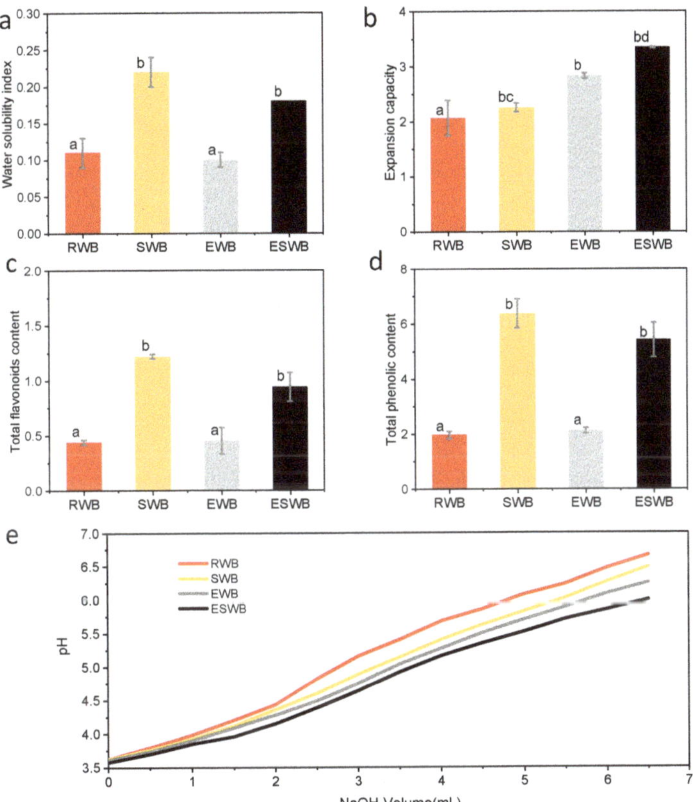

Figure 4. Water solubility index (**a**), expansion capacity (**b**), total flavonoid (**c**) and phenolic (**d**) contents, as well as cation exchange capacity (**e**) of raw wheat bran (RWB), steam-exploded wheat bran (SWB), extruded wheat bran (EWB), and extruded steam-exploded wheat bran (ESWB); means that are significantly different ($p < 0.05$) are indicated by different letters.

Due to its side chain groups, such as carboxyl, hydroxyl, and amino groups, dietary fiber can reversibly exchange cations, lowering blood pressure [35,36]. As the content of NaOH in the system increased, its pH was raised (Figure 4b). From Equation (1), it can be seen that the larger volume of sodium hydroxide consumed the higher cation exchange capacity of the sample as the target pH was reached. After extrusion treatment, the steam-exploded wheat bran exhibited the highest sodium hydroxide demand, indicating its high cation exchange capacity. The enhanced cation exchange capacity of wheat bran may be due to a tremendous number of functional groups being exposed after the dual treatment of steam explosion and extrusion [37]. The above results indicate that steam explosion improves the puffing performance of wheat bran and enhances the functionality of the puffed product, thus promoting the functionalization of wheat bran.

4. Conclusions

In this study, steam explosion pretreatment was applied to improve the extrusion ratio of wheat bran. The steam explosion can promote the conversion of insoluble dietary fiber to soluble dietary fiber, increase the content of water-extractable arabinoxylan, as well as break the dense structure of the cell wall of wheat bran to facilitate the dissolution of flavonoid and phenolic compounds. Steam-exploded wheat bran exhibited a higher extrusion capacity than untreated samples. Furthermore, the grindability of the samples was further evidenced; the extrusion ratio of steam-exploded wheat bran was superior to that of raw wheat bran. Physicochemical properties such as the water solubility index, expansion capacity, and cation exchange capacity of the steam-exploded wheat bran after extrusion were enhanced compared to the other samples. In summary, this work demonstrates that steam explosion pretreatment upgrades the extrusion capacity of wheat bran, broadening the approach to the processing of wheat bran and boosting the potential of wheat bran applications in foods.

Author Contributions: Conceptualization, L.W. and H.C.; methodology, F.K.; software, T.P.; validation, F.K.; investigation, F.K.; resources, H.C.; data curation, F.K.; writing—original draft preparation, T.P.; writing—review and editing, L.W. and H.C.; visualization, T.P.; supervision, L.W. and H.C.; project administration, L.W. and H.C.; funding acquisition, L.W. All authors have read and agreed to the published version of the manuscript.

Funding: This study was financially supported by the National Key Research and Development Program of China (Grant 2019YFB1503800) and Transformational Technologies for Clean Energy and Demonstration (Strategic Priority Research Program of the Chinese Academy of Sciences, Grant No. XDA 21060300).

Institutional Review Board Statement: Not applicable.

Informed Consent Statement: Not applicable.

Data Availability Statement: The data presented in this study.

Conflicts of Interest: The authors declare no conflict of interest.

References

1. Zhao, Z.M.; Wang, L.; Chen, H.Z. A novel steam explosion sterilization improving solid-state fermentation performance. *Bioresour. Technol.* **2015**, *192*, 547–555. [CrossRef] [PubMed]
2. Nada, A.-A.M.A.; Yousef, M.A.; Shaffei, K.A.; Salah, A.M. Infrared spectroscopy of some treated lignins. *Polym. Degrad. Stab.* **1998**, *62*, 157–163. [CrossRef]
3. Hashimoto, S.; Shogren, M.D.; Pomeranz, Y. Cereal pentosans: Their estimation and significance. I. Pentosans in wheat and milled wheat products. *Cereal Chem.* **1987**, *64*, 30–34.
4. Wu, L.; Huang, Z.; Qin, P.; Ren, G. Effects of processing on phytochemical profiles and biological activities for production of sorghum tea. *Food Res. Int.* **2013**, *53*, 678–685. [CrossRef]
5. Li, Y.; Sun, Y.; Zhong, M.; Xie, F.; Wang, H.; Li, L.; Qi, B.; Zhang, S. Digestibility, textural and sensory characteristics of cookies made from residues of enzyme-assisted aqueous extraction of soybeans. *Sci. Rep.* **2020**, *10*, 4222. [CrossRef] [PubMed]
6. Zhishen, J.; Mengcheng, T.; Jianming, W. The determination of flavonoid contents in mulberry and their scavenging effects on superoxide radicals. *Food Chem.* **1999**, *64*, 555–559. [CrossRef]
7. Zhu, L.-J.; Shukri, R.; de Mesa-Stonestreet, N.J.; Alavi, S.; Dogan, H.; Shi, Y.-C. Mechanical and microstructural properties of soy protein–high amylose corn starch extrudates in relation to physiochemical changes of starch during extrusion. *J. Food Eng.* **2010**, *100*, 232–238. [CrossRef]
8. He, S.; Tang, M.; Sun, H.; Ye, Y.; Cao, X.; Wang, J. Potential of water dropwort (*Oenanthe javanica* DC.) powder as an ingredient in beverage: Functional, thermal, dissolution and dispersion properties after superfine grinding. *Powder Technol.* **2019**, *353*, 516–525. [CrossRef]
9. Xiao, W.; Zhang, Y.; Fan, C.; Han, L. A method for producing superfine black tea powder with enhanced infusion and dispersion property. *Food Chem.* **2017**, *214*, 242–247. [CrossRef]
10. Raghavendra, S.N.; Rastogi, N.K.; Raghavarao, K.S.M.S.; Tharanathan, R.N. Dietary fiber from coconut residue: Effects of different treatments and particle size on the hydration properties. *Eur. Food Res. Technol.* **2004**, *218*, 563–567. [CrossRef]
11. Jafari, S.M.; Ghalegi Ghalenoei, M.; Dehnad, D. Influence of spray drying on water solubility index, apparent density, and anthocyanin content of pomegranate juice powder. *Powder Technol.* **2017**, *311*, 59–65. [CrossRef]
12. Ralet, M.C.; Della Valle, G.; Thibault, J.F. Raw and extruded fibre from pea hulls. Part I: Composition and physico-chemical properties. *Carbohydr. Polym.* **1993**, *20*, 17–23. [CrossRef]
13. Wang, L.P.; Shen, Q.R.; Yu, G.H.; Ran, W.; Xu, Y.C. Fate of biopolymers during rapeseed meal and wheat bran composting as studied by two-dimensional correlation spectroscopy in combination with multiple fluorescence labeling techniques. *Bioresour. Technol.* **2012**, *105*, 88–94. [CrossRef] [PubMed]
14. Paiva, M.C.; Ammar, I.; Campos, A.R.; Cheikh, R.B.; Cunha, A.M. Alfa fibres: Mechanical, morphological and interfacial characterization. *Compos. Sci. Technol.* **2007**, *67*, 1132–1138. [CrossRef]
15. Honců, I.; Sluková, M.; Vaculová, K.; Sedláčková, I.; Wiege, B.; Fehling, E. The effects of extrusion on the content and properties of dietary fibre components in various barley cultivars. *J. Cereal Sci.* **2016**, *68*, 132–139. [CrossRef]

16. Zhang, M.Y.; Liao, A.M.; Thakur, K.; Huang, J.H.; Zhang, J.G.; Wei, Z.J. Modification of wheat bran insoluble dietary fiber with carboxymethylation, complex enzymatic hydrolysis and ultrafine comminution. *Food Chem.* **2019**, *297*, 124983. [CrossRef]

17. Kong, F.; Wang, L.; Gao, H.; Chen, H. Process of steam explosion assisted superfine grinding on particle size, chemical composition and physico-chemical properties of wheat bran powder. *Powder Technol.* **2020**, *371*, 154–160. [CrossRef]

18. Chen, H. Gas Explosion Equipments. In *Gas Explosion Technology and Biomass Refinery*; Chen, H., Ed.; Springer: Dordrecht, The Netherlands, 2015; pp. 87–143. [CrossRef]

19. Alam, S.A.; Järvinen, J.; Kirjoranta, S.; Jouppila, K.; Poutanen, K.; Sozer, N. Influence of Particle Size Reduction on Structural and Mechanical Properties of Extruded Rye Bran. *Food Bioprocess Technol.* **2013**, *7*, 2121–2133. [CrossRef]

20. Liu, W.; Mohanty, A.K.; Drzal, L.T.; Askel, P.; Misra, M. Effects of alkali treatment on the structure, morphology and thermal properties of native grass fibers as reinforcements for polymer matrix composites. *J. Mater. Sci.* **2004**, *39*, 1051–1054. [CrossRef]

21. Liu, Z.-H.; Qin, L.; Pang, F.; Jin, M.-J.; Li, B.-Z.; Kang, Y.; Dale, B.E.; Yuan, Y.-J. Effects of biomass particle size on steam explosion pretreatment performance for improving the enzyme digestibility of corn stover. *Ind. Crops Prod.* **2013**, *44*, 176–184. [CrossRef]

22. Zheng, Y.; Li, Y. Physicochemical and functional properties of coconut (*Cocos nucifera* L.) cake dietary fibres: Effects of cellulase hydrolysis, acid treatment and particle size distribution. *Food Chem.* **2018**, *257*, 135–142. [CrossRef] [PubMed]

23. Robin, F.; Dubois, C.; Pineau, N.; Schuchmann, H.P.; Palzer, S. Expansion mechanism of extruded foams supplemented with wheat bran. *J. Food Eng.* **2011**, *107*, 80–89. [CrossRef]

24. Robin, F.; Schuchmann, H.P.; Palzer, S. Dietary fiber in extruded cereals: Limitations and opportunities. *Trends Food Sci. Technol.* **2012**, *28*, 23–32. [CrossRef]

25. Cao, L.; Liu, X.; Qian, T.; Sun, G.; Guo, Y.; Chang, F.; Zhou, S.; Sun, X. Antitumor and immunomodulatory activity of arabinoxylans: A major constituent of wheat bran. *Int. J. Biol. Macromol.* **2011**, *48*, 160–164. [CrossRef] [PubMed]

26. Peyron, S.; Chaurand, M.; Rouau, X.; Abecassis, J. Relationship between Bran Mechanical Properties and Milling Behaviour of Durum Wheat (*Triticum durum* Desf.). Influence of Tissue Thickness and Cell Wall Structure. *J. Cereal Sci.* **2002**, *36*, 377–386. [CrossRef]

27. Arte, E.; Rizzello, C.G.; Verni, M.; Nordlund, E.; Katina, K.; Coda, R. Impact of Enzymatic and Microbial Bioprocessing on Protein Modification and Nutritional Properties of Wheat Bran. *J. Agric. Food Chem.* **2015**, *63*, 8685–8693. [CrossRef] [PubMed]

28. Aktas-Akyildiz, E.; Mattila, O.; Sozer, N.; Poutanen, K.; Koksel, H.; Nordlund, E. Effect of steam explosion on enzymatic hydrolysis and baking quality of wheat bran. *J. Cereal Sci.* **2017**, *78*, 25–32. [CrossRef]

29. Wang, P.; Hou, C.; Zhao, X.; Tian, M.; Gu, Z.; Yang, R. Molecular characterization of water-extractable arabinoxylan from wheat bran and its effect on the heat-induced polymerization of gluten and steamed bread quality. *Food Hydrocoll.* **2019**, *87*, 570–581. [CrossRef]

30. Singkhornart, S.; Edou-ondo, S.; Ryu, G.H. Influence of germination and extrusion with CO_2 injection on physicochemical properties of wheat extrudates. *Food Chem.* **2014**, *143*, 122–131. [CrossRef]

31. Ridzuan, M.J.M.; Abdul Majid, M.S.; Afendi, M.; Aqmariah Kanafiah, S.N.; Zahri, J.M.; Gibson, A.G. Characterisation of natural cellulosic fibre from Pennisetum purpureum stem as potential reinforcement of polymer composites. *Mater. Des.* **2016**, *89*, 839–847. [CrossRef]

32. Chen, Y.; Shan, S.; Cao, D.; Tang, D. Steam flash explosion pretreatment enhances soybean seed coat phenolic profiles and antioxidant activity. *Food Chem.* **2020**, *319*, 126552. [CrossRef] [PubMed]

33. Chen, Y.; Zhang, R.; Liu, C.; Zheng, X.; Liu, B. Enhancing antioxidant activity and antiproliferation of wheat bran through steam flash explosion. *J. Food Sci. Technol.* **2016**, *53*, 3028–3034. [CrossRef]

34. Liu, B.; Chen, Y.; Mo, H.; Ma, H.; Zhao, J. Catapult steam explosion significantly increases cellular antioxidant and anti-proliferative activities of *Adinandra nitida* leaves. *J. Funct. Foods* **2016**, *23*, 423–431. [CrossRef]

35. Wang, T.; Sun, X.; Zhou, Z.; Chen, G. Effects of microfluidization process on physicochemical properties of wheat bran. *Food Res. Int.* **2012**, *48*, 742–747. [CrossRef]

36. Wang, N.; Wang, C.; Chen, H.; Bai, L.; Wang, W.; Yang, H.; Wei, D.; Yang, L. Facile fabrication of a controlled polymer brush-type functional nanoprobe for highly sensitive determination of alpha fetoprotein. *Anal. Methods* **2020**, *12*, 4438–4446. [CrossRef]

37. Chau, C.-F.; Huang, Y.-L. Comparison of the chemical composition and physicochemical properties of different fibers prepared from the peel of *Citrus sinensis* L. Cv. Liucheng. *J. Agric. Food Chem.* **2003**, *51*, 2615–2618. [CrossRef]

Review

Principle and Application of Steam Explosion Technology in Modification of Food Fiber

Chao Ma [1,2], Liying Ni [2], Zebin Guo [1], Hongliang Zeng [1], Maoyu Wu [2], Ming Zhang [2] and Baodong Zheng [1,*]

[1] Department of Food Science, Fujian Agriculture and Forestry University, Fuzhou 350002, China
[2] Jinan Fruit Research Institute All China Federation of Supply and Marketing Co-Operatives, Jinan 250014, China
*   Correspondence: zbdfst@163.com

Abstract: Steam explosion is a widely used hydrothermal pretreatment method, also known as autohydrolysis, which has become a popular pretreatment method due to its lower energy consumption and lower chemical usage. In this review, we summarized the technical principle of steam explosion, and its definition, modification and application in dietary fiber, which have been explored by researchers in recent years. The principle and application of steam explosion technology in the modification of food dietary fiber were analyzed. The change in dietary fiber structure; physical, chemical, and functional characteristics; the advantages and disadvantages of the method; and future development trends were discussed, with the aim to strengthen the economic value and utilization of plants with high dietary fiber content and their byproducts.

Keywords: steam explosion; dietary fiber; modification; promote dissolution; application

Citation: Ma, C.; Ni, L.; Guo, Z.; Zeng, H.; Wu, M.; Zhang, M.; Zheng, B. Principle and Application of Steam Explosion Technology in Modification of Food Fiber. *Foods* **2022**, *11*, 3370. https://doi.org/10.3390/foods11213370

Academic Editors: Fuping Lu and Wenjie Sui

Received: 22 September 2022
Accepted: 19 October 2022
Published: 26 October 2022

Publisher's Note: MDPI stays neutral with regard to jurisdictional claims in published maps and institutional affiliations.

1. Introduction

With the development of living and economic standards, people have placed more importance on healthy diet requirements by eating more plant-based food like fruits and vegetables, which helps to maintain a healthy lifestyle. Cereals, fruits, and vegetables and other plant-derived foods are rich in functional compounds such as vitamins and dietary fiber. However, at present, more than one-third of the fruits and vegetables and their processing byproducts are not fully processed and utilized. These plant resources are usually treated by animal feed, landfill or incineration, and producers usually pay a certain price for it [1,2]. Thus, this inefficient processing of plant-derived food potentially has negative impacts on the environment.

Dietary fiber (DF) is a class of carbohydrate polymers that are neither digested nor absorbed in the small intestine, comprised of cellulose, hemicellulose, pectin, algae, and lignin. Dietary fiber mainly exists in the tough wall layer of plant cells. According to relative water solubility, dietary fiber can be divided into soluble and insoluble forms. These substances make up the hulls of cereal and wheat; and the roots, skin, stems and leaves of vegetables and fruits, which play an important beneficial physiological role. Over the years, research regarding the potential health benefits of dietary fiber in disease prevention has received considerable attention; obesity, type II diabetes, and cardiovascular diseases have been extensively studied [3]. Many previous studies found that there is an inverse relationship between dietary fiber intake and change in body weight. Koh-Banerjee et al. [4] supported this statement in a study that for every 40 g/d increase in whole grain intake, weight gain decreased by 1.1 lbs. A majority of studies show a positive correlation between high glycemic foods and type II diabetes. However, Meyer et al. [5] found that there was a strong inverse relationship between dietary fiber intake and diabetes. In the study, women who consumed an average of 26 g/d of dietary fiber had a 22% decreased risk of getting diabetes than women who consumed just 13 g/d. Recent studies suggest that increasing levels of dietary fiber may improve carbohydrate metabolism in a non–pharmacological way, resultingly having a positive effect on weight control and diabetes prevention. High quality DF should include more than 10% SDF and have favorable processing properties, as well as physiological activity and a healthcare function. Although the amount of total dietary fiber is fairly high in many plants, the content of SDF is only approximately 3–4% of the total dietary fiber. Insoluble dietary fiber may have some outstanding problems, such as rough taste, poor water-holding capacity, poor swelling power and weak functional activity; consequently, the utilization of this kind of raw material is not high [6]. It is urgent to carry out cost-effective industrial treatment of these resources and transform as much insoluble dietary fiber into soluble fiber as possible, which means modifying the dietary fiber raw materials in an efficient green way, in order to make better use of these resources.

As for the high-value utilization of raw materials rich in dietary fiber, pretreatment is needed to destroy their relatively dense physical structural barrier, so as to promote the efficient extraction, transformation, and utilization of DF [7]. At present, dietary fiber modification strategies may be classified into four types: physical, chemical, biological and combined method [7]. The glycosidic bond of DF is melted or broken using a physical method that makes use of high temperature, high pressure, immediate pressure decrease, explosion, high-speed impact, and shearing. This accomplishes the goal of modification, such as in steam explosion (SE) [8], ultrafine comminution (UC) [9], ultrasound [10], microwave [11], etc. The chemical method alters the structural and functional characteristics of DF by chemical reactions, such as alkaline hydrogen peroxide (AHP) treatment, alkali treatment, acid treatment, and Na_2HPO_4 treatment [12,13]. Compared with the physical method, the chemical method shows a short processing time and can react at room temperature. However, after the chemical method treatment, the modified DF exhibited low purity, and was prone to produce harmful components [14]. Utilizing certain enzymes or microorganisms to enzymatically hydrolyze or ferment raw materials with the goal of changing the content and bioactivity of DF is known as the biological method [15,16]. Biological methods are widely used in the modification of DF due to their high specificity, and benefits of milder processing conditions and environmental friendliness. The disadvantages of biological methods might be the high cost of enzyme purification and strain breeding [7]. Combining two or more ways to modify the DF is referred to as a combined modification method. Since the biological method requires a mild environment, the conditions of chemical methods are very harsh. Overall, the physical method is the main choice to combine with the other three methods [17,18]. Among these modification methods of DF, the physical methods are frequently used in the modification of dietary fiber because their advantages of low cost, short time consumption, simple operation, and lack of toxic waste generation [7].

Steam explosion (SE) technology, as a new type of physicochemical modification technology of food raw materials, has increasingly been studied by more and more researchers in recent years. SE is a method that presses high-pressure and high-temperature steam into cell walls and plant tissues, applying the thermochemical action of high-temperature cooking coupled with the physical tearing action of instantaneous blasting [8,19,20]. The physicochemical properties of the macromolecules of the fibrous raw materials are changed, thereby promoting the subsequent separation and conversion of solid-phase multi-component materials, which entails heating the biomass under pressurized steam to explode the cellulose fibrils and depolymerize lignin. At present, the steam explosion pretreatment has been used for ethanol production from straw materials, such as switchgrass and sugarcane bagasse [21], corn stover [22] and sunflower stalks [23]. In addition, the wall-breaking effect of SE on natural products has been widely studied. To sum up, this method has been gradually applied to the processing and modification of dietary fiber raw materials due to its low energy consumption and chemical usage [24,25].

In this review, we summarize the principle of structural modification of dietary fiber by steam explosion technology explored by researchers in recent years. The change in dietary fiber structure; physical, chemical, and functional characteristics; advantages and disadvantages of the method; and future development trends are discussed, with the aim to strengthen the economic value and utilization of plants with high dietary fiber content and their byproducts.

2. Source and Classification of Dietary Fiber in Food

In the 1950s, Eben Hipsley [26] first proposed the definition of dietary fiber (DF), which described the food components from the cell walls of plants, and many scholars have since carried out research on DF. As suggested by Trowell et al. [27], DF was characterized as lignin-resistant and plant polysaccharides to hydrolyze human digestive enzymes. The American Society of Cereal Chemists includes oligosaccharides (DP 3–9) in the DF definition, as these substances have some physiological characteristics with the majority of DF. According to the AACC, DF is the edible section of plants, or carbohydrate-like substances, that are difficult for humans to digest and absorb in the small intestine; but are fully or partially fermented in the large intestine [28]. Polysaccharides, lignin, and associated plant matter are all included in DF. This was the first time that DF-related substances, such as phenolic compounds, were included in the definition of DF [28]. In 2009, the Codex Alimentarius Commission defined DF as 10 or more monomeric units of carbohydrate polymers (whether to include 3 or 9 monomer chain carbohydrates is determined by the national authority) that are not hydrolyzed by the endogenous enzymes in the human small intestine and fall into one of the following categories: (1) edible carbohydrate polymers naturally occurring in the food as consumed; (2) synthetic carbohydrate polymers or carbohydrate polymers that have been obtained from food raw material by physical, enzymatic, or chemical means and which have been shown to have a physiological effect or benefit to health as demonstrated by generally accepted scientific evidence to competent authorities [29]. With the development of nutrition and other disciplines, dietary fiber (DF) is known as the seventh most important nutrient by the nutrition community, and plays an important role in human health. DF exists in most natural plants, such as fruits (16.74–91.24%), vegetables (6.53–85.19%), grains (9.76–69.20%) and so on [30]. Indeed, dietary fiber is a heterogeneous complex of components with different physical, chemical, and physiological properties, which complicates direct analytical measurements [31]. DF is an important component of a healthy diet; the positive correlation with human health has been established by the scientific community [32]. More than 50% of functional foods in the market contain DF as the active ingredient [33]. Additionally, DF has many biologically active substances that are beneficial for our health, such as improving the intestinal flora, lowering blood glucose, decreasing the probability of obesity and cardiovascular disease, reducing the risk of some cancers, increasing fecal volume, promoting bowel movements, etc. [26,34–37].

The Chinese Nutrition Society defines DF as a carbohydrate polymer that is not easily digested by digestive enzymes, mainly from the plant cell wall, and comprised of cellulose, hemicelluloses, lignin, pectin and resin; more specifically defined as follows: (1) Cellulose is a long-chain polymer composed of glucose linked by β-1,4 glycosidic bonds; (2) Hemicellulose is a polymer composed of a mixture of monosaccharides such as arabinose, galactose, and xylose; (3) Lignin is not a polysaccharide, but a polymer composed of phenylpropane units; (4) Pectin is a polymer composed of uronic acid residues with rhamnose and contains neutral sugar branched chains; (5) Mucus and gums are mostly hemicelluloses. The existence of bioactive substances linked to the cell wall has a significant impact on the physicochemical characteristics of DF and also affects its physiological properties in humans. Additionally, according to the DF definition supplied by the AACC, phenolic compounds were included in the definition of DF. These components may play a significant role in DF properties, even though the properties are generally attributed to those of polysaccharides [33]. Goñi et al. [38] found that polyphenols appear in both fractions of fiber, but in a greater proportion in the insoluble portion. Consequently, polyphenols are an essential component of DF, and the definition of DF restricted to non-digestible polysaccharides and lignin could be extended to include polyphenols [39]. According to the solution properties of DF, they can be divided in two types: insoluble dietary fiber (IDF) and soluble dietary fiber (SDF) [40]. Additionally, the degree of fermentation in the colon can be divided into complete fermentation fibers (mostly SDF) and partial fermentation fibers (mostly IDF). SDF is mainly fermented by bacteria in the ileum and ascending colon, while IDF is primarily fermented in the distal colon. Many kinds of natural DF tend to have high IDF content and low SDF content, resulting in the rough taste of raw materials; poor functional properties such as water-holding capacity (WHC), oil-holding capacity (OHC), and swelling capacity (SC); and low physiological activity, which have seriously limited the development and utilization of dietary fiber resources [7]. Thus, finding the most appropriate modification method to enhance the yield and functional properties of DF is becoming the hot topic in the food processing research fields.

3. DF Modification Methods

The current DF modification methods can be divided into four types: the physical method, chemical method, biological method, and combined method [7]. The above various modification methods have their own advantages and disadvantages. Due to their low cost, straightforward operation, and lack of harmful waste formation, physical methods are frequently used in the modification of DF. But, some physical methods require a large operating area and involve high levels of danger. Compared with physical and biological methods, chemical methods can create DF quickly and at room temperature, but they also have a tendency to produce a lot of hazardous byproducts. The products prepared by the biological method have the advantages of high purity and fewer byproducts. However, the only disadvantage of the biological method may be the high cost of enzymes and long operation cycles [7]. Based on the advantages and disadvantages of the above methods, it is urgent to find a DF modification method with low cost, high efficiency, a short reaction time, lower toxicity, fewer harmful byproducts, large processing capacity, and easy industrialization.

4. Steam Explosion Technology Process and Principle

Steam explosion (SE), also known as autohydrolysis, is an atypical hydrothermal pretreatment. According to the SE principle, the raw materials are placed into a cylinder at a high temperature and high pressure. The steam that is generated then permeates into the interior of the materials and fills the tissue pores with steam. Finally, the high pressure created by the saturated steam is immediately released within milliseconds [41]. The water contained in the substrate evaporates and rapidly expands, which causes cell wall rupture to form pores, so that small molecular material is released from the cells [42]. This also causes the reduction of cellulose crystallinity, delignification [43], and hydrolyzation of the hemicelluloses [44]. This process was widely used in the physicochemical pretreatment of lignocellulosic biomass, since there was no need to add chemicals in the process [45], basically eliminating three waste emissions, and therefore was an effective environmental friendly treatment method [46,47]. The SE was first invented by M.H. Mason in the United States in 1928, and was mainly used in the production of artificial fiberboard [48]. Initially, saturated steam from 7 to 8 MPa was used as a medium, which was not well promoted due to the high blasting pressure [48]. In the 1970s, the technology was widely used in animal feed processing, and extraction of ethanol and special chemicals from wood fiber. SE was mainly produced by an intermittent method. That is to say, after the material input, it was sequentially subjected to high temperature and high pressure, and instantaneously blasted in a closed reactor, during which time no feeding was carried out. The method for the separation of lignocellulose components was first seen in Delong's patent [49]. After the 1980s, this technology received renewed attention and rapidly developed. Continuous SE production technology and equipment have emerged, and the application fields have been further expanded, including the fermentation of feed and ethanol [50], municipal waste and sewage, sludge processing [51], tobacco processing [52], and applications in the food and pharmaceutical fields.

Steam explosion technology, as a new green food processing technology, has attracted extensive attention through research and applications in the food field in recent years with its advantages of strong applicability, short-term high efficiency, no pollution, and industrial amplification. Biomass was heated at high temperatures between 160 and 280 °C for 10 to 30 min under high pressure (i.e., 0.69–4.83 MPa) during the steam explosion. Most of the early SE methods were of the thermal spray type and screw extrusion type [53]. Due to the long pressure relief time and low energy conversion efficiency of the above technologies, the process of "explosion" could not be reflected. In view of this, the instantaneous ejection steam explosion technology came into being. The equipment is mainly composed of three parts: steam generator, steam explosion chamber, and material receiving bin. The key innovation of the technology is the introduction of a piston valve drive system. When the holding pressure ends, the piston bursts out of the cylinder causing an instantaneous blasting motion to release the sealed state of the cavity, and ejects the material and steam together within milliseconds [43]. According to its function characteristics, the steam explosion process is mainly divided into two stages: the high temperature cooking stage and instantaneous decompression stage. In the high-temperature cooking stage, the raw material was maintained under a saturated steam of pressure and temperature for a period of time, and the hemicelluloses and other parts of the plant raw materials were hydrolyzed to form soluble carbohydrates. At the same time, the lignin in the composite intercellular layer was softened and partially degraded, which reduced the lateral bonding strength of the fibers; and the cell pores were filled with high-pressure water vapor, which became soft and plastic. During the instantaneous decompression stage, the pore gas undergoes sharp expansion due to the sudden decompression, resulting in an explosion. The fiber material was split into small fiber bundles, which resulted in modification of the physical structure and redistribution of partial components. By employing SE, it is possible to destroy the crystal structure of cellulose and make hemicelluloses easier to use in subsequent processes. In the steam explosion, there following processes were mainly contained: acid hydrolysis, thermal degradation, mechanical fracture, hydrogen bond destruction, and structural rearrangement [54,55]. Two different SE pretreatment models are shown in Figures 1 and 2.

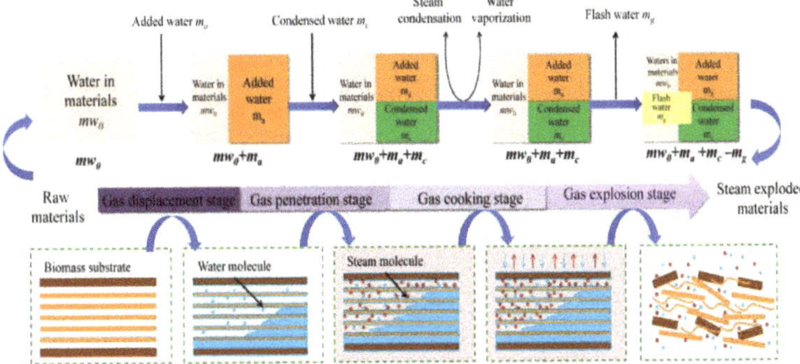

Figure 1. Schematic diagram of steam explosion process [56]. Reproduced with permission from Wenjie Sui and Hongzhang, Industrial Crops and Products; published by Elsevier, 2015.

As an industrialized technology, SE has economical and eco-friendly advantages in processing large quantities of food materials [57,58]. The technology of steam explosion pretreatment has been investigated for food production from a wide range of aspects, including wall-breaking extraction of natural products, oil extraction, hydrothermal conversion of active ingredients, comprehensive utilization of raw materials and DF modification. Several studies and applications of steam explosion processing of medicinal and edible fiber show that: on the one hand, the physical explosion of steam explosion can break the multi-scale anti-extraction barrier structure at the level of the plant tissue-cell-cell wall, improving the wall-breaking and solubilizing effect of active ingredients. It has been applied to the extraction and strengthening process of various active ingredients in various raw materials, such as *Radix astragali*, *Rhus chinensis* mill [59], and wheat bran, resulting in the extraction yield and efficiency being significantly improved. On the other hand, the thermochemical action of steam explosion promotes hydrolysis reactions of the intrinsic structural components and active ingredients of plant materials. It has a deglycosylation effect on some glycoside active ingredients, which separates the sugar group from the aglycone and promotes high activity. Effective utilization of aglycone components was demomstrated'. There have been studies using steam explosion to hydrolyze rutin glycosidic bonds to prepare quercetin, process turmeric to extract and prepare diosgenin, and combine steam explosion and solid-state fermentation to convert *Polygonum cuspidatum* resveratrol glycosides to prepare resveratrol. In recent years, the technology of steam explosion pretreatment has been investigated for DF modification from a wide range of feedstocks, including wheat bran, grapefruit peel, okara, sugarcane bagasse, pineapple peel, citrus pomace, and vegetable waste [60–66].

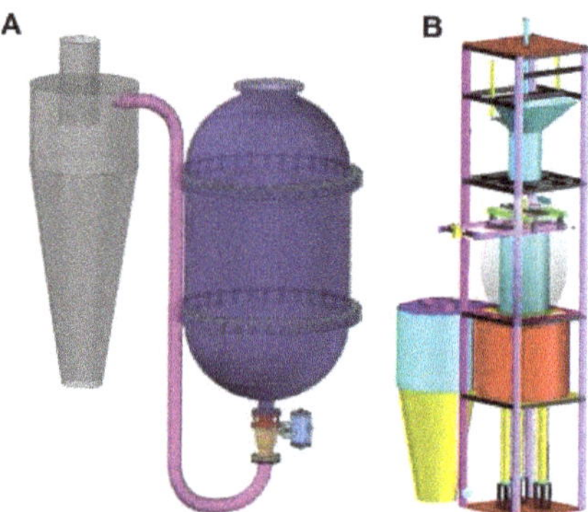

Figure 2. The structure diagram of two SE pretreatment models. (**A**) A model in valve blow mode. (**B**) A model in catapult explosion mode [43]. Reproduced with permission from Zhengdao Yu, Bailiang Zhang, Fuqiang Yu, Guizhuan Xu and Andong Song, Bioresource Technology; published by Elsevier, 2012.

5. Effect of SE on the Soluble Modification of DF

In accordance with the principle of SE technology, this technology is very suitable for DF modification processing. As shown in Figure 2, SE technology is commonly used for the dietary fiber modification of agricultural products such as fruits and vegetables, grains, edible fungi, and aquatic products. Its typical characteristic is that the IDF in the materials will be moderately reduced, but the SDF content increased. With the further increase of steam explosion pressure and pressure holding time, the macromolecular polysaccharides will be degraded or depolymerized to different degrees [63]. Thus, the content of DF will increase first and then decrease. The modification conditions of SE technology in grains, fruits, and vegetables and yield changes in DF are shown in Table 1. When the SE strength was 1.5 MPa for 30 s, the SDF content from okara increased to 36.28%, 26 times higher compared with the control [63]. Additionally, with appropriate steam explosion conditions, the yield of SDF in raw materials, orange peel, soybean hulls, and sweet potato residue will significantly increase [67].

Steam explosion has the characteristics of both the mechanical modification and thermal modification of fibrous materials. The structure of high-fiber raw material was modified by the effects of acid-like hydrolysis, thermal degradation, mechanical rupture, hydrogen bond rupture, and structural rearrangement generated in the SE process. On the one hand, the interlaced lignin, cellulose, and hemicelluloses, staggered distributed in lignocellulosic biomass structures, were separated to a maximum extent by SE. Through SE technology, the physical structure was destroyed, loose microstructure and honeycomb-like porous structure obtained, and the specific surface area increased, thereby improving the extraction efficiency of SDF. On the other hand, due to the destruction of the lignin, cellulose and hemicelluloses structures, the cell wall structure was disintegrated, resulting in improving the yield of SDF through thermochemical degradation that converts IDF to SDF. This is mainly achieved because steam explosion can remove a large amount of hemicelluloses from fibrous raw materials [68]. In addition, the high-pressure thermoacidic environment during the steam explosion process led to massive hydrolysis of hemicelluloses on the cell wall, and the lignin wrapped around cellulose were hydrolyzed and softened. Additionally, SE was able to readily break the hydrogen bonds of the starch structure, leading the starch to partially degrade and gelatinize; the advanced spatial structure of the protein shifted to denaturation, all of which resulted in the conversion of IDF to SDF [58]. With the increase in steam explosion strength, the conversion trends become more obvious. The results of steam explosion of lignocellulosic biomass materials show that the removal rate of hemicelluloses can reach more than 85%. The principle is shown in Figure 3 [69].

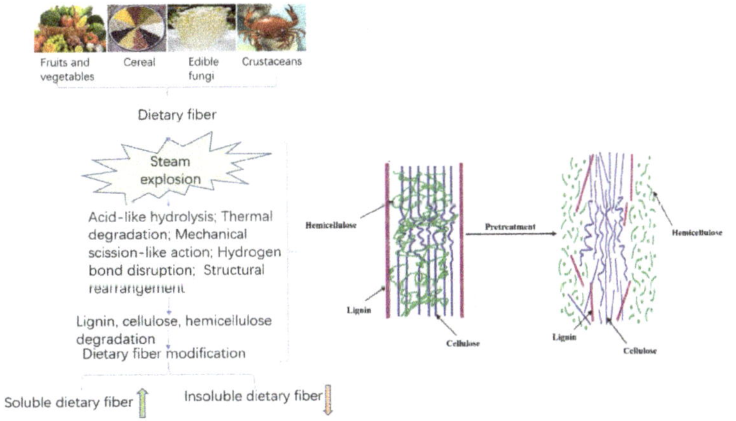

Figure 3. Schematic diagram of the modification mechanism of dietary fiber in Steam explosion.

Table 1. Effect of Steam explosion on the dietary fiber content of different food raw materials.

Material Category	Steam Explosion Conditions	DF Content Changes	Reference Documentation
Wheat bran	160 °C, 5 min	Water extractable arabinoxylan content was increased from 0.75% to 2.06%.	[70]
	0.8 MPa, 5 min	The SDF content increased to 9.62 g/100 g, which is 2.08-fold higher than that of the untreated.	[58]
Okara	1.5 MPa, 30 s	SDF content increased to 36.28%, higher by 26 times compared with the control okara.	[63]
Soybean hulls	1.2 MPa, 180 s	The SDF yield rate was 11.12%.	[67]
Sweet potato residue	0.35 MPa, 122 s	The SDF yield rate was increased from $3.81 \pm 0.62\%$ to $22.59 \pm 0.35\%$.	[71]
Rosa roxburghii pomace	0.87 MPa, 97 s	The SDF content was increased from $9.31 \pm 0.07\%$ to $15.82 \pm 0.31\%$.	[72]
Grapefruit peel	0.8 MPa, 90 s	The pectin yield rate reached 17.5%.	[62]
Orange peel	0.8 MPa, 7 min, combined with 0.8% sulfuric-acid soaking	SDF was increased from 8.04% to 33.74% in comparison with the control.	[17]
Apple pomace	0.51 MPa, 168 s, sieving mesh size, 60.	The SDF yield from apple pomace after SE was 29.85%, which is 4.76 times the yield of SDF (6.27%) in untreated apple pomace.	[73]
Okra seed	1.5 MPa, 5 min	The SDF content reached 6.5%.	[74]
Ampelopsis grossedentata	0.4 MPa, 4 min	The yield of crude polysaccharides reached $5.35 \pm 0.12\%$, which is an increase of 2.2 times over that of the materials without SE pretreatment.	[75]

6. Effect of the SE on the Structure of DF

6.1. Effect of the SE on the Apparent Morphology of DF

DF is a major component of the cell wall. Steam explosion (SE) is a type of physical method used in the DF modification process in recent years. The sample material was placed into a closed environment with high temperature and high pressure, and the pores of the material filled with steam. The cell wall structure and crystal structure of cellulose were torn and destroyed by the mechanical and thermochemical action. When the high pressure is instantly released, the superheated steam in the pores will vaporize rapidly and the volume will expand sharply, leading to the "explosion" of cells. The cell walls broke out to become porous, and the chemical bond between lignin-cellulose-hemicelluloses was destroyed. Meanwhile, the lignin and cellulose structures were loosened and softened, and the hemicelluloses partially hydrolyzed, resulting in the release of low molecular weight substances in cells. In the process of SE treatment, cellulose and hemicelluloses will be degraded into SDF by acid-like hydrolysis, thermal degradation, mechanical fracture, and hydrogen bond destruction. The action principle of steam explosion is shown in Figure 4 [76].

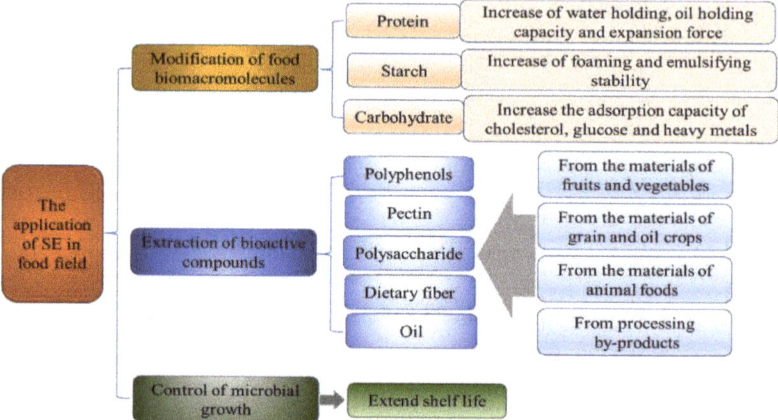

Figure 4. Schematic diagram of action principle of steam explosion [76]. Reproduced with permission from Fachun Wan, Chengfeng Feng, Kaiyun Luo, Wenyu Cui, Zhihui Xia and Anwei Cheng, Trends in Food Science & Technology; published by Elsevier, 2022.

Studies have shown that after SE treatment, the DF raw material is decomposed into fine fiber bundles, which reduces the toughness and is easier to grind. Aktas et al. found that through SE treatment, the particle size of the 50% bran was reduced from 420 to 177 μm, and the particle diameter of the 90% sample was decreased from 902 to 410 μm [70]. After SE treatment, the structure of okara significantly changed. Under weak SE strength, the layered structure of okara was destroyed into scattered small fragments, while the internal structure was relatively intact. When the strength of SE was further increased, the nucleated structure inside okara was gradually destroyed and cracked, forming small particles. Due to the destruction of the loose sheet structure on the surface of okara, the physical properties, including water and oil-holding capacities, and the swelling capacity significantly decreased. In addition, the water solubility of okara increased as the large lamellar structure cracked into small fragments and the internal nucleated structure disintegrated into small particles under high temperature and pressure [63]. Kong et al. [77] indicated that SE greatly disrupted the structure of wheat bran. The thin rod-like fibers of bran are crimped and broken, expanded into a porous structure, and a concave pore-like structure appears on the surface. This indicated that the SE seemed to have an effect on decreasing the toughness and increasing the brittleness of wheat bran. Therefore, wheat bran was easily crushed by a grinder. In another study, Sui, et al. studied the effect of SE treatment on the apparent morphology and microstructure of wheat bran. The results showed that the surface of raw bran was intact and smooth with no apparent fractures or debris. After SE, some obvious changes of bran were observed: the surface began to lose its luster and became porous, rough, and even honeycombed (as shown in Figure 5) [58].

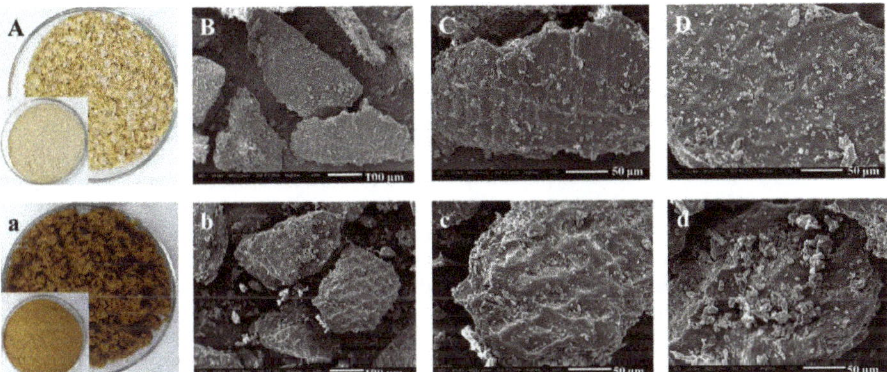

Figure 5. Photographs and SEM micrographs of raw bran and SE bran. (**A,a**): photographs of raw bran and SE bran; (**B** **D**): SEM micrographs of raw bran powders through a 60 mesh screen (**B**): 200×; (**C,D**): 500×; (**b–d**): SEM micrographs of SE bran powders through a 60 mesh screen (**b**): 200×; (**c,d**): 500×.

Above all, steam explosion leads to changes in the morphology and structure of DF, which usually shows the honeycomb porous structure formed inside, the number of cracks and roughness increase, and the granulation is obvious, leading to the structure and physical properties being completely changed after the steam explosion. Additionally, with the SE strength increased, the melting of the particle surface was intensified. The specific reason is that during the high-temperature cooking process of SE, the fiber structure is softened, and some structural components are depolymerized and degraded, thereby the surface is in a molten state. Meanwhile, in the subsequent instantaneous explosion process, the liquid water in the porous structure gasified and the water vapor adiabatically expanded, which destroyed the stereoscopic structure and exposed the internal structure, resulting in increasing the complexity of the system.

6.2. Effects of SE on the Molecular Structure and Properties of DF

Rapid pressure release during SE treatment causes intricate mechanical action. Meanwhile, the residence time of the materials with the high-pressure steam produces various chemical reactions. The typical results of this treatment includes significant disintegration of the lignocellulosic structure, hydrolysis of the hemicellulosic fraction, and depolymerization of the lignin components, changing the molecular structure of DF and its monosaccharide composition.

Li et al. [63] assessed the impact of steam explosion (SE) treatment on the dietary fiber, and physiochemical and protein properties of okara. The findings demonstrated that as SE strength rose, the molecular weight of the okara polysaccharide reduced and the proportion of low molecular weight fractions increased. Additionally, the molecular weight distribution range decreased. Okara polysaccharide, which has a molecular weight of 1.5 KDa, had a symmetrical peak at 2.0 MPa for 60 s and 120 s, showing that certain high molecular weight polysaccharides were gradually degraded into low molecular weight polysaccharides and oligosaccharides, increasing the amount of soluble dietary fiber. Furthermore, as SE intensity increased, so did the degree of polysaccharides breakdown and content of SDF. When the SE strength was too high, the polysaccharides were excessively degraded to convert into monosaccharides or oligosaccharides, which cannot be precipitated by 75% ethanol, leading to the decrease of contents of TDF and SDF. Another study discovered that pretreatment with SE reduced the molecular weight of *A. grossedentata* polysaccharide (AGP), implying that some high molecular weight polysaccharides were gradually degraded into low molecular weight polysaccharides and oligosaccharides, resulting in an increase in soluble dietary fiber. The results of monosaccharide composition analysis showed that AGP was composed of mannose, glucuronic acid, galacturonic acid, galactose, rhamnose, xylose, glucose, and arabinose with a molar ratio of 1.00:1.02:1.24:1.30:3.90:6.44:20.46:24.85. Meanwhile, AGP-SE had eight monosaccharide species, which was consistent with AGP. However, there was a difference in monosaccharide molar ratios between the two polysaccharides. As a result, the SE pretreatment had no discernible influence on the polysaccharide monosaccharide composition [75].

He et al. used SE to improve the physicochemical and functional properties of tartary buckwheat dietary fiber, and found that after SE pretreatment the surface structure changed from a dense intact to loose multilayer structure, and a loose multilayer honeycomb-like network structure was observed in the treated SDF. Meanwhile, the content of D-galacturonate and glucuronose in SDF significantly reduced by 40.78% and 15.07% compared with that of the untreated. However, the content of D-mannose and D-glucose were notably increased by 126.36% and 59.23%, respectively [78].

Many studies showed that the molecular weight of SDF after steam explosion treatment was reduced, and the steam explosion treatment destroyed the molecular chain of SDF. During the SE process, the hydrogen bonds and ordered structure of cellulose were broken with the high temperature and high pressure. Furthermore, the stable hydroxyl groups disordered and the macromolecular crystal was destroyed, leading to degradation of the polysaccharide from macromolecules into low molecular weight polysaccharides. Therefore, cellulose and hemicelluloses are degraded into soluble sugar such as oligosaccharides, monosaccharides, furfural, and hydroxymethylfurfural with SE treatment [79]. However, if the SE strength further increased, this soluble sugar will be further degraded into lower molecular weight carboxylic acids such as acetic acid and levulinic acid, or polymerized again [80]. Additionally, the steam explosion method causes glycosidic linkages in hemicelluloses and celluloses to hydrolyze. Furthermore, it causes lignin to depolymerize by cleaving β-O-4 ether and other acid-labile bonds. The SE technique can also release insoluble bound phenolic compounds by degrading the ester bonds between lignin and phenolic acids [81].

6.3. Effect of SE on the Crystallinity of DF

The crystalline state of the cellulose samples was determined by the X-ray diffraction isothermal method. Furthermore, the ICDD PDF-2 (2008) database of XRD patterns was used for analyzing the XRD patterns. According to Segal's method, the crystallinity of each sample was determined by the intensity of the crystal peak at 2θ of ~22° and the minimum intensity of the crystal peak between 16.5 and 22.6°, corresponding to the amorphous area [82]. This method was recently used by Tanpichai et al. [83] on cellulose microfibers isolated from pineapple leaves using SE. The authors found that all treated pineapple leaf fiber samples exhibited the outstanding peaks located at 2θ of 14.8, 16.5, and 22.6°, corresponding to (101), (101⁻) and (002) lattice planes of the typical cellulose I structure, respectively, while the peaks located at 2θ of 12, 20 and 22° would be determined from the cellulose II structure, which means that the crystal structure of pineapple leaf fiber after SE treatment cannot be changed. However, with the increase of the steam explosion treatment cycle, the crystallinity index increased. Similar improvement in the crystallinity of *Opuntia ficus indica* fibers treated with three green chemistry methods was found by Marwa et al. [84]. With the steam treatment assisted by lemon juice to extract cactus rackets fibers, the crystallinity was improved with the destruction of the amorphous phase of the sample and retention of the crystalline phase. In other research, the authors used SE to treated chitin and systematically investigated the crystallinity using X-ray diffraction.

The results indicated that with the powerful seepage force of the steam during SE, the crystallinity index of chitin decreased 10.2% in the (110) plane and 13.3% in the (020) plane, indicating that the crystallinity of the chitin was destroyed by SE treatment [85]. Therefore, a longer time and higher pressure in SE treatment will decrease the crystallinity of the treated fibers due to the cellulose degradation.

6.4. Effect of SE on the Thermal Stability of DF

The thermal stability of the biomass is analyzed by using thermal gravimetric analysis (TGA), a widely adopted technique to determine the thermal degradation of plant biomass [86,87]. According to the research, the thermal degradation reaction of cellulose can be divided into four stages. The first stage, from room temperature to 160 °C involves the evaporation of extractives or water. The second stage (160–285 °C) is the thermal depolymerization of hemicelluloses or pectin. The major degradation occurred in the third stage (285–380 °C), which is attributed to cellulose degradation. The fourth stage (380–550 °C) corresponds to lignin degradation [88,89]. Liu et al. [90] used SE pretreatment to process sweet potato vine with different moisture content, and found that the hemicelluloses shoulder of manually dried feedstock samples' derivative thermal gravimetric (DTG) curve almost disappeared after SE pretreatment, and the height of the maximum peak increased 12.5% and moved to a higher temperature due to the degradation of hemicelluloses and the related lifting of cellulose. In the study of Sui et al. [66], the authors investigated SE processing on the deconstruction of wasted broccoli stems. The results showed that the SE process led to the removal of hemicelluloses, re-condensation of lignin, degradation of the cellulosic amorphous region, and enhancement of the thermal stability of broccoli wastes. Furthermore, Tanpichai et al. [83] found similar results for thermal properties from pineapple leaves fibers after steam explosion treatment. However, Jacquet et al. [91] investigated the effect of intensity of a steam explosion on the thermal degradation of a cellulose. The results indicated that the thermal stability of the steam explosion samples decreased with the intensity of the treatment, which was contrary to the above findings. This phenomenon may be because of the high intensity of SE and the glucose ring cracking reaction. Thus, SE treatment has an obvious impact on the thermal stability of DF.

7. Effect of the SE on the Physicochemical Properties of DF

The capacity of DF to interact with water via physical adsorption is a significant aspect of its physiological activity [72]. The results of swelling capacity can be used to determine how much water and volume DF can hold [92]. Through changing the structure of DF raw materials, SE treatment increases the content of SDF and soluble small molecular substances, reducing IDF. Moreover, the microstructure of the material changes from dense to lose and porous, resulting in changes in the functional properties of DF, such as WHC, SC, OHC, and other physicochemical activities. Based on the modification of the conditions of steam explosion, the property changes of DF are shown in Table 2.

In much research, SE pretreatment can effectively improve the WHC and water-retention capacity (WRC) of DF. Moderate steam explosion treatments can increase the water-retention capacity coupled to an increase of the overall crystallinity [93]. In the study of Zhao et al. [61], the outcomes demonstrate that SE pretreatment can raise the WHC of wheat bran. The increase in soluble dietary fiber causes a decrease in the WRC of wheat bran, which could be ascribed to the exposure of hydroxyl and carboxyl groups bound to water groups by the instantaneous shear stress of SE. In addition, Zhai et al. investigated the effect of SE technology on the modification of DF from *Rosa roxburghii pomace*. The results revealed that the physicochemical properties of SDF and IDF were improved after SE treatment. Compared with without SE treatment, the hydration capacity of IDF was significantly increased from 6.77 ± 0.33 g/g to 7.92 ± 0.27 g/g. This is possibly because the molecular morphology of SE-IDF was changed as the specific surface area and rougher surface structure of the material increased, exposing more binding sites for water molecules and so enhancing the ability of DF to capture water molecules [72].

Table 2. Effect of Steam explosion on the physicochemical properties of dietary fiber.

Material	Modification Conditions	Property Changes	Reference Documentation
Wheat bran	0.1~1.5 MPa, 5 min	WHC↑, WRC↓, cellulose content↓	[70]
	0.8 MPa, 5 min	SDF content↑, IDF↓, WHC↑, SC↑, OHC↑	[70]
Okara	1.5 MPa, 30 s	SDF content↑, WHC↓, SC↓, OHC↓	[63]
Sweet potato residue	0.35 MPa, 122 s	SDF content↑, WHC↑, SC↑, OHC↑	[71]
Rosa roxburghii pomace	0.87 MPa, 97 s	SDF content↑, IDF↓, WHC↑, SC↑, OHC↑	[72]
Orange peel	0.8 MPa, 7 min, combined with 0.8% sulfuric-acid soaking	SDF content↑, WS↑, WHC↑, SC↑, OHC↑, emulsion stability↑	[17]
Apple pomace	0.51 MPa, 168 s, sieving mesh size, 60	SDF content↑, WHC↑, SC↑, OHC↑	[73]
Cellulose fibers	1.0~4.0 MPa, 5 min	crystallinity↑, Water retention values↑	[74]

"↑" means this characteristic was improved after SE treatment. "↓" means this characteristic was reduced after SE treatment.

Oil-holding capacity is one of the important indexes of physical and chemical properties of DF [30]. Research indicates that the OHC mainly relates to the protein structure and absorption property. SE treatment caused the depolymerization of proteins, which might lead to the decrease of oil holding capacity. A similar result was observed in Li's study [63], that is, the oil-holding capacity of okara decreased with the increase of SE strength. At SE strengths of 2.0 MPa for 120 s, the oil-holding capacity decreased 60.9%, compared with the control. In the study of Zhai et al. [72], the results showed that the OHC of the modified IDF was 1.3 times higher than that of IDF before modification by SE, and the OHC of SDF increased from 1.53 ± 0.51 g/g (0-SDF) to 5.96 ± 0.49 g/g (SE-SDF) [94]. It is possible that SE treatment led to the porous structure and large specific surface area, which are beneficial for the absorption property of dietary fiber.

Dietary fiber is a general term for non-starch polysaccharide and lignin. Due to the molecular structure, non-starch polysaccharide contains many hydrophilic groups, which play the role of hydrophilic hydroxyl groups, and has certain hydrophilic properties and exhibits certain swelling capacity (SC). Wang et al. [71] found that SE treatment increased SC values of DF from sweet potato by 37.79%. In another study, the authors used SE treatment conditions to optimize SDF extraction from apple pomace [73]; the results illustrated that the SC was increased, which may be attributed to the increase in the surface area of SDF after SE. Additionally, it was applied by Wang et al. to prepare SDF from orange peel assisted by steam explosion and dilute acid soaking, and found that the SC of SDF from orange peel treated by SE-SAS increased from 4.83 ± 0.52 mL/g db to 6.28 ± 0.73 mL/g db. The SC was dependent on the characteristics of individual components and the physical structure (porosity and crystallinity) of the fiber matrix [95]; thereby, the increase in SC might be attributed to a rise in the amount of short-chains and surface area of dietary fiber induced by SE.

8. Effect of SE on the Functional Activities of DF

SE treatment not only changed the structural composition of DF, but also improved the biological activity of DF. After SE treatment, the specific surface area of DF was increased, and more groups were exposed. Meanwhile, a part of the hydrogen bonds inside the cellulose were broken to form free hydroxyl groups. The uronic acid content in SDF was significantly increased after SE treatment, which had a positive effect on the adsorption of excess cholesterol and toxic cations [96].

Wang et al. [17] found that the binding capacity of SDF from orange peel treated by SE–SAS to three toxic cations (Pb, As and Cu) was significantly improved. Furthermore, compared with the control group, the main weight-average molecular weight of SDF in orange peel treated by SE-SAS was obviously smaller and the thermal stability of SDF was higher. In another study, the authors illustrated that SE can increase the cation exchange capacity (CEC) of *Polygonatum odoratum* DF [97]. Three possible mechanisms were proposed for the binding capacity of dietary fibers to heavy metals: chemical adsorption, physical adsorption, and mechanical adsorption [98]. Chemical adsorption is related to the existence of ophenolic groups in lignin and carboxyl groups in uronic acid. When the pH value was increased, the carboxyl groups were dissociated into carboxyl anions (RCOO·), which showed stronger interaction with the toxic cations, resulting in higher binding capacity of the dietary fibers with the cations. Physical adsorption was caused by van der Waals' force, which was connected with temperature, while mechanical adsorption depended on the degree of porosity of DF and its ability to trap the substances in its spatial structure.

According to the research, it is shown that the bran constituent of cereal is rich in phenolic compounds, which has a positive effect on oxidative stress. Phenolic acids can be divided into three types: free, soluble esters, or conjugates and insoluble combined forms. Insoluble bound phenolics are abundant and combined with ester and ether bonds of cell wall components such as arabinoxylans and lignin, showing significantly higher antioxidant capacity compared with soluble free and conjugated phenolics [99–101]. The bound phenolic acids (BPA) can be released by steam explosion-assisted extraction. Meanwhile, the content of phenolic acids and its antioxidant properties were significantly improved [102]. Kong et al. [103] used SE treatment to process wheat bran; the results showed that the phenolic content and total flavonoids were increased by 83% and 198%, respectively, at SE strengths of 0.8 MPa for 5 min. Li et al. [104] found that with SE treatment of 1.5 MPa for 60 s, the amount of free and bound phenolics in the tartary buckwheat bran were increased from 23.62 and 0.36 mg gallic acid per gram dry weight (mg GAE per g DW) of tartary buckwheat bran to 27.34 and 0.99 mg GAE per g DW, which increased by 15.7% and 175%, respectively. Moreover, the biological activity tests indicated that SE effectively increased the oxygen radical absorbance capacity (ORAC) in vitro of the bound phenolics by 270%. It also enhanced the cellular antioxidant activity (CAA) in vitro of free phenolics by 215%. In another study, after SE treatment at 220 °C for 60 s, the yields of free and conjugate ferulic acid increased by about 59.0 and 8.45 times, respectively. Meanwhile, the corresponding increases of p-coumaric acid were 47.6 and 7.25 times. With the residence time prolonged to 120 s, the total soluble phenolic content from barley bran reached 1686.4 gallic acid equivalents mg/100 g dry weight, which was about 9.83 times higher than that of the untreated sample. Therefore, the extraction rate of free phenolic acids and soluble conjugates increased sharply after steam explosion, and some bound phenolics like ferulic acid and p-coumaric acid in the cell walls might be released. 2,2-Azinobis (3-ethylbenzothiazoline-6-sulfonic acid (ABTS) and ferric-reducing antioxidant power (FRAP) assays indicated that SE can obviously improve the antioxidant capacity of the Tibetan hull-less barley bran extract [81].

In addition, DF possess several biological properties such as hypoglycemic activity, cholesterol-lowering activity, and prebiotic activity. The results obtained from Chen et al. [105] indicated that the content of SDF was increased from $2.6 \pm 0.3\%$ to $30.1 \pm 0.6\%$ from soybean residues treated by SE combined with extrusion treatment (BEP). Moreover, in vivo experiment results indicated that after BEP processing, the SDF significantly reduces the concentrations of total cholesterol (TC), low-density lipoprotein cholesterol (LDL-C), and triglyceride (TG) while increasing the concentration of high-density lipoprotein cholesterol (HDL-C). In other research, Liu et al. [75] observed that SE pretreatment can increase the content of polysaccharide from *Ampelopsis grossedentata* and slightly change the physicochemical properties of AGP (i.e., molecular weight, monosaccharide composition, and glycoside bonding). Furthermore, the polysaccharides pretreated with SE showed significantly higher α-glucosidase inhibitory activities than those from the without SE-pretreatment sample, which might serve as an efficient α-glucosidase inhibitor. Additionally, SE treatment can obtain fiber components with potential prebiotic activity and high antioxidant activity from asparagus byproducts [106].

9. Conclusions

SE technology has been widely used in the treatment of lignocellulosic materials. With the development of equipment and technology, it has been gradually applied in the food industry, pharmaceutical industry, biological energy, chemical materials, environmental protection, and other fields. This technology has its significant advantages in the processing of DF materials, but also has many disadvantages.

9.1. Analysis of Technical Advantages and Disadvantages

As shown in Table 3, SE treatment has obvious technical advantages, but also has defects. SE technology is widely used in the modification of DF because of its low cost, simple operation only needing to pass high temperature and high pressure steam, no chemical additives and low pollution of the environment [107]. Meanwhile, SE treatment can destroy the connection of cellulose, promote the release of functional components in materials, and also modify DF to enhance the corresponding functional properties. With short time consumption, large-scale production, and relatively low energy consumption, SE is capable of improving the comprehensive utilization and added value of materials.

Table 3. Comparison of the advantages and disadvantages of steam explosion and conventional treatment methods.

Subjects	Advantages	Disadvantages	Reference
Industrial application cost	Short time consumption, Large-scale production, low cost	Equipment cost investment is large in the early stage.	[7]
Operation convenience	Higher automation and lower operation intensity	Continuous production cannot be realized.	
Operation safety	Better equipment safety protection, less potential safety hazard in operation	The equipment requires high temperature and high pressure, which will generate huge noises during work. It needs a muffler device, and an exhaust gas absorption or waste heat recovery device.	
Product safety	Except high-pressure steam, no other chemicals are brought into the process, which will not pollute the materials.	High temperature and high-pressure process is prone to adverse reactions such as Maillard reaction and denaturation of nutrients.	

However, there are also some deficiencies in SE treatment. Due to its complex physical and chemical processes, it is difficult to accurately control the strength and uniformity of treatment, which makes it easier to cause the degradation of other effective components and occurrence of Maillard reaction, and the treatment cannot be continuous.

9.2. Development Trends

(1) Further study the change process and mechanism of DF during SE treatment, and further clarify the exact effect of SE treatment on food safety and its influence on nutritional value;

(2) Further clarify the physical and chemical changes in DF materials during SE treatment, to clarify the structure-activity relationship between its chemical components and functional activities, and reveal the molecular mechanism of the biological activity change in DF materials in SE treatment;

(3) Further improve the research and upgrade of industrialized SE processing equipment and facilities; develop in the direction of large-scale, continuous, automatic and precise control; and explore combined process technologies that combine multiple pretreatment methods.

SE treatment technology is an efficient modification technology of fiber material, which has the advantages of low investment, low energy consumption, short time consumption, and environmental friendliness. However, the current research mainly focuses on the industrial application of wood fibers. Limited information is available on the modification of DF in food by SE treatment, mainly in the grain bran, and fruit and vegetable byproducts. After SE treatment, under the dual action of high temperature and high-pressure steam, the physical and chemical properties of the material significantly changed. After process optimization, the physical and chemical quality and nutritional value of DF have been markedly improved. Therefore, finding a suitable method that is more conducive to industrialization and improves the yield and functional properties of DF will continue to be a hot topic in the food industry.

Author Contributions: Conceptualization, methodology, validation, writing—original draft preparation, C.M. and L.N.; investigation, Visualization, Methodology, H.Z. and M.Z.; Conceptualization, Writing—review and editing, Supervision, Z.G. and B.Z.; Project administration, funding acquisition, M.W. All authors have read and agreed to the published version of the manuscript.

Funding: This work was supported by the Key Research and Development Program of Shandong Province (LJNY202105).

Institutional Review Board Statement: Not applicable.

Informed Consent Statement: Not applicable.

Data Availability Statement: Not applicable.

Conflicts of Interest: The authors declare no conflict of interest.

Abbreviations

DF	Dietary fiber
SDF	Soluble dietary fiber
IDF	Insoluble dietary fiber
TDF	Total dietary fiber
SE	Steam explosion
OHC	Oil-holding capacity
WHC	Water-holding capacity
WRC	Water-retention capacity
SC	Swelling capacity

References

1. Angulo, J.; Mahecha, L.; Yepes, S.A.; Yepes, A.M.; Bustamante, G.; Jaramillo, H.; Valencia, E.; Villamil, T.; Gallo, J. Nutritional evaluation of fruit and vegetable waste as feedstuff for diets of lactating Holstein cows. *J. Environ. Manag.* **2012**, *95*, S210–S214. [CrossRef]
2. Leroy, B.; Bommele, L.; Reheul, D.; Moens, M.; Neve, S.D. The application of vegetable, fruit and garden waste (VFG) compost in addition to cattle slurry in a silage maize monoculture: Effects on soil fauna and yield. *Eur. J. Soil Biol.* **2007**, *43*, 91–100. [CrossRef]
3. Telrandhe, U.B.; Kurmi, R.; Uplanchiwar, V.; Mansoori, M.H.; Jain, S.K. Nutraceuticals—A Phenomenal Resource in Modern Medicine. *Int. J. Univers. Pharm. Life Sci.* **2012**, *2*, 179–195.
4. Koh-Banerjee, P.; Franz, M.; Sampson, L.; Liu, S.; Jacobs, D.R.; Spiegelman, D.; Willett, W.; Rimm, E. Changes in whole-grain, bran, and cereal fiber consumption in relation to 8-y weight gain among men. *Am. J. Clin. Nutr.* **2004**, *80*, 1237–1245. [CrossRef]
5. Meyer, K.A.; Kushi, L.H.; Jacobs, D.R.; Slavin, J.; Sellers, T.A.; Folsom, A.R. Carbohydrates, dietary fiber, and incident type 2 diabetes in older women. *Am. J. Clin. Nutr.* **2000**, *71*, 921–930. [CrossRef]
6. Liu, M.; Jiang, Y.; Zhang, L.; Wang, X.; Lin, L. Effect of Steam Explosion Modification and in Vitro Simulated Digestion on Antioxidant Capacity of Dietary Fiber of Pineapple Peel, IOP conference series. *Earth Environ. Sci.* **2019**, *330*, 42055.
7. Gan, J.; Xie, L.; Peng, G.; Xie, J.; Chen, Y.; Yu, Q. Systematic review on modification methods of dietary fiber. *Food Hydrocoll.* **2021**, *119*, 106872. [CrossRef]
8. Chandra, R.P.; Bura, R.; Mabee, W.E.; Berlin, A.; Saddler, J.N. Substrate Pretreatment: The Key to Effective Enzymatic Hydrolysis of Lignocellulosics. *Adv. Biochem. Eng. Biotechnol.* **2007**, *108*, 67–93.
9. Guoyong, Y.; Jia, B.; Jing, Z.; Quanhong, L.; Chen, C. Modification of carrot (*Daucus carota* Linn. var. *Sativa Hoffm.*) pomace insoluble dietary fiber with complex enzyme method, ultrafine comminution, and high hydrostatic pressure. *Food Chem.* **2018**, *257*, 333–340.
10. Liurong, H.; Xiaona, D.; Yunshu, Z.; Yuxiang, L.; Haile, M. Modification of insoluble dietary fiber from garlic straw with ultrasonic treatment. *J. Food Process Pres.* **2018**, *42*, e13399.
11. Enwei, W.; Rui, Y.; Hepeng, Z.; Penghui, W.; Suqing, Z.; Wanchen, Z.; Yang, Z.; Hongli, Z. Microwave-assisted extraction releases the antioxidant polysaccharides from seabuckthorn (*Hippophae rhamnoides* L.) berries. *Int. J. Biol. Macromol.* **2019**, *123*, 280–290.

12. Yuge, N.; Na, L.; Qi, X.; Yuwei, H.; Guannan, X. Comparisons of three modifications on structural, rheological and functional properties of soluble dietary fibers from tomato peels. *LWT* **2018**, *88*, 56–63.
13. Kunli, W.; Mo, L.; Yuxiao, W.; Zihao, L.; Yuanying, N. Effects of extraction methods on the structural characteristics and functional properties of dietary fiber extracted from kiwifruit (*Actinidia deliciosa*). *Food Hydrocoll.* **2021**, *110*, 106162.
14. Suya, H.; Yawen, H.; Yanping, Z.; Zhuang, L. Modification of insoluble dietary fibres in soya bean okara and their physicochemical properties. *Int. J. Food Sci. Technol.* **2015**, *50*, 2606–2613.
15. Chaofan, W.; Rongzhen, S.; Siqing, W.; Wenliang, W.; Feng, L.; Xiaozhen, T.; Ningyang, L. Modification of insoluble dietary fiber from ginger residue through enzymatic treatments to improve its bioactive properties. *LWT* **2020**, *125*, 109220.
16. Mengyun, J.; Jiajun, C.; Xiaozhen, L.; Mingyong, X.; Shaoping, N.; Yi, C.; Jianhua, X.; Qiang, Y. Structural characteristics and functional properties of soluble dietary fiber from defatted rice bran obtained through Trichoderma viride fermentation. *Food Hydrocoll.* **2019**, *94*, 468–471.
17. Lei, W.; Honggao, X.; Fang, Y.; Rui, F.; Yanxiang, G. Preparation and physicochemical properties of soluble dietary fiber from orange peel assisted by steam explosion and dilute acid soaking. *Food Chem.* **2015**, *185*, 90–98.
18. Jiapan, G.; Ziyan, H.; Qiang, Y.; Guanyi, P.; Yi, C.; Jianhua, X.; Shaoping, N.; Mingyong, X. Microwave assisted extraction with three modifications on structural and functional properties of soluble dietary fibers from grapefruit peel. *Food Hydrocoll.* **2020**, *101*, 105549.
19. Auxenfans, T.; Cr Nier, D.; Chabbert, B.; Pa, S.G. Understanding the structural and chemical changes of plant biomass following steam explosion pretreatment. *Biotechnol. Biofuels* **2017**, *10*, 36. [CrossRef]
20. Sarker, T.R.; Pattnaik, F.; Nanda, S.; Dalai, A.K.; Meda, V.; Naik, S. Hydrothermal pretreatment technologies for lignocellulosic biomass: A review of steam explosion and subcritical water hydrolysis. *Chemosphere* **2021**, *284*, 131372. [CrossRef]
21. Shannon, E.; Renata, B. The effect of biomass moisture content on bioethanol yields from steam pretreated switchgrass and sugarcane bagasse. *Bioresour. Technol.* **2011**, *102*, 2651–2658.
22. Yanling, Y.; Yujie, F.; Chen, X.; Jia, L.; Dongmei, L. Onsite bio-detoxification of steam-exploded corn stover for cellulosic ethanol production. *Bioresour. Technol.* **2011**, *102*, 5123–5128.
23. Vaithanomsat, P.; Chuichulcherm, S.; Apiwatanapiwat, W. Bioethanol production from enzymatically saccharified sunflower stalks using steam explosion as pretreatment. *World Acad. Sci. Eng. Technol.* **2009**, *49*, 140–143.
24. Balat, M. Production of bioethanol from lignocellulosic materials via the biochemical pathway: A review. *Energ. Convers. Manag.* **2011**, *52*, 858–875. [CrossRef]
25. Maria, C.D.A.W.; Carlos, M.; George, J.D.M.R.; Ester, R.G. Increase in ethanol production from sugarcane bagasse based on combined pretreatments and fed-batch enzymatic hydrolysis. *Bioresour. Technol.* **2013**, *128*, 448–453.
26. Brownlee, I.A. The physiological roles of dietary fibre. *Food Hydrocoll.* **2011**, *25*, 238–250. [CrossRef]
27. Trowell, H.; Southgate, D.A.; Wolever, T.M.; Leeds, A.R.; Jenkins, D.J. Letter: Dietary fibre redefined. *Lancet* **1976**, *307*, 967. [CrossRef]
28. Paola, V.; Aurora, N.; Vincenzo, F. Cereal dietary fibre: A natural functional ingredient to deliver phenolic compounds into the gut. *Trends Food Sci. Tech.* **2008**, *19*, 451–463.
29. Lupton, J.R.; Betteridge, V.A.; Pijls, L.J. Codex final definition of dietary fibre: Issues of implementation. *Qual. Assur. Saf. Crops Foods* **2009**, *1*, 206–212. [CrossRef]
30. Yang, H.; Bixiang, W.; Liankui, W.; Fengzhong, W.; Hansong, Y.; Dongxia, C.; Xin, S.; Chi, Z. Effects of dietary fiber on human health. *Food Sci. Hum. Wellness* **2022**, *11*, 1–10.
31. Susanne, W.; Kommer, B.; Jan-Willem, V.D.K. Dietary fibre: Challenges in production and use of food composition data. *Food Chem.* **2013**, *140*, 562–567.
32. Melissa, M.K.; Michael, J.M.; Gregory, G.F. The health benefits of dietary fiber: Beyond the usual suspects of type 2 diabetes mellitus, cardiovascular disease and colon cancer. *Metabolism* **2012**, *61*, 1058–1066.
33. Macagnan, F.T.; Da Silva, L.P.; Hecktheuer, L.H. Dietary fibre: The scientific search for an ideal definition and methodology of analysis, and its physiological importance as a carrier of bioactive compounds. *Food Res. Int.* **2016**, *85*, 144–154. [CrossRef]
34. Jinjin, C.; Qingsheng, Z.; Liwei, W.; Shenghua, Z.; Lijun, Z.; Bing, Z. Physicochemical and functional properties of dietary fiber from maca (*Lepidium meyenii* Walp.) liquor residue. *Carbohyd. Polym.* **2015**, *132*, 509–512.
35. Bader, U.A.H.; Saeed, F.; Khan, M.A.; Niaz, B.; Rohi, M.; Nasir, M.A.; Tufail, T.; Anbreen, F.; Anjum, F.M. Modification of barley dietary fiber through thermal treatments. *Food Sci. Nutr.* **2019**, *7*, 1816–1820. [CrossRef]
36. Shuai, L.; Mengyun, J.; Jiajun, C.; Haisheng, W.; Ruihong, D.; Shaoping, N.; Mingyong, X.; Qiang, Y. Removal of bound polyphenols and its effect on antioxidant and prebiotics properties of carrot dietary fiber. *Food Hydrocoll.* **2019**, *93*, 284–292.
37. Ji, Y.S.; Young-Min, K.; Byung-Hoo, L.; Sang-Ho, Y. Increasing the dietary fiber contents in isomaltooligosaccharides by dextran-sucrase reaction with sucrose as a glucosyl donor. *Carbohyd. Polym.* **2020**, *230*, 115607.
38. Isabel, G.; Díaz-Rubio, M.E.; Jara, P.; Fulgencio, S. Towards an updated methodology for measurement of dietary fiber, including associated polyphenols, in food and beverages. *Food Res. Int.* **2009**, *42*, 840–846.
39. Saura-Calixto, F. Concept and health-related properties of nonextractable polyphenols: The missing dietary polyphenols. *J. Agr. Food Chem.* **2012**, *60*, 11195–11200. [CrossRef]
40. Xiaoran, D.; Suyan, G.; Shoumin, J.; Litian, Z.; Ruihuan, D.U.; Aili, X.; Biao, Q.I.; Lei, W. Research Status and Progress on Modification of Dietary Fiber at Home and Abroad. *Agric. Biotechnol.* **2019**, *8*, 131–135.

41. Marzieh, S.; Maryam, M.K.; Hamid, Z.; Ilona, S.H.; Keikhosro, K. Techno-economical study of biogas production improved by steam explosion pretreatment. *Bioresour. Technol.* **2013**, *148*, 53–60.

42. Sheldon, J.B.D.; William, D.M. Bioconversion of forest products industry waste cellulosics to fuel ethanol: A review. *Bioresour. Technol.* **1996**, *55*, 1–33.

43. Yu, Z.; Zhang, B.; Yu, F.; Xu, G.; Song, A. A real explosion: The requirement of steam explosion pretreatment. *Bioresour. Technol.* **2012**, *121*, 335–341. [CrossRef]

44. Jung, M.L.; Hasan, J.; Richard, A.V. A comparison of the autohydrolysis and ammonia fiber explosion (AFEX) pretreatments on the subsequent enzymatic hydrolysis of coastal Bermuda grass. *Bioresour. Technol.* **2010**, *101*, 5449–5458.

45. Itziar, E.; Cristina, S.; Iñaki, M.; Jalel, L. Effect of alkaline and autohydrolysis processes on the purity of obtained hemicelluloses from corn stalks. *Bioresour. Technol.* **2012**, *103*, 239–248.

46. Yibo, Z. Enhanced saccharification of steam explosion pretreated corn stover by the supplementation of thermoacidophilic β-glucosidase from a newly isolated strain, Tolypocladium cylindrosporum syzx4. *Afr. J. Microbiol. Res.* **2011**, *5*, 2413–2421.

47. Saleha, S.; Umi, K.M.S.; Huzairi, Z.; Suraini, A.; Siti, M.M.K.; Yoshihito, S.; Mohd, A.H. Effect of steam pretreatment on oil palm empty fruit bunch for the production of sugars. *Biomass Bioenergy.* **2012**, *36*, 280–288.

48. Manzanares, P.; Ballesteros, I.; Negro, M.J.; Oliva, J.M.; Gonzalez, A.; Ballesteros, M. Biological conversion of forage sorghum biomass to ethanol by steam explosion pretreatment and simultaneous hydrolysis and fermentation at high solid content. *Biomass Convers. Biorefinery* **2012**, *2*, 123–132. [CrossRef]

49. Delong, E.A. Method of Rendering Lignin Separable from Cellulose and Hemicellulose in Lignocellulosic Material and the Product so Produced. Canadian Patent No1096374, 24 February 1981.

50. Sipos, B.; Kreuger, E.; Suensson, S.E.; Reczey, K.; Bjoernsson, L.; Zacchi, G. Steam pretreatment of dry and ensiled industrial hemp for ethanol production. *Biomass Bioenerg.* **2010**, *34*, 1721–1731. [CrossRef]

51. Peitao, Z.; Shifu, G.; Kunio, Y. An orthogonal experimental study on solid fuel production from sewage sludge by employing steam explosion. *Appl. Energ.* **2013**, *112*, 1213–1221.

52. Song, G.F.; Dong-Liang, L.I.; Gang, L.I.; Li, L.I.; Zhu, L.J. Study on reduction of ammonia delivery in mainstream cigarette smoke by steam explosion. *Acta Tab. Sin.* **2012**, *24*, 37–53.

53. Turn, S.Q.; Kinoshita, C.M.; Kaar, W.E.; Ishimura, D.M. Measurements of gas phase carbon in steam explosion of biomass. *Bioresour. Technol.* **1998**, *64*, 71–75. [CrossRef]

54. Chen, H.; Sui, W. *Steam Explosion as a Hydrothermal Pretreatment in the Biorefinery Concept, Hydrothermal Processing in Biorefineries*; Springer: Cham, Switzerland, 2017; Chapter 12; pp. 317–332.

55. Chen, H. *Gas Explosion Technology and Biomass Refinery*; Springer: Dordrecht, The Netherlands, 2015.

56. Wenjie, S.; Hongzhang, C. Water transfer in steam explosion process of corn stalk. *Ind. Crops Prod.* **2015**, *76*, 977–986.

57. Sui, W.; Chen, H. Extraction enhancing mechanism of steam exploded Radix Astragali. *Process Biochem.* **2014**, *49*, 2181–2190. [CrossRef]

58. Sui, W.; Xie, X.; Liu, R.; Wu, T.; Zhang, M. Effect of wheat bran modification by steam explosion on structural characteristics and rheological properties of wheat flour dough. *Food Hydrocoll.* **2018**, *84*, 571–580. [CrossRef]

59. Guozhong, C.; Hongzhang, C. Extraction and deglycosylation of flavonoids from sumac fruits using steam explosion. *Food Chem.* **2011**, *126*, 1934–1938.

60. Wu, T.; Li, Z.; Liu, R.; Sui, W.; Zhang, M. Effect of Extrusion, Steam Explosion and Enzymatic Hydrolysis on Functional Properties of Wheat Bran. *Food Sci. Technol. Res.* **2018**, *24*, 591–598. [CrossRef]

61. Zhao, G.; Gao, Q.; Hadiatullah, H.; Zhang, J.; Zhang, A.; Yao, Y. Effect of wheat bran steam explosion pretreatment on flavors of nonenzymatic browning products. *LWT* **2021**, *135*, 110026. [CrossRef]

62. Zhao, B.; Wang, C.; Zeng, Z.; Xu, Y. Study on the extraction process of pectin from grapefruit peel based on steam explosion technology. *IOP Conf. Ser. Earth Environ. Sci.* **2021**, *792*, 12008. [CrossRef]

63. Li, B.; Yang, W.; Nie, Y.; Kang, F.; Goff, H.D.; Cui, S.W. Effect of steam explosion on dietary fiber, polysaccharide, protein and physicochemical properties of okara. *Food Hydrocoll.* **2019**, *94*, 48–56. [CrossRef]

64. Golbaghi, L.; Khamforoush, M.; Hatami, T. Carboxymethyl cellulose production from sugarcane bagasse with steam explosion pulping: Experimental, modeling, and optimization. *Carbohyd. Polym.* **2017**, *174*, 780–788. [CrossRef]

65. Dorado, C.; Cameron, R.G.; Manthey, J.A. Study of Static Steam Explosion of Citrus sinensis Juice Processing Waste for the Isolation of Sugars, Pectic Hydrocolloids, Flavonoids, and Peel Oil. *Food Bioprocess Tech.* **2019**, *12*, 1293–1303. [CrossRef]

66. Sui, W.; Li, S.; Zhou, X.; Dou, Z.; Liu, R.; Wu, T.; Jia, H.; Wang, G.; Zhang, M. Potential Hydrothermal-Humification of Vegetable Wastes by Steam Explosion and Structural Characteristics of Humified Fractions. *Molecules* **2021**, *26*, 3841. [CrossRef]

67. Zhu, L.; Yu, B.; Chen, H.; Yu, J.; Yan, H.; Luo, Y.; He, J.; Huang, Z.; Zheng, P.; Mao, X.; et al. Comparisons of the micronization, steam explosion, and gamma irradiation treatment on chemical composition, structure, physicochemical properties, and in vitro digestibility of dietary fiber from soybean hulls. *Food Chem.* **2021**, *366*, 130618. [CrossRef]

68. Glasser, W.G.; Wright, R.S. Steam-assisted biomass fractionation. II. fractionation behavior of various biomass resources. *Biomass Bioenerg.* **1998**, *14*, 219–235. [CrossRef]

69. Kumar, P.; Barrett, D.M.; Delwiche, M.J.; Stroeve, P. Methods for Pretreatment of Lignocellulosic Biomass for Efficient Hydrolysis and Biofuel Production. *Ind. Eng. Chem. Res.* **2009**, *48*, 3713–3729. [CrossRef]

70. Aktas-Akyildiz, E.; Mattila, O.; Sozer, N.; Poutanen, K.; Koksel, H.; Nordlund, E. Effect of steam explosion on enzymatic hydrolysis and baking quality of wheat bran. *J. Cereal. Sci.* **2017**, *1*, 25–32. [CrossRef]

71. Wang, T.; Liang, X.; Ran, J.; Sun, J.; Jiao, Z.; Mo, H. Response surface methodology for optimisation of soluble dietary fibre extraction from sweet potato residue modified by steam explosion. *Int. J. Food Sci. Technol.* **2017**, *52*, 741–747. [CrossRef]
72. Zhai, X.; Ao, H.; Liu, W.; Zheng, J.; Li, X.; Ren, D. Physicochemical and structural properties of dietary fiber from Rosa roxburghii pomace by steam explosion. *J. Food Sci. Technol.* **2022**, *59*, 2381–2391. [CrossRef]
73. Liang, X.; Ran, J.; Sun, J.; Wang, T.; Jiao, Z.; He, H.; Zhu, M. Steam-explosion-modified optimization of soluble dietary fiber extraction from apple pomace using response surface methodology. *CyTA J. Food* **2018**, *16*, 20–26. [CrossRef]
74. Hu, L.; Guo, J.; Zhu, X.; Liu, R.; Wu, T.; Sui, W.; Zhang, M. Effect of steam explosion on nutritional composition and antioxidative activities of okra seed and its application in gluten-free cookies. *Food Sci. Nutr.* **2020**, *8*, 4409–4421. [CrossRef]
75. Liu, C.; Sun, Y.; Jia, Y.; Geng, X.; Pan, L.; Jiang, W.; Xie, B.; Zhu, Z. Effect of steam explosion pretreatment on the structure and bioactivity of Ampelopsis grossedentata polysaccharides. *Int. J. Biol. Macromol.* **2021**, *185*, 194–205. [CrossRef]
76. Wan, F.; Feng, C.; Luo, K.; Cui, W.; Xia, Z.; Cheng, A. Effect of steam explosion on phenolics and antioxidant activity in plants: A review. *Trends Food Sci. Tech.* **2022**, *124*, 13–24. [CrossRef]
77. Kong, F.; Wang, L.; Gao, H.; Chen, H. Process of steam explosion assisted superfine grinding on particle size, chemical composition and physico-chemical properties of wheat bran powder. *Powder Technol.* **2020**, *371*, 154–160. [CrossRef]
78. He, X.; Liu, X.; Li, W.; Zhao, J.; Li, F.; Ming, J. Modification of Dietary Fiber from Tartary Buckwheat Bran by Steam Explosion. *Food Sci.* **2021**, *42*, 46–54. (In Chinese)
79. Persson, T.; Ren, J.L.; Joelsson, E.; Jönsson, A.S. Fractionation of wheat and barley straw to access high-molecular-mass hemicelluloses prior to ethanol production. *Bioresour. Technol.* **2009**, *100*, 3906–3913. [CrossRef]
80. Sun, X.F.; Xu, F.; Sun, R.C.; Geng, Z.C.; Fowler, P.; Baird, M.S. Characteristics of degraded hemicellulosic polymers obtained from steam exploded wheat straw. *Carbohydr. Polym.* **2005**, *60*, 15–26. [CrossRef]
81. Gong, L.; Huang, L.; Zhang, Y. Effect of Steam Explosion Treatment on Barley Bran Phenolic Compounds and Antioxidant Capacity. *J. Agric. Food Chem.* **2012**, *60*, 7177–7184. [CrossRef]
82. Segal, L.; Creely, J.J.; Martin, A.E., Jr.; Conrad, C.M. An Empirical Method for Estimating the Degree of Crystallinity of Native Cel-lulose Using the X-Ray Diffractometer. *Text. Res. J.* **1959**, *29*, 786–794. [CrossRef]
83. Supachok, T.; Suteera, W.; Anyaporn, B. Study on structural and thermal properties of cellulose microfibers isolated from pineapple leaves using steam explosion. *J. Environ. Chem. Eng.* **2018**, *7*, 102836.
84. Marwa, C.R.; Souhir, A.; Sabu, T.; Hamadi, A.; Dorra, G. Use of green chemistry methods in the extraction of dietary fibers from cactus rackets (*Opuntia ficus* indica): Structural and microstructural studies. *Int. J. Biol. Macromol.* **2018**, *116*, 901–910.
85. Zhiqing, T.; Shikui, W.; Xuefang, H.; Zhimin, Z.; Liang, L. Crystalline reduction, surface area enlargement and pore generation of chitin by instant catapult steam explosion. *Carbohyd. Polym.* **2018**, *200*, 255–261.
86. Marion, C.; Anne, L.; Dominique, D.; Jean-Michel, L.; Frédérique, H.; François, C.; Cyril, A. Thermogravimetric analysis as a new method to determine the lignocellulosic composition of biomass. *Biomass Bioenergy* **2010**, *35*, 298–307.
87. Sanchez-Silva, L.; López-González, D.; Villaseñor, J.; Sánchez, P.; Valverde, J.L. Thermogravimetric–mass spectrometric analysis of lignocellulosic and marine biomass pyrolysis. *Bioresour. Technol.* **2012**, *109*, 163–172. [CrossRef]
88. Yong, G.S.; Yulong, M.; Zheng, W.; Junkang, Y. Evaluating and optimizing pretreatment technique for catalytic hydrogenolysis conversion of corn stalk into polyol. *Bioresour. Technol.* **2014**, *158*, 307–312.
89. Zhangbing, Z.; Zhidan, L.; Yuanhui, Z.; Baoming, L.; Haifeng, L.; Na, D.; Buchun, S.; Ruixia, S.; Jianwen, L. Recovery of reducing sugars and volatile fatty acids from cornstalk at different hydrothermal treatment severity. *Bioresour. Technol.* **2016**, *199*, 220–227.
90. Li-Yang, L.; Jin-Cheng, Q.; Kai, L.; Muhammad, A.M.; Chen-Guang, L. Impact of moisture content on instant catapult steam explosion pretreatment of sweet potato vine. *Bioresour. Bioprocess.* **2017**, *4*, 49.
91. Jacquet, N.; Quiévy, N.; Vanderghem, C.; Janas, S.; Blecker, C.; Wathelet, B.; Devaux, J.; Paquot, M. Influence of steam explosion on the thermal stability of cellulose fibres. *Polym. Degrad. Stabil.* **2011**, *96*, 1582–1588. [CrossRef]
92. Nadia, B.; Mouna, K.; Radhouane, K. Extraction and characterization of three polysaccharides extracted from Opuntia ficus indica cladodes. *Int. J. Biol. Macromol.* **2016**, *92*, 441–450.
93. Jacquetab, N.; Vanderghema, C.; Danthine, S.; Quiévy, N.; Blecker, C.; Devaux, J.; Paquot, M. Influence of steam explosion on physicochemical properties and hy-drolysis rate of pure cellulose fibers. *Bioresour. Technol.* **2012**, *121*, 221–227. [CrossRef]
94. Qi, M. Sunflower Stalk Pith Fibre: Investigation on Oil Holding Capacity, Oil-Fibre Interaction, and Related Application in Food. Master's Thesis, University of Guelph, Guelph, ON, Canada, 2017.
95. Raghavendra, S.N.; Ramachandra Swamy, S.R.; Rastogi, N.K.; Raghavarao, K.S.M.S.; Kumar, S.; Tharanathan, R.N. Grinding characteristics and hydration properties of coconut residue: A source of dietary fiber. *J. Food Eng.* **2004**, *72*, 281–286. [CrossRef]
96. Uskoković, V. Composites comprising cholesterol and carboxymethyl cellulose. *Colloids Surf. B Biointerfaces* **2008**, *61*, 250–261. [CrossRef]
97. Gaoshuang, L.; Haixia, C.; Shuhan, C.; Jingge, T. Chemical composition and physicochemical properties of dietary fiber from Polygonatum odoratum as affected by different processing methods. *Food Res. Int.* **2012**, *49*, 406–410.
98. Guohua, H.; Shaohua, H.; Hao, C.; Fei, W. Binding of four heavy metals to hemicelluloses from rice bran. *Food Res. Int.* **2009**, *43*, 203–206.
99. Robbins, R.J. Phenolic acids in foods: An overview of analytical methodology. *J. Agr. Food Chem.* **2003**, *51*, 2866–2887. [CrossRef]
100. Liyana-Pathirana, C.M.; Shahidi, F. Importance of insoluble-bound phenolics to antioxidant properties of wheat. *J. Agr. Food Chem.* **2006**, *54*, 1256–1264. [CrossRef]
101. Chandrasekara, A.; Shahidi, F. Content of insoluble bound phenolics in millets and their contribution to antioxidant capacity. *J. Agr. Food Chem.* **2010**, *58*, 6706–6714. [CrossRef]

102. Liu, L.; Zhao, M.; Liu, X.; Zhong, K.; Tong, L.; Zhou, X.; Zhou, S. Effect of steam explosion-assisted extraction on phenolic acid profiles and antioxidant properties of wheat bran. *J. Sci. Food Agr.* **2016**, *96*, 3484–3491. [CrossRef]
103. Kong, F.; Wang, L.; Chen, H.; Zhao, X. Improving storage property of wheat bran by steam explosion. *Int. J. Food Sci. Technol.* **2020**, *56*, 287–292. [CrossRef]
104. Li, W.; Zhang, X.; He, X.; Li, F.; Zhao, J.; Yin, R.; Ming, J. Effects of steam explosion pretreatment on the composition and biological activities of tartary buckwheat bran phenolics. *Food Funct.* **2020**, *11*, 4648–4658. [CrossRef]
105. Ye, C.; Ran, Y.; Luo, Y.; Ning, Z. Novel blasting extrusion processing improved the physicochemical properties of soluble dietary fiber from soybean residue and in vivo evaluation. *J. Food Eng.* **2014**, *120*, 1–8.
106. Sara, J.; Rocío, R.; Rafael, G.; Ana, J. Hydrothermal treatments enhance the solubility and antioxidant characteristics of dietary fiber from asparagus by-products. *Food Bioprod. Process.* **2019**, *114*, 175–184.
107. Alvira, P.; Tomás-Pejó, E.; Ballesteros, M.; Negro, M.J. Pretreatment technologies for an efficient bioethanol production process based on enzymatic hydrolysis: A review. *Bioresour. Technol.* **2009**, *101*, 4851–4861. [CrossRef] [PubMed]

MDPI
St. Alban-Anlage 66
4052 Basel
Switzerland
Tel. +41 61 683 77 34
Fax +41 61 302 89 18
www.mdpi.com

Foods Editorial Office
E-mail: foods@mdpi.com
www.mdpi.com/journal/foods